incremental motion control

volume I

DC Motors and Control Systems

Edited by

Benjamin C. Kuo

University of Illinois at Urbana-Champaign

Jacob Tal

The University of Utah

SRL Publishing Company, *Champaign, Illinois*

Library of Congress Catalog Card No. 78-53480
ISBN 0-918152-02-X

Printed in the United States of America

Contents

Preface

This is the first of a two-volume book on incremental motion control systems and devices. The present volume covers the subject on dc motors and controls. Volume II which is scheduled to be published in 1979 will be devoted to step motors and controls.

Incremental motion controls using dc motors and step motors have attracted a great deal of attention in the industry in recent years. The Symposium on Incremental Motion Control Systems and Devices sponsored annually by the Department of Electrical Engineering of the University of Illinois at Urbana-Champaign and the Warner Electric Brake and Clutch Company since 1972 has been attended by many engineers and scientists from all over the world. The advent of the microprocessor age has further enhanced the applications of high-performance dc motors and step motors to complex motion control systems.

The main objective of this book is to familiarize the practicing engineer on how to apply dc motors as system components for the control of motion. The latter need not be incremental at all times, although some of the most challenging control problems stem from the highly critical performance requirements in the numerous start-stop motions found in the industry. A good portion of this volume contains material that was presented at the past Symposiums on Incremental Motion Control Systems and Devices. The editors believe that the subject matters selected are of fundamental importance, and the chapter arrangements form a continuous and logical flow of information. Most of these chapters have been rewritten by the individual authors and then edited. The chapter on microprocessors and their application to dc motor control is new and represents approximately one fourth of the entire book.

The editors are indebted to the many contributing authors of this volume. The names and the affiliations of these authors are listed elsewhere in the front material of the book.

Special thanks should go to Jane Carlton who typeset the entire manuscript and contributed in many ways in the editing efforts.

Benjamin C. Kuo
Jacob Tal
May 1978

CONTRIBUTORS

D. K. Gehmlich	*Department of Electrical Engineering, The University of Utah*
B. C. Kuo	*Department of Electrical Engineering, University of Illinois at Urbana-Champaign*
F. G. Moritz	*Mosciatti, Foley, and Moritz, Inc.*
E. K. Persson	*Electro-Craft Corporation*
S. H. Pollack	*Department of Electrical Engineering, University of Illinois at Urbana-Champaign*
C. K. Rushforth	*Department of Electrical Engineering, The University of Utah*
D. C. Schlick	*Burroughs Corporation*
R. Schmidt	*Data Card Corporation*
J. Tal	*Department of Electrical Engineering, The University of Utah*
H. Waagen	*PMI Motors, Division of Kollmorgen Corporation*

1 Introduction

B. C. KUO

Incremental motion control systems are found in numerous industrial processes. Because of the advances made in microelectronics and microprocessor technology in recent years, motion control systems design and analysis have entered a new and important era.

In general, incremental motion, as the name implies, is concerned with the control of motions that are characterized in a step-by-step fashion. Of course, incremental motion is not limited only to start-stop actions. Often, when the speed is relatively high, the motion becomes continuous with respect to time. Thus, the subjects covered in this book are quite broad, and they encompass essentially all control systems that contain dc motors as prime movers.

The problem of incremental motion control is not new. There were many old-time applications in process and system controls that required step-by-step or point-to-point control of a certain load. However, without high-performance prime movers with low electrical inductance and low mechanical inertia, fast electronic switching devices, high-precision and high-resolution encoders, the speed and the overall performance of the traditional incremental motion control systems are quite limited.

Historically, the prime movers used in motion control systems have been electric motors, hydraulic actuators and brake-and-clutch packages. In recent years the advances made in the technology on permanent magnets have made dc motors an important actuator in control systems of all types. The advent of the microprocessor has created new applications and popularity for step motors. This volume of the two-volume book on Incremental Motion Control is devoted to DC Motors and Control Systems. Volume II, edited by B. C. Kuo, is devoted to Step Motors and Controls.

Traditionally, two-phase induction motors have been used in many feedback control systems. This type of motor has the inherent advantage of being rugged in construction and requiring very little maintenance. DC motors, on the other hand, were relatively more expensive; they require two energy sources with the wound-field construction, and have brushes to maintain. However, in recent years, the permanent-magnet technology, particularly in the area of rare-earth magnets, has made the permanent-magnet dc motor highly efficient, and for all practical purposes under normal conditions very little brush maintenance is necessary.

Because dc motors are analog actuators with no preferential stopping positions, practically all dc motor control systems are closed-loop systems. In a closed-loop system the output position and/or speed are/is fed back and compared with the reference input to achieve the desired performance.

Two of the most important areas of incremental motion control systems are in the computer peripheral systems and numerical control and machine tool controls. The following list represents some of the typical incremental motion control applications found in the computer industries:

- random-access disk memories
- printwheel control in word processors
- carriage control in word processors
- tape drives
- card punches and readers
- tape readers and punches
- incremental tape recorders

Some of the other typical applications of dc motor control systems are:

- automatic typesetting for printing industry
- banking terminals
- coil-winding machines
- wire-wrapping machines
- photo-copying machines
- X-Y plotters
- X-Y microfiche access mechanisms
- sun-tracking control for solar panels
- strip chart recorders
- numerically-controlled machine tool drives
- medical equipment
- packaging machines
- agricultural machines
- photographic and optical equipment
- robots and translating machines
- textile machines
- aircraft control systems

and many others.

The basic components of a typical dc-motor control system are shown by the block diagram of Fig. 1-1. We shall use this diagram to illustrate the material covered in this book.

The dc motor shown in Fig. 1-1 is typically a high-performance permanent-magnet dc motor with low armature inductance and low rotor inertia. The principle of operation of this type of dc motor is described in Chapter 5. Chapter 6 gives the mathematical modeling of dc motors.

Depending on the specific application of the dc-motor control system, the encoder shown in Fig. 1-1 may be one of many types. For positional control a simple potentiometer may be used. With digital control an incremental photoelectric encoder is often used. For speed control, one may use a tachometer or derive the speed information from the incremental encoder pulses. The various types of encoders used in incremental motion control systems are discussed in Chapter 3, and dc tachometers are discussed in Chapter 8.

The power amplifier shown in the block diagram of Fig. 1-1 in general can be a linear dc servo amplifier (Chapter 10) or a pulsewidth-modulated amplifier (Chapter 11). In a linear amplifier the voltages vary continuously. This type of amplifier is relatively simple and is capable of wide bandwidth control applications. The disadvantage is in the relatively high power dissipation in the output transistors. The pulsewidth-modulated amplifier has the characteristics that the transistors are turned on and off like a switch. When the transistor is switched on, the voltage across it is negligible; when the transistor is switched off, the voltage across it is high but the current is zero. This reduces the power dissipated in the output transistors.

In order to conduct a successful analysis and design of a control system, one has to have an adequate background on control systems theory. Chapters 2 and 4 cover the fundamental principles of incremental motion control and feedback control theory. The specific designs of velocity, positional and phase-lock-loop control systems are discussed in Chapters 12, 13 and 14, respectively. Chapter 15 covers the design of incremental motion control systems using dc motors.

The controller block shown in Fig. 1-1 can assume a variety of forms. Using the conventional design techniques, the controller is often described mathematically by a transfer function. The transfer function of the controller can be realized by a passive electric network or by a digital network. For more complex functions the controller can be realized by a microcomputer. Chapter 16 is a fairly extensive chapter which covers first the fundamentals of microprocessor programming, and then the application of microprocessors to the control of dc motor systems. The last two chapters provide some practical examples of the applications of dc motors in incremental control systems.

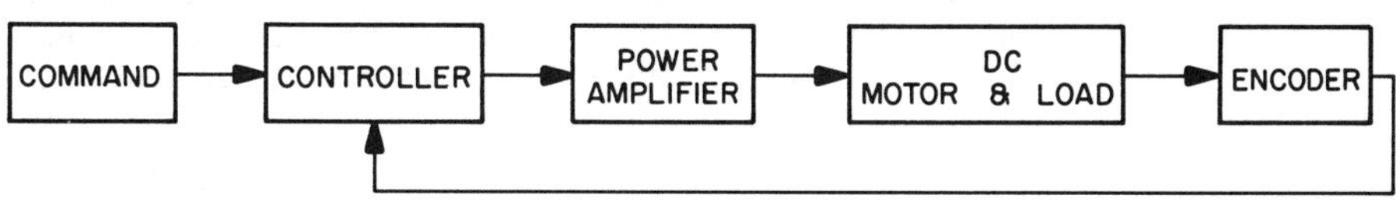

Fig. 1-1. DC motor control system.

2 Incremental Motion Control

B. C. KUO

2.1 INTRODUCTION

The purpose of this chapter is to establish some basic foundation on the mathematical description of elements of control systems. Emphasis will be placed on the modeling of mechanical elements and the treatment of system responses of incremental motion control systems.

2.2 MATHEMATICAL MODELING OF MECHANICAL SYSTEM ELEMENTS

Since practically all incremental motion control systems involve some types of mechanical linkages and elements at the output, it is essential to mathematically describe these components. For instance, a majority of the loads which are driven by a step motor or a clutch-brake combination can be regarded as an inertial load. Sometimes the load is subject to frictional torque or force. Often, the load is coupled to the prime mover through cables, a flexible shaft, or a gear train.

In general, the motion involving mechanical elements can be described as translational, rotational, or a combination of both. The equations governing the motions of mechanical system elements are often directly or indirectly formulated from Newton's law of motion.

Translational Motion

The motion of translation is defined as a motion which takes place along a straight line. The variables which are used to describe translational motion are: acceleration, velocity and displacement.

Newton's law of motion states that *the algebraic sum of forces acting on a rigid body in a given direction is equal to the product of the mass of the body and its acceleration in the same direction.* The law can be expressed as

$$\Sigma \text{ forces} = Ma \tag{2-1}$$

where M denotes the mass and a is the acceleration in the direction considered.

For translational motion, the following elements are usually involved:

Mass: Mass is considered as an indication of the property of an element which stores the kinetic energy of translational motion. It is analogous to inductance of electrical networks. If W denotes the weight of a body, then M is given by

$$M = W/g \tag{2-2}$$

where g is the acceleration of the body due to gravity, or the acceleration of free fall. Three consistent sets of units for the elements in Eqs. (2-1) and (2-2) are given below.

Units	Mass M	Weight W	acceleration	force
MKS	Kg = newton/m/sec^2	newton	m/sec^2	newton
CGS	g = dyne/cm/sec^2	dyne	cm/sec^2	dyne
British	Slug = lb/ft/sec^2	lb	ft/sec^2	lb

Figure 2-1 illustrates the situation where a force is acting on a body with mass M. The force equation is written

$$f(t) = Ma(t) = M\frac{d^2y(t)}{dt^2} = M\frac{dv(t)}{dt} \tag{2-3}$$

where y(t) represents displacement, v(t) the velocity, and a(t) is the acceleration, all referenced in the direction of the applied force.

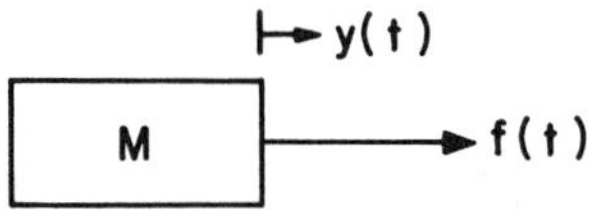

Fig. 2-1. A force-mass system.

Linear Spring: A linear spring in practice may be an actual spring or the compliance of a cable or a belt. In general, a spring is considered as an element which stores potential energy. It is analogous to a capacitor in electric networks. In practice, all springs are nonlinear to some extent. However, if the deformation of a spring is small, its behavior may be approximated by a linear relationship,

$$f(t) = Ky(t) \tag{2-4}$$

where K is the spring constant, or simply, stiffness. The three unit systems for the spring constant are:

Units	K
MKS	newton/m
CGS	dyne/cm
British	lb/ft

Equation (2-4) implies that the force acting on the spring is directly proportional to the displacement (deformation) of the spring. The model representing a linear spring element is shown in Fig. 2-2.

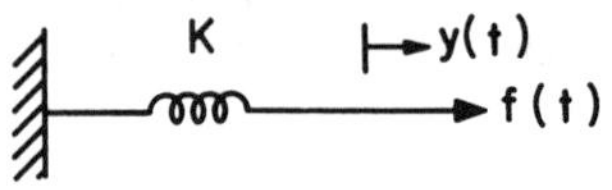

Fig. 2-2. A force-spring system.

If the spring is preloaded with a preload tension of T, then Eq. (2-4) should be modified to

$$f(t) - T = Ky(t) \tag{2-5}$$

Friction For Translational Motion

Whenever there is motion or tendency of motion between two elements, frictional forces exist. The frictional forces encountered in physical systems are usually of a nonlinear nature. The characteristics of the frictional forces between two contacting surfaces often depends on such factors as the composition of the surfaces, the pressure between the surfaces, their relative velocity, and others, so that an exact mathematical description of the frictional force is difficult. However, for practical purposes, frictional forces can be divided into three basic categories: viscous friction, static friction, and Coulomb friction. These are discussed separately in detail in the following.

Viscous Friction: Viscous friction represents a retarding force which is proportional to velocity. The schematic diagram element for friction is often represented by a dashpot such as that shown in Fig. 2-3. The mathematical expression of viscous friction is

$$f(t) = B\frac{dy(t)}{dt} \tag{2-6}$$

where B is the viscous frictional coefficient. The dimensions of B in the three unit systems are:

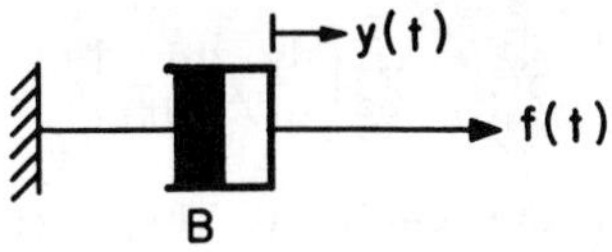

Fig. 2-3. A force-viscous friction system.

Units	B
MKS	$\frac{\text{newton}}{\text{m/sec}}$
CGS	$\frac{\text{dyne}}{\text{cm/sec}}$
British	$\frac{\text{lb}}{\text{ft/sec}}$

Figure 2-4(a) shows the functional relation between the viscous frictional force and velocity.

Static Friction: Static friction represents a retarding force which tends to prevent motion from beginning. The static frictional force can be represented by the following expression:

$$f(t) = \pm(F_s)_{\dot{y}=0} \tag{2-7}$$

where $(F_s)_{\dot{y}=0}$ is defined as the static frictional force which exists only when the body is stationary but has a tendency of moving. The sign of the friction depends on the direction of motion or the initial direction of velocity. The force-velocity relation of static friction is illustrated in Fig. 2-4(b). Notice that once motion begins, the static frictional force vanishes, and other frictions take over.

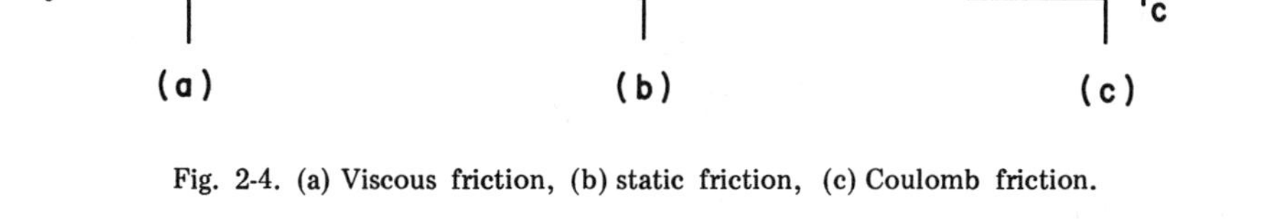

Fig. 2-4. (a) Viscous friction, (b) static friction, (c) Coulomb friction.

Coulomb Friction: Coulomb friction is a retarding force which has a constant amplitude with respect to the change in velocity, but the sign of the frictional force changes with the reversal of the direction of velocity. The mathematical relation for the Coulomb friction is given by

$$f(t) = F_c \left[\frac{dy}{dt} \Big/ \left| \frac{dy}{dt} \right| \right] \tag{2-8}$$

where F_c is the Coulomb friction coefficient. The functional description of the friction to velocity relation is shown in Fig. 2-4(c).

Rotational Motion

The rotational motion of a body may be defined as motion about a fixed axis. The variables generally used to describe the motion of rotation are torque, angular acceleration, α, angular velocity, ω, and angular displacement, θ. The following elements are usually involved with the rotational motion:

Inertia: Inertia, J, is considered as an indication of the property of an element which stores the kinetic energy of rotational motion. The inertia of a given element depends on the geometric composition about the axis of rotation and its density.

For instance, the inertia of a circular disk or a circular shaft about its geometric axis is given by

$$J = \frac{1}{2} Mr^2 \tag{2-9}$$

where M is the mass of the disk or shaft and r is its radius.

Example 2-1.

Given a disk which is 1 inch in diameter, 0.25 inch thick, and weighing 5 ounces, its inertia is

$$J = \frac{1}{2} \frac{Wr^2}{g} = \frac{1}{2} \frac{(5 \text{ oz})(1 \text{ in})^2}{386 \text{ in/sec}^2}$$

$$= 0.00647 \text{ oz-in-sec}^2 \tag{2-10}$$

Usually the density of the material is given in weight per unit volume. Then, for a circular disk or shaft it can be shown that the inertia is proportional to the fourth power of the radius and the first power of the thickness or length. Therefore, if the weight W is expressed as

$$W = \rho(\pi r^2 h) \tag{2-11}$$

where ρ is the density in weight per unit volume, r is the radius, and h is the thickness or length, then Eq. (2-9) is written

$$J = \frac{1}{2} \frac{\rho \pi h r^4}{g} = 0.00406 \, \rho h r^4 \tag{2-12}$$

where h and r are in inches.

For steel, ρ is 4.53 oz/in^3; Eq. (2-12) becomes

$$J = 0.0184 \, hr^4 \tag{2-13}$$

For aluminum, ρ is 1.56 oz/in^3; Eq. (2-12) becomes

$$J = 0.00636\ hr^4 \tag{2-14}$$

When a torque is applied to a body with inertia J, as shown in Fig. 2-5, the torque equation is written

$$T(t) = J\alpha(t) = J\frac{d\omega(t)}{dt} = J\frac{d^2\theta(t)}{dt^2} \tag{2-15}$$

The three generally used sets of units for the quantities in Eq. (2-15) are tabulated below:

Units	Inertia	Torque	Angular displacement
MKS	Kg-m-sec^2	newton-meter	radian
CGS	gm-cm-sec^2	dyne-cm	radian
English	slug-ft^2 or lb-ft-sec^2	pound-ft	radian
	oz-in-sec^2	ounce-in	

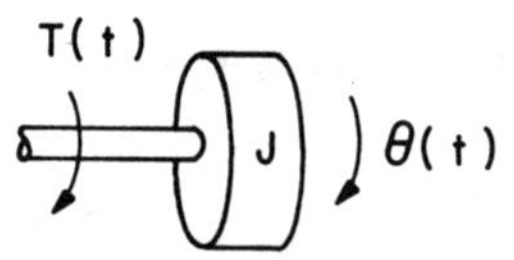

Fig. 2-5. A torque-inertia system.

The following conversion factors are often found useful:

Angular Displacement

1 radian = $180/\pi$ = 57.3 degrees

Angular Velocity

1 RPM = $2\pi/60$ = 0.1047 rad/sec

1 RPM = 6 degrees/sec

Torque

1 gm-cm = 0.0139 oz-in

1 lb-ft = 192 oz-in

1 oz-in = 0.00521 lb-ft

Inertia

$1 \text{ gm-cm}^2 = 1.417 \times 10^{-5} \text{ oz-in-sec}^2$

$1 \text{ lb-ft-sec}^2 = 192 \text{ oz-in-sec}^2$

$1 \text{ oz-in-sec}^2 = 386 \text{ oz-in}^2$

$1 \text{ gm-cm-sec}^2 = 980 \text{ gm-cm}^2$

$1 \text{ lb-ft-sec}^2 = 32.2 \text{ lb-ft}^2$

Torsional Spring: As with the linear spring for translational motion, a torsional spring constant, K, in torque per unit angular displacement can be devised to represent the compliance of a rod, or a shaft when it is subject to an applied torque. Figure 2-6 illustrates a simple torque-spring system which can be represented by the following equation:

$$T(t) = K\theta(t) \tag{2-16}$$

The dimension for K is given in the following units:

Units	K
MKS	newton-m/rad
CGS	dyne-cm/rad
British	oz-in-rad

If the torsional spring is preloaded by a preload torque of TP, then Eq. (2-16) is modified to

$$T - TP = K\theta(t) \tag{2-17}$$

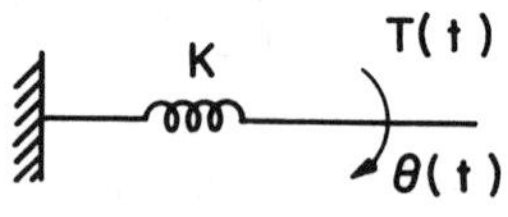

Fig. 2-6. A torque-torsional spring system.

Friction For Rotational Motion

The three types of frictions described for translational motion can be carried over to the motion of rotation. Therefore, Eqs. (2-6), (2-7), and (2-8) can respectively be replaced by their counterparts as follows:

$$T(t) = B\frac{d\theta(t)}{dt} \tag{2-18}$$

$$T(t) = \pm(F_s)_{\dot{\theta}=0} \tag{2-19}$$

$$T(t) = F_c\left[\frac{d\theta}{dt}\Big/\left|\frac{d\theta}{dt}\right|\right] \tag{2-20}$$

where B is the viscous frictional coefficient in torque per unit angular velocity, $(F_s)_{\dot{\theta}=0}$ is the static friction, and F_c is the Coulomb friction coefficient.

Relation Between Translational And Rotational Motions

In motion control problems it is often necessary to convert rotational motion into a translational one. For instance, a load may be controlled to move along a straight line through a rotary motor and screw assembly, such as that shown in Fig. 2-7. Figure 2-8 shows a similar situation in which a rack-and-pinion is used as the mechanical linkage. Another common system in motion control is the control of a mass through a pulley by a rotary prime mover, such as that shown in Fig. 2-9. The systems shown in Figs. 2-7, 2-8, and 2-9 can all be represented by a simple system with an equivalent inertia connected directly to the drive motor. For instance, the mass in Fig. 2-9 can be regarded as a point mass which moves about the pulley which has a radius r. Disregarding the inertia of the pulley, the equivalent inertia which the motor sees is

$$J = Mr^2 = \frac{W}{g} r^2 \tag{2-21}$$

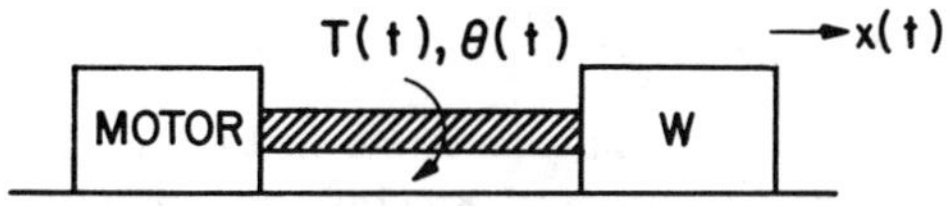

Fig. 2-7. A motor-lead screw system.

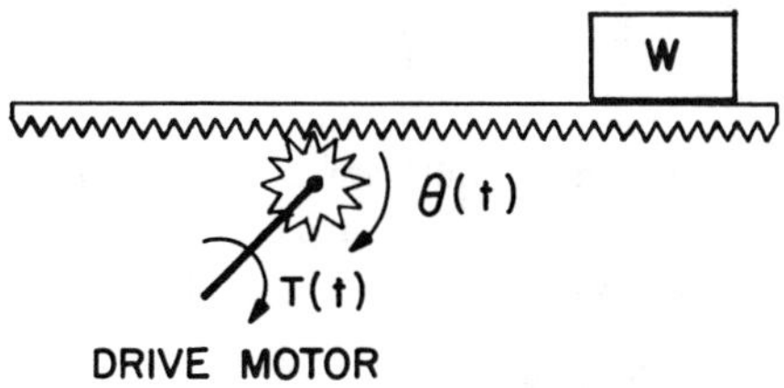

Fig. 2-8. A rack-and-pinion system.

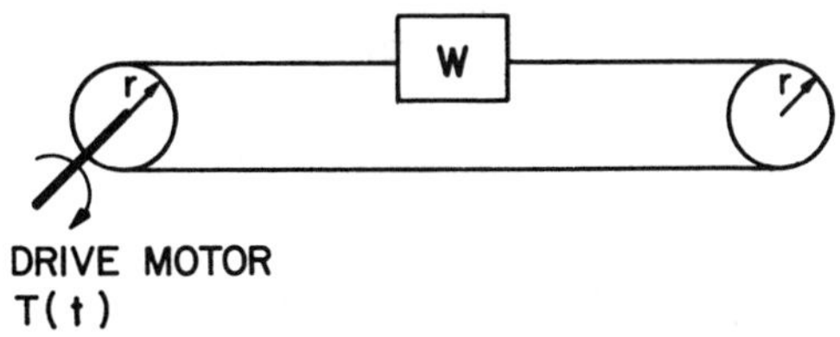

Fig. 2-9. A belt-and-pulley system.

If the radius of the pinion in Fig. 2-8 is r, the equivalent inertia which the motor sees is also given by Eq. (2-21).

Now consider the system of Fig. 2-7. The lead of the screw, L, is defined as the linear distance which the mass travels per revolution of the screw. Therefore, in principle, the two systems in Figs. 2-8 and 2-9 are equivalent. In Fig. 2-8, the distance travelled by the mass per revolution of the pinion is $2\pi r$. Therefore, using Eq. (2-21) as the equivalent inertia for the system of Fig. 2-9, we have

$$J = \frac{W}{g}\left[\frac{L}{2\pi}\right]^2 \tag{2-22}$$

where in the British system,

J = inertia (oz-in-sec^2)

W = weight (oz)

L = screw lead (in)

G = gravitational force (386.4 in/sec^2)

Mechanical Energy And Power

Energy and power play an important role in the design of electromechanical systems. Stored energy in the form of kinetic and potential energy controls the dynamics of the system, whereas dissipative energy usually is spent in the form of heat which must be closely controlled.

The mass or inertia of a body indicates its ability to store kinetic energy. The kinetic energy of a moving mass with a velocity v is

$$W_k = \frac{1}{2} Mv^2 \tag{2-23}$$

The following consistent sets of units are given for the kinetic energy relation.

Units	Energy	Mass	Velocity
MKS	Joule = newton-m	Kg	m/sec
CGS	erg = dyne-cm	gram	cm/sec
British	ft-lb	lb/ft/sec^2 (slug)	ft/sec

For a rotational system, the kinetic energy relation is written

$$W_k = \frac{1}{2} J\omega^2 \tag{2-24}$$

where J is the moment of inertia, and ω is the angular velocity. The following units are given for the rotational kinetic energy.

Units	Energy	Inertia	Angular velocity
MKS	Joule = newton-m	kg-m-sec^2	rad/sec
CGS	erg = dyne-cm	gm-cm-sec^2	rad/sec
British	oz-in	oz-in-sec^2	rad/sec

Potential energy stored in a mechanical element represents the amount of work required to change the configuration. For a linear spring which is deformed by y in length, the potential energy stored in the spring is

$$W_p = \frac{1}{2} Ky^2 \tag{2-25}$$

where K is the spring constant. For a torsional spring, the potential energy stored is given by

$$W_p = \frac{1}{2} K\theta^2 \tag{2-26}$$

When dealing with a frictional element, the form of energy differs from the previous two cases in that the energy represents a loss or dissipation by the system in overcoming the frictional force. Power is the time rate of doing work. Therefore, the power dissipated in a frictional element is the product of force and velocity; i.e.,

$$P = fv \tag{2-27}$$

Since f = Bv, where B is the frictional coefficient, Eq. (2-27) becomes

$$P = Bv^2 \tag{2-28}$$

The MKS unit for power is in newton-m/sec or watt; for the CGS system it is dyne-cm/sec. In the British unit system, power is represented in ft-lb/sec or horsepower (hp). Furthermore,

$$\begin{aligned} 1 \text{ hp} &= 746 \text{ watts} \\ &= 550 \text{ ft-lb/sec} \end{aligned} \tag{2-29}$$

Since power is the rate at which energy is being dissipated, the energy dissipated in a frictional element is

$$W_d = B\int v^2 dt \tag{2-30}$$

Gear Trains, Levers, And Timing Belts

A gear train, a lever, or a timing belt over pulleys is a mechanical device that transmits energy from one part of a system to another in such a way that force, torque, speed, and displacement are altered. These devices may also be regarded as matching devices which are used to attain maximum power transfer.

Two gears are shown coupled together in Fig. 2-10. The inertia and friction of the gears are neglected in the ideal case considered.

The relationships between the torques, T_1 and T_2, angular displacements, θ_1 and θ_2, and the teeth numbers, N_1 and N_2, of the gear train are derived from the following facts:

1. The number of teeth on the surface of the gears is proportional to the radius r_1 and r_2 of the gears; that is,

$$r_1 N_2 = r_2 N_1 \tag{2-31}$$

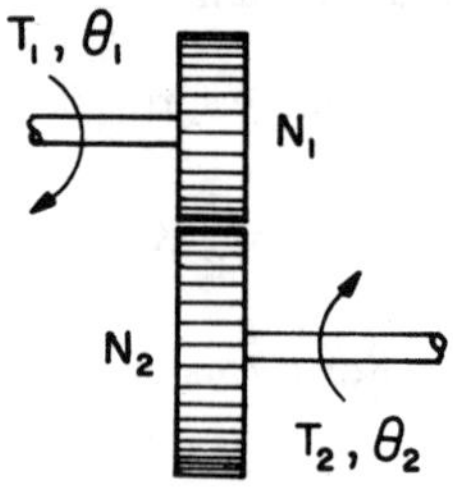

Fig. 2-10. Coupled gear system.

2. The distance travelled along the surface of each gear is the same. Therefore,

$$\theta_1 r_1 = \theta_2 r_2 \tag{2-32}$$

3. The work done by one gear is equal to that of the other since there is assumed to be no losses. Thus,

$$T_1 \theta_1 = T_2 \theta_2 \tag{2-33}$$

If the angular velocities of the two gears, ω_1 and ω_2, are brought into the picture, the last three equations lead to

$$\frac{T_1}{T_2} = \frac{\theta_2}{\theta_1} = \frac{N_1}{N_2} = \frac{\omega_2}{\omega_1} = \frac{r_1}{r_2} \tag{2-34}$$

In practice, real gears do have inertia and friction between the coupled gear teeth which often cannot be neglected. An equivalent representation of a gear train with viscous friction, Coulomb friction, and inertia considered as lumped elements, is shown in Fig. 2-11. The following variables and parameters are defined for the gear train:

T = applied torque

θ_1, θ_2 = angular displacements

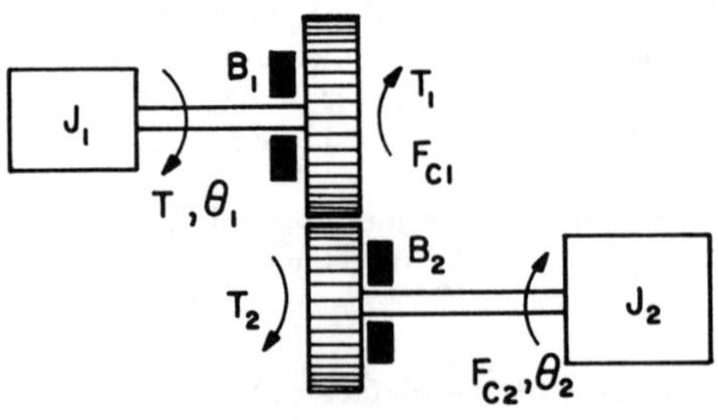

Fig. 2-11. Coupled gears with inertias.

T_1, T_2 = torque transmitted to gears

J_1, J_2 = inertia of gears

N_1, N_2 = numbers of teeth

F_{c1}, F_{c2} = Coulomb friction coefficients

B_1, B_2 = viscous frictional coefficients

The torque equation for gear 2 is written

$$T_2(t) = J_2 \frac{d^2\theta_2(t)}{dt^2} + B_2 \frac{d\theta_2(t)}{dt} + F_{c2} \frac{\dot{\theta}_2}{|\dot{\theta}_2|} \tag{2-35}$$

The torque equation on the side of gear 1 is

$$T(t) = J_1 \frac{d^2\theta_1(t)}{dt^2} + B_1 \frac{d\theta_1(t)}{dt} + F_{c1} \frac{\dot{\theta}_1}{|\dot{\theta}_1|} + T_1(t) \tag{2-36}$$

By the use of Eq. (2-34), Eq. (2-35) is converted to

$$T_1(t) = \frac{N_1}{N_2} T_2(t) = \left(\frac{N_1}{N_2}\right)^2 J_2 \frac{d^2\theta_1(t)}{dt^2} + \left(\frac{N_1}{N_2}\right)^2 B_2 \frac{d\theta_1(t)}{dt} + \left(\frac{N_1}{N_2}\right) F_{c2} \frac{\dot{\theta}_2}{|\dot{\theta}_2|} \tag{2-37}$$

This last equation indicates that it is possible to reflect inertia, friction, (and compliance), torque, speed, and displacement from one side of a gear train to the other.

Therefore, the following quantities are obtained when reflecting from gear 2 to gear 1:

Inertia: $(N_1/N_2)^2 J_2$

Viscous frictional coefficient: $(N_1/N_2)^2 B_2$

Torque: $(N_1/N_2) T_2$

Angular displacement: $(N_1/N_2)\theta_2$

Angular velocity: $(N_1/N_2)\omega_2$

Coulomb frictional torque: $(N_1/N_2) F_{c2}(\omega_2/|\omega_2|)$

If torsional spring effect were present, the spring constant is also multiplied by $(N_1/N_2)^2$ in reflecting from gear 2 to gear 1.

Now substituting Eq. (2-37) into Eq. (2-36), we get

$$T(t) = J_{1e} \frac{d^2\theta_1(t)}{dt^2} + B_{1e} \frac{d\theta_1(t)}{dt} + T_F \tag{2-38}$$

where

$$J_{1e} = J_1 + \left(\frac{N_1}{N_2}\right)^2 J_2 \tag{2-39}$$

$$B_{1e} = B_1 + \left(\frac{N_1}{N_2}\right)^2 B_2 \tag{2-40}$$

$$T_F = F_{c1} \frac{\dot{\theta}_1}{|\dot{\theta}_1|} + \left(\frac{N_1}{N_2}\right) F_{c2} \frac{\dot{\theta}_2}{|\dot{\theta}_2|} \tag{2-41}$$

Example 2-2

Given a load which has inertia of 0.05 oz-in-sec^2 and a Coulomb friction torque of 2 oz-in, find the inertia and frictional torque reflected through a 1:5 gear train ($N_1/N_2 = 5$, with N_2 on the load side). The reflected inertia on the side of N_1 is $(1/5)^2\ 0.05 = 0.002$ oz-in-sec^2. The reflected Coulomb friction is $(1/5)2 = 0.4$ oz-in.

Timing belts and chain drives serve the same purposes as the gear train except that they allow the transfer of energy over a longer distance without using an excessive number of gears. Figure 2-12 shows the diagram of a belt or chain drive between two pulleys. Assuming that there is no slippage between the belt and the pulleys, it is easy to see that Eq. (2-34) still applies to this case. In fact, the reflection or transmittance of torque, inertia, friction, etc., is similar to that of a gear train.

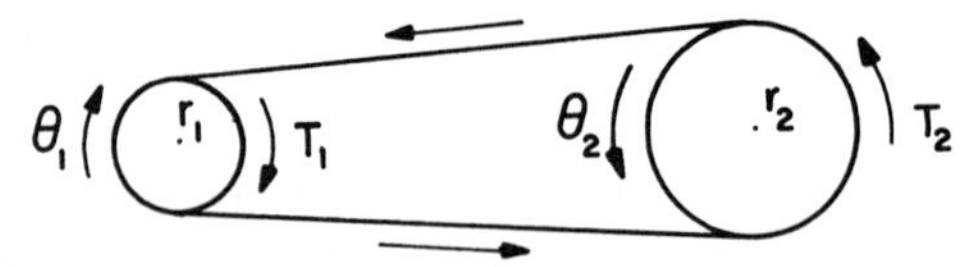

Fig. 2-12. Belt-and-capstan system.

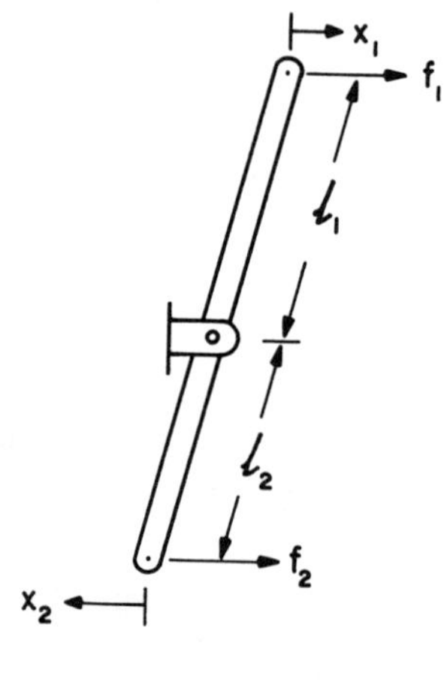

Fig. 2-13. A lever system.

The lever system shown in Fig. 2-13 transmits translational motion and force in the same way that gear trains transmit rotational motion. The relation between the forces and distances is

$$\frac{f_1}{f_2} = \frac{l_2}{l_1} = \frac{x_2}{x_1} \tag{2-42}$$

Backlash And Dead Zone

Backlash and dead zone usually play an important role in gear trains or other similar mechanical linkages. In a great majority of the situations backlash may give rise to undesirable oscillations and instability in control systems. In addition, it has a tendency of wearing out the mechanical elements. Regardless of the actual mechanical elements, a physical model of backlash or dead zone between an input and an output member is shown in Fig. 2-14. The model can be used for a rotational system as well as a translational system. The amount of backlash is b/2 on either side of the reference position.

In general, the dynamics of the mechanical linkage with backlash depend upon the relative inertia-to-friction ratio of the output member. If the inertia of the output member is very small compared with that of the input member, the motion is controlled predominantly by friction. This means that the output member will not coast whenever there is no contact between the two members. When the output is driven by the input the two members will travel together until the input member reverses its direction; then the output member will stand still until the backlash is taken up on the other side, at which time it is assumed that the output member instantaneously takes on the velocity of the input member. The transfer characteristic between the input and the output displacements of a backlash element with negligible output inertia is shown in Fig. 2-15. To illustrate the relative motion between the in-

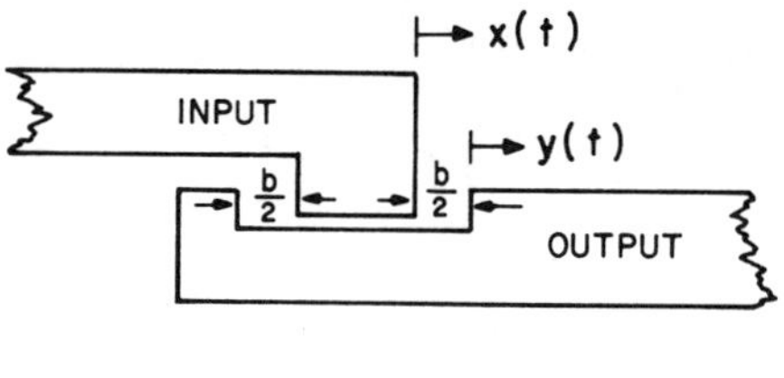

Fig. 2-14. Backlash or deadplay.

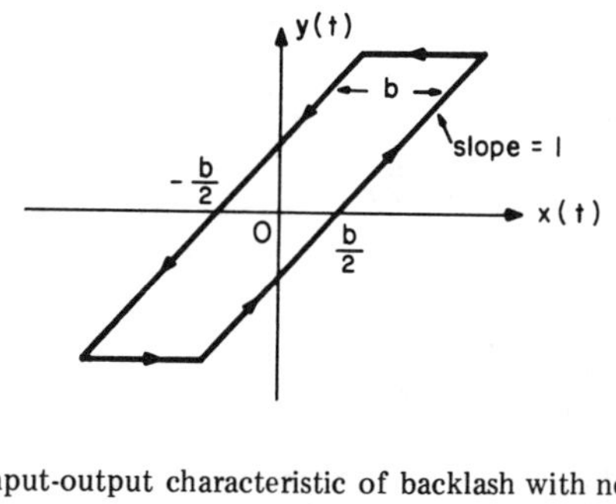

Fig. 2-15. Input-output characteristic of backlash with negligible output inertia.

put and the output members, let us assume that the input displacement is driven sinusoidally with respect to time. The displacements and velocities of the input and output members are illustrated as shown in Fig. 2-16. Note that the reference position of the two members is taken to be that of Fig. 2-14; that is, with the input member starting at the center of the total backlash. For Fig. 2-16, it is assumed that when motion begins the input member is in contact with the output member on the right, so that $x(0) = 0$ and $y(0) = -b/2$. In the other extreme, if the friction on the output member is so small that it may be neglected, then the inertia of the output member becomes the controlling factor. Under such condition, the output member remains in contact with the input member so long as the acceleration is in the direction to keep the two members together. When the acceleration of the input member becomes zero, the output member does not stop immediately, but leaves the input member and coasts at a constant velocity which is equal to the maximum velocity which the input member had attained. When the output member has traversed a distance, relative to the input member, which is equal to the full width of the backlash, it will be restrained by the opposite side of the input member. At this time, the output member will again assume the velocity of the input member. The transfer characteristic between the input and the output displacement of a backlash element with negligible output friction is shown in Fig. 2-17. The displacement, velocity, and acceleration

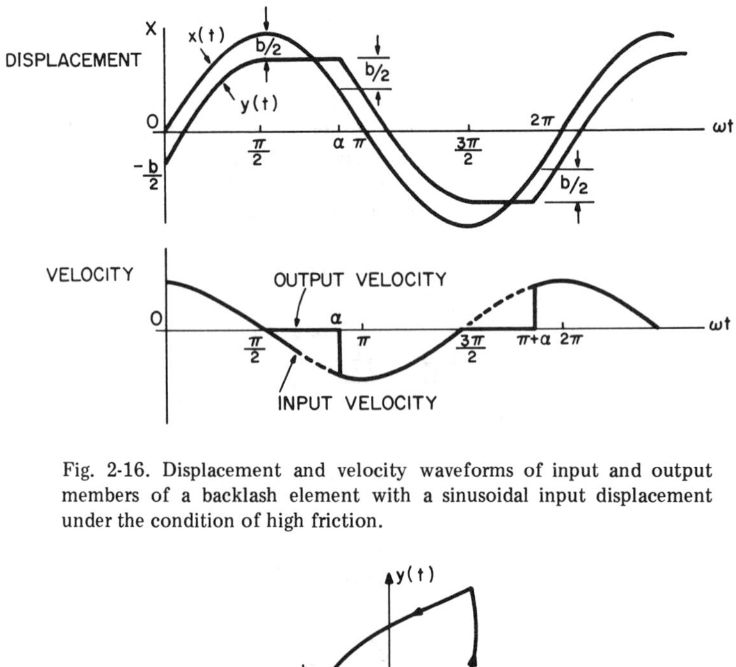

Fig. 2-16. Displacement and velocity waveforms of input and output members of a backlash element with a sinusoidal input displacement under the condition of high friction.

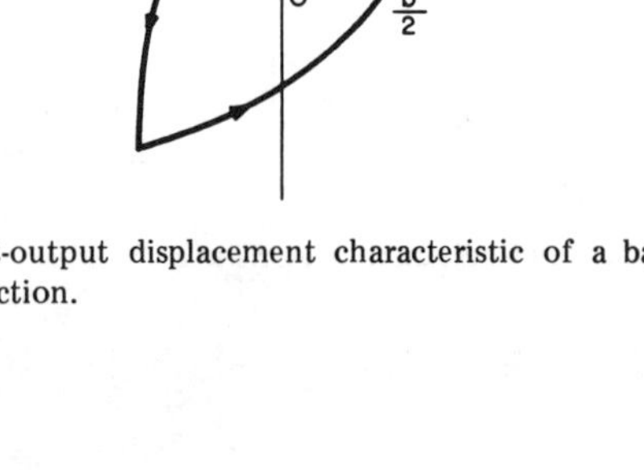

Fig. 2-17. Input-output displacement characteristic of a backlash element without friction.

waveforms of the input and output members, when the input displacement is driven sinusoidally, is shown in Fig. 2-18.

In practice, of course, the output member of a mechanical linkage with backlash usually has friction as well as inertia. Then the output waveforms in response to a sinusoidally driven input displacement should lie between those of Figs. 2-16 and 2-18.

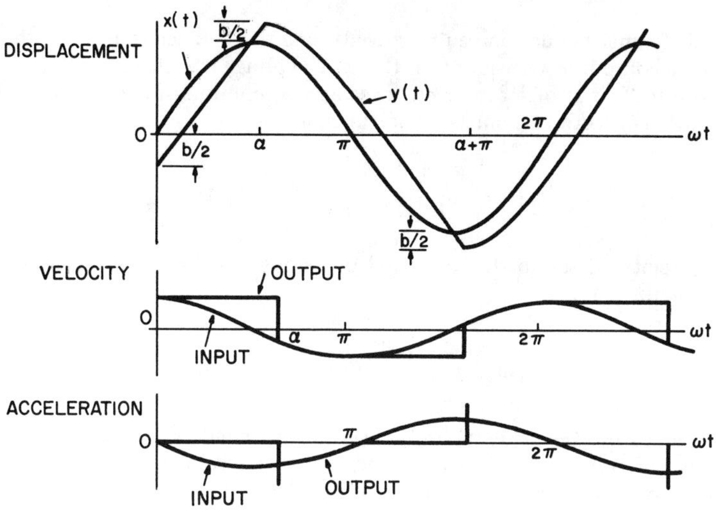

Fig. 2-18. Displacement, velocity, and acceleration waveforms of input and output members of a backlash element when the input displacement is driven sinusoidally, and under the condition of low friction.

2.3 ANALYSIS OF MECHANICAL SYSTEMS

In this section we shall consider the analysis of a typical motion control system which is modeled by the mechanical elements discussed in the last section.

Figure 2-19 illustrates a mechanical load with inertia J and viscous friction B. The load is connected to a motor through a shaft with a torsional spring constant K. The torque developed by the motor is T(t). The torque equations of the system are written from the free-body diagram of Fig. 2-19.

$$T(t) = K[\theta_1(t) - \theta_2(t)] \tag{2-43}$$

$$K[\theta_1(t) - \theta_2(t)] = J\ddot{\theta}_2(t) + B\dot{\theta}_2(t) \tag{2-44}$$

When the shaft is considered to be rigid, $K = \infty$, $\theta_1(t) = \theta_2(t)$; then the torque equation simply becomes

$$T(t) = J\ddot{\theta}_2(t) + B\dot{\theta}_2(t) \tag{2-45}$$

Let us assume that the motor is ideal so that its torque is proportional to the applied voltage

$$T(t) = K_t e(t) \tag{2-46}$$

where K_t is the torque constant.

Therefore, Eq. (2-45) is written

$$K_t e(t) = J\ddot{\theta}_2(t) + B\dot{\theta}_2(t) \tag{2-47}$$

This differential equation can be solved using the Laplace transform, the classical method, or a computer. Consider that the input voltage is constant with an amplitude of E for $t > 0$. This is often referred to as a step function of magnitude E. The solution of Eq. (2-47), with zero initial conditions, is

$$\theta_2(t) = \frac{K_m E}{B}\left[t - \frac{J}{B} + \frac{J}{B} e^{-Bt/J}\right] \tag{2-48}$$

The velocity of the load is obtained by taking the derivative of Eq. (2-48) with respect to time;

$$\omega_2(t) = \frac{K_m E}{B}\left(1 - e^{-Bt/J}\right) \tag{2-49}$$

Typical waveforms of $\theta_2(t)$ and $\omega_2(t)$ are sketched as shown in Fig. 2-20. The term J/B is defined as the *time constant*, T, of the velocity response. Notice that $\omega_2(t)$ is an exponential rise function. When $t = T = J/B$

$$\omega_2(T) = \frac{K_m E}{B}\left(1 - e^{-1}\right) = 0.632\,\frac{K_m E}{B} \tag{2-50}$$

Therefore, the time constant of the velocity response is the time required for the system to reach 63.2 percent of its final value.

Now let us consider the same system in Fig. 2-19, except that the shaft between the motor and the load is no longer rigid. Assuming that the initial conditions of the system are zero, and the motor displacement, $\theta_1(t)$, is suddenly changed by one unit at $t = 0$, the displacement $\theta_2(t)$ is to be determined. Knowing $\theta_1(t)$, $\theta_2(t)$ can be determined from Eq. (2-44). Since $\theta_1(t)$ is given to be unity for $t > 0$, Eq. (2-44) is written

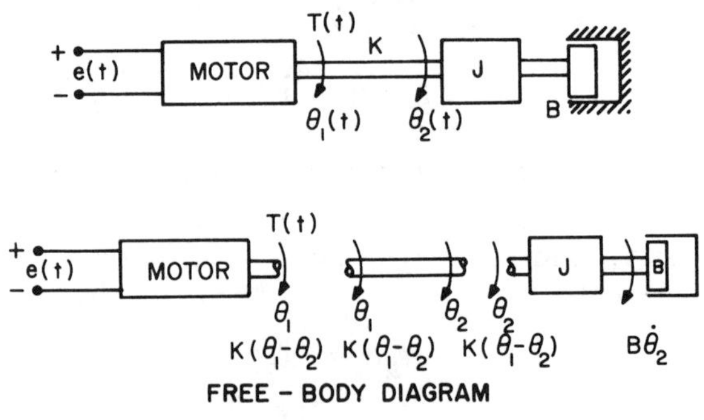

Fig. 2-19. A motor-load system and its free-body diagram.

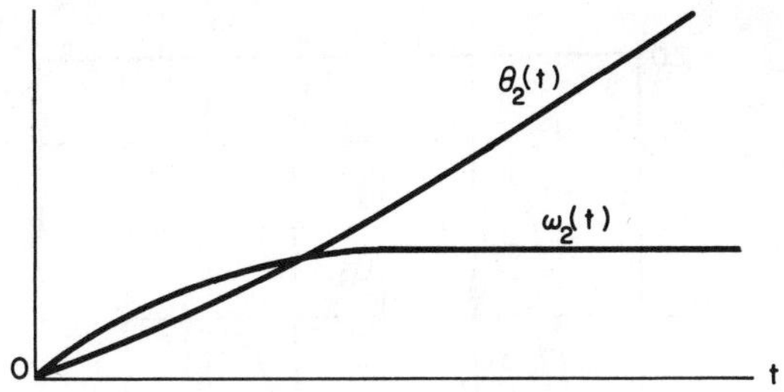

Fig. 2-20. Displacement and velocity responses of the system in Fig. 2-19.

$$J\ddot{\theta}_2(t) + B\dot{\theta}_2(t) + K\theta_2(t) = K \tag{2-51}$$

The solution of the last equation is carried out by taking the Laplace transform on both sides of the equation. Thus,

$$\Theta_2(s) = \frac{K}{s(Js^2 + Bs + K)} \tag{2-52}$$

or

$$\Theta_2(s) = \frac{\omega_n^2}{s(s^2 + 2\zeta\omega_n s + \omega_n^2)} \tag{2-53}$$

where

$$\omega_n = \sqrt{K/J} \tag{2-54}$$

$$\zeta = \frac{B}{2\sqrt{KJ}} \tag{2-55}$$

Equation (2-53) represents the Laplace transform of the output of a typical second-order system. The constant, ζ, is defined as the *damping ratio*, and ω_n is defined as the *natural undamped frequency*, of the system. The solution for $\theta_2(t)$ is obtained by taking the inverse Laplace transform of Eq. (2-53). We have

$$\theta_2(t) = 1 + \frac{e^{-\zeta\omega_n t}}{\sqrt{1-\zeta^2}} \sin[\omega_n\sqrt{1-\zeta^2}t - \tan^{-1}\frac{\sqrt{1-\zeta^2}}{-\zeta}] \tag{2-56}$$

Figure 2-21 gives the plots of $\theta_2(t)$ as a function of the normalized time $\omega_n t$ for various values of damping ratio ζ. It is seen that the response becomes more oscillatory as ζ decreases. When $\zeta > 1$, the response is nonoscillatory and it never exceeds the value of the input which is unity. When ζ lies between zero and one, the response oscillates before settling down to its final value. When $\zeta = 0$, Eq. (2-56) becomes

$$\theta_2(t) = 1 + \sin\omega_n t \tag{2-57}$$

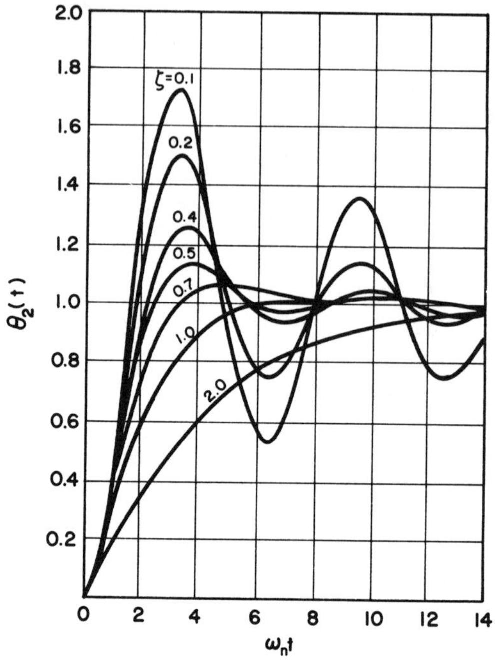

Fig. 2-21. Step responses of $\theta_2(t)$ for various damping ratios.

which is an undamped sine wave.

The difference between the maximum value and the final value of $\theta_2(t)$ is defined as the *peak overshoot*. The maximum value of $\theta_2(t)$ is found by taking the derivative of Eq. (2-56) with respect to time and then setting it equal to zero.

$$\frac{d\theta_2(t)}{dt} = \frac{e^{-\zeta\omega_n t}}{\sqrt{1-\zeta^2}}\left[\omega_n\sqrt{1-\zeta^2}\cos(\omega_n\sqrt{1-\zeta^2}\,t-\phi) - \omega_n \sin(\omega_n\sqrt{1-\zeta^2}\,t-\phi)\right] = 0 \qquad (2\text{-}58)$$

where

$$\phi = \tan^{-1}\frac{\sqrt{1-\zeta^2}}{-\zeta} \qquad (2\text{-}59)$$

Simplifying, Eq. (2-58) gives

$$\tan(\omega_n\sqrt{1-\zeta^2}\,t-\phi) = \frac{\sqrt{1-\zeta^2}}{\zeta} \qquad (2\text{-}60)$$

Thus, the time at which the peak overshoot occurs is

$$t_{max} = \frac{\pi}{\omega_n\sqrt{1-\zeta^2}} \tag{2-61}$$

Substituting Eq. (2-61) into Eq. (2-56), we get the maximum value of $\theta_2(t)$ as

$$\theta_2(t)\big|_{max} = 1 + e^{-\zeta\pi/\sqrt{1-\zeta^2}} \tag{2-62}$$

Therefore, the peak overshoot is

$$\text{Peak overshoot} = e^{-\zeta\pi/\sqrt{1-\zeta^2}} \tag{2-63}$$

Often the peak overshoot is represented as a percentage of the final value of the response; that is,

$$\text{Percent overshoot} = 100e^{-\zeta\pi/\sqrt{1-\zeta^2}} \tag{2-64}$$

2.4 INCREMENTAL MOTION

In this section we shall investigate the responses of systems which are subject to start-stop or incremental motion control. Some of the important characteristics of typical incremental motion control will be illustrated.

Inertial Load

Consider the simple motor-inertia system shown in Fig. 2-22. It is assumed that the total inertia of the load and motor is represented by J. There is no friction and no compliance in the coupling shaft. The motor is assumed to be ideal, so that the torque developed is directly proportional to the applied voltage. The equations of the system are written

$$T_m(t) = K_m e(t) \tag{2-65}$$

$$T_m(t) = J\ddot{\theta}_m(t) \tag{2-66}$$

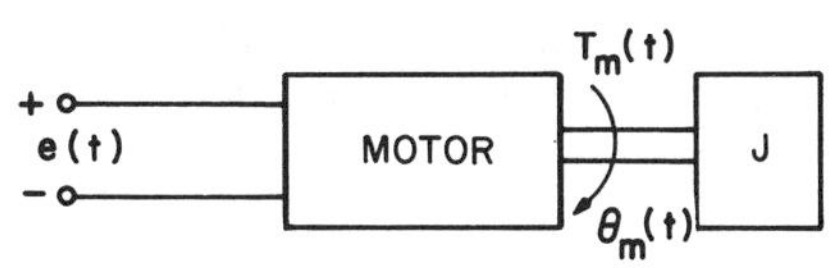

Fig. 2-22. A motor-inertia system.

For incremental start-stop motion, let the control voltage of the motor be that of Fig. 2-23. If this system is to represent a printer mechanism, it is expected that the load is to be stopped once every two seconds, and the printing operation takes place for a one second duration.

Since e(t) of Fig. 2-23 is a periodic function, for the present system it is necessary only to consider the first period which is from 0 to 3 seconds. For this duration, e(t) can be expressed as

$$e(t) = E[u(t) - 2u(t-1) + u(t-2)] \qquad 0 \leqslant t \leqslant 3 \tag{2-67}$$

where u(t) is a unit-step function.

Combining Eq. (2-65) with Eq. (2-66), and taking the Laplace transforms, we get

$$\Theta_m(s) = \frac{K_m E}{Js^3} [1 - 2e^{-s} + e^{-2s}] \tag{2-68}$$

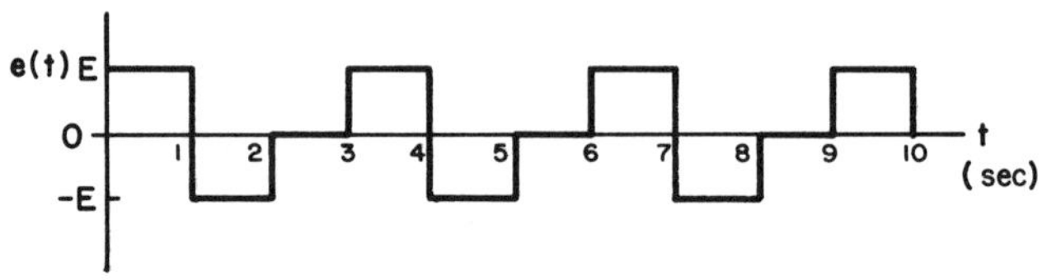

Fig. 2-23. Input control voltage of motor for incremental motion.

where it is assumed that the initial conditions are zero. Taking the inverse Laplace transform on both sides of Eq. (2-68) gives the displacement of the load from $0 \leqslant t \leqslant 3$. Therefore,

$$\theta_m(t) = \frac{K_m E}{J} [\frac{t^2}{J} u(t) - (t-1)^2 u(t-1) + \frac{1}{2} (t-2)^2 u(t-2)] \tag{2-69}$$

The Laplace transformed expression for the velocity of the load is

$$\Omega_m(s) = \frac{K_m E}{Js^2} [1 - 2e^{-s} + e^{-2s}] \tag{2-70}$$

Therefore, the velocity of the load is

$$\omega_m(t) = \frac{K_m E}{J} [tu(t) - 2(t-1)u(t-1) + (t-2)u(t-2)] \tag{2-71}$$

Since the velocity of the load is zero, and stays at zero for $2 \leqslant t < 3$, the same velocity response is obtained for subsequent periodic inputs starting from the initial value at the beginning of the new period. The waveforms for $\theta_m(t)$ and $\omega_m(t)$ are sketched as shown in Fig 2-24.

Viscous Friction-Inertia Load

Now consider that the system shown in Fig. 2-22 has viscous friction at the output shaft. The torque equation of Eq. (2-66) becomes

$$T_m(t) = J\ddot{\theta}_m(t) + B\dot{\theta}_m(t) \tag{2-72}$$

Let us assume that the waveform of the input voltage to the motor is as shown in Fig. 2-25. However, theoretically, when viscous friction exists in the system, the motor will not come to rest completely after the input voltage becomes zero. Therefore, in this case a mechanical detent mechanism is activated every 2 seconds to hold the load at the desired position.

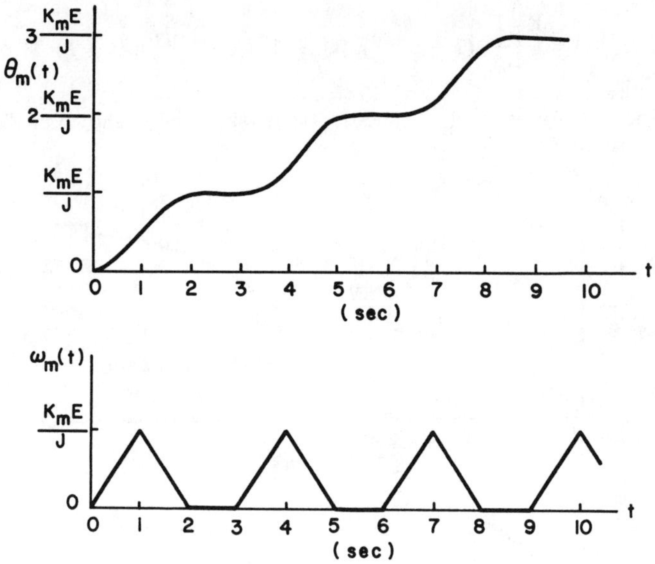

Fig. 2-24. Displacement and velocity waveforms of the system in Fig. 2-22.

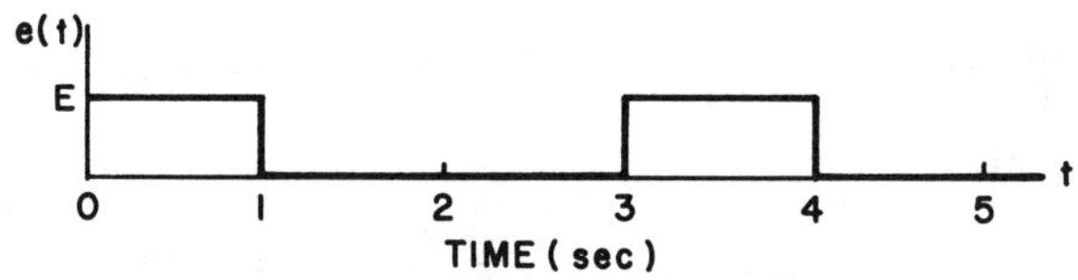

Fig. 2-25. Input control voltage of motor for incremental motion.

The Laplace transform of the output displacement is

$$\Theta_m(s) = \frac{K_m E}{s^2(Js + B)}(1 - e^{-s}) \tag{2-73}$$

for $0 \leqslant t < 2$ sec, and $\Theta_m(s) = 0$ for $2 \leqslant t < 3$, if the mechanical detent operates perfectly.

The output displacement for the first period is obtained by taking the inverse Laplace transform of Eq. (2-73). Thus,

$$\theta_m(t) = \frac{K_m E}{B}\ [t - T(1 - e^{-t/T})]\,u(t) - [(t - 1) - T(1 - e^{-(t-1)/T})]\,u(t - 1) \tag{2-74}$$

where $T = J/B$ is the time constant. The velocity of the load over the same time interval is

$$\omega_m(t) = \frac{K_m E}{B} (1 - e^{-t/T})u(t) - (1 - e^{-(t-1)/T})u(t - T) \tag{2-75}$$

$0 \leqslant t < 2$. The waveforms for $\theta_m(t)$ and $\omega_m(t)$ are sketched as shown in Fig. 2-26 for $t \geqslant 0$.

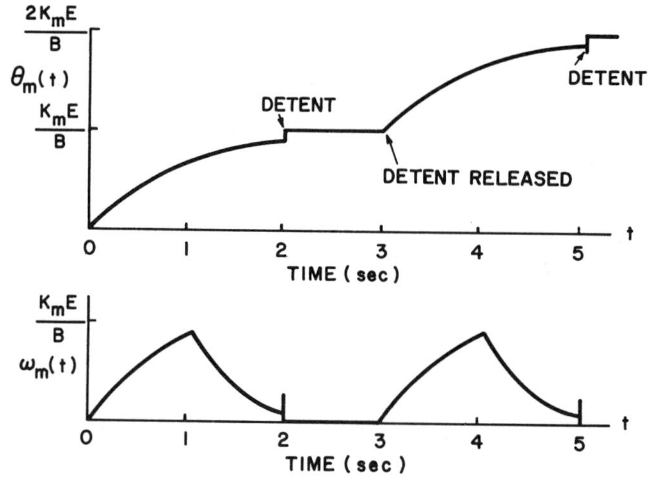

Fig. 2-26. Load displacement and velocity waveforms of the system in Fig. 2-22.

An Incremental Motion Control System With Feedback

Feedback is often used in control systems to reduce the error in the output, among other objectives. Figure 2-27 illustrates a system which has a potentiometer coupled to the output shaft for the purpose of feeding back a signal which is proportional to the angular displacement of the output. The feedback signal is compared with a reference input, and the difference is termed the error. If the system parameters are selected properly, the output displacement should be driven to the desired value according to the reference input in some optimal fashion.

Consider that the motor is ideal and the load is characterized by inertia J and viscous frictional coefficient B. The transfer relation of the potentiometer is

$$f(t) = K_p \theta_m(t) \tag{2-76}$$

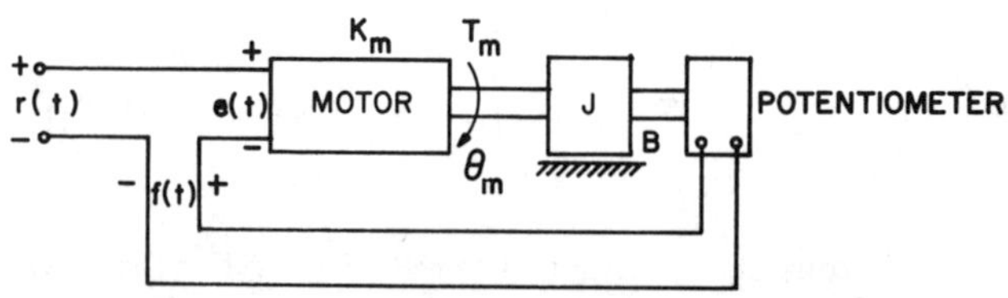

Fig. 2-27. A feedback control system.

where K_p is in volts per radian. The dynamic equations of the system are written,

$$e(t) = r(t) - f(t) \tag{2-77}$$

$$T_m(t) = K_m e(t) \tag{2-78}$$

$$T_m(t) = J\ddot{\theta}_m(t) + B\dot{\theta}_m(t) \tag{2-79}$$

Taking the Laplace transform on both sides of the last three equations, and solving for the relationship between $\Theta_m(s)$ and $R(s)$, we have

$$\frac{K_p \Theta_m(s)}{R(s)} = \frac{K_m K_p}{Js^2 + Bs + K_m K_p} \tag{2-80}$$

Since this equation has the same form as that of Eq. (2-52), it is apparent that if the reference input is a unit step function, the output displacement should reach $1/K_p$ radians after the transient has died out. Furthermore, it is recognized that the damping ratio of the system is given by

$$\zeta = \frac{B}{\sqrt{2JK_m K_p}} \tag{2-81}$$

and the natural undamped frequency is

$$\omega_n = \sqrt{K_m K_p / J} \tag{2-82}$$

With a step input as r(t) various forms of output can be obtained depending upon the values of ζ and ω_n. Although in principle the output displacement will always reach the desired value in the steady state, sometimes it takes time for the transient to die out completely, especially if the load inertia is very large. Therefore, for incremental motion control where the load is to follow a start-stop motion, the design objective is to have the transient response damp out as quickly as possible.

The reference input signal which is necessary for controlling the output in a start-stop motion is shown in Fig. 2-28. Figures 2-29(a) and 2-29(b) illustrate the typical displacement and velocity of the system, respectively. It is assumed that the output has essentially reached its steady-state value before the next step input is applied.

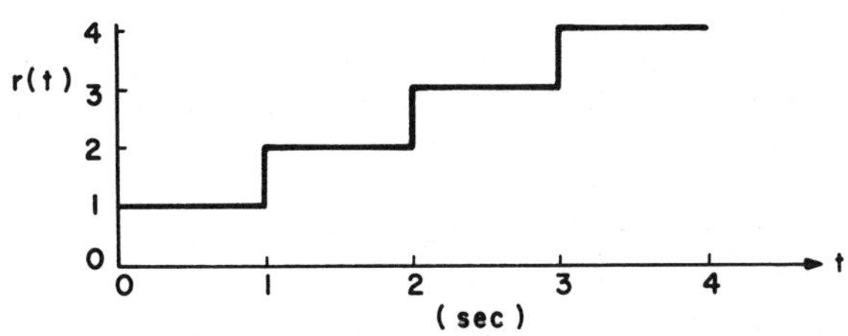

Fig. 2-28. Reference input to the system in Fig. 2-27.

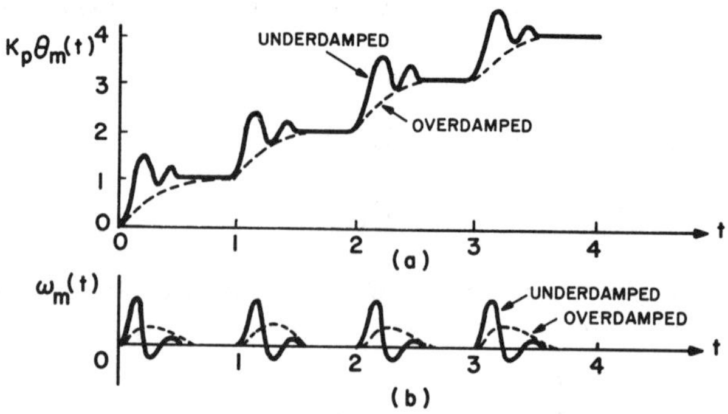

Fig. 2-29. (a) Output displacement of the system in Fig. 2-27.
(b) Output velocity of the system in Fig. 2-27.

Effect Of Coulomb Friction

Coulomb friction has been described previously in this chapter as a constant frictional torque which is independent of velocity. In practice, Coulomb friction is encountered quite frequently in incremental motion control systems.

Let us consider that the system shown in Fig. 2-27 now has Coulomb friction at the output, in addition to the viscous friction. The torque equation of the system is

$$T_m(t) = J_m \ddot{\theta}_m(t) + B\dot{\theta}_m(t) + T_c \frac{\dot{\theta}_m}{|\dot{\theta}_m|} \tag{2-83}$$

where T_c is the Coulomb friction torque. Combining Eq. (2-83) with Eqs. (2-77) and (2-78), we have

$$K_m r(t) - T_c \frac{\dot{\theta}_m}{|\dot{\theta}_m|} = J_m \ddot{\theta}_m(t) + B\dot{\theta}_m(t) + K_m K_p \theta_m(t) \tag{2-84}$$

Since T_c is a constant, the solution of the last equation is similar to that of the previous system when the Coulomb friction was absent, except that the input step amplitude is effectively increased or decreased by T_c/K_m depending on whether $\dot{\theta}_m$ is positive or negative.

Consider that the system is initially at standstill and a unit step input is applied at t = 0. Since once motion begins the output velocity is positive, the solution of Eq. (2-84) is

$$\theta_m(t) = \frac{1}{K_p}(1 - \frac{T_c}{K_m})[1 + \frac{e^{-\zeta\omega_n t}}{\sqrt{1-\zeta^2}} \sin(\omega_n\sqrt{1-\zeta^2}\, t - \phi)] \tag{2-85}$$

where $\phi = \tan^{-1}(\sqrt{1-\zeta^2}/-\zeta)$, ζ and ω_n are given by Eqs. (2-81) and (2-82), respectively. Equation (2-85) clearly shows that the effect of the Coulomb friction is to reduce the effective magnitude of the input by the amount T_c/K_m. For a critically damped or overdamped system, $\zeta \geqslant 1$, the velocity of the output is always positive, and Eq. (2-85) is valid for all $t > 0$. The steady-state response of $\theta_m(t)$ is $(1 - T_c/K_m)/K_p$. However, if ζ is less than unity, the system is underdamped; the response of $\theta_m(t)$ will overshoot. When $\theta_m(t)$ reaches its peak value, the velocity, $\dot{\theta}_m(t)$, will reverse its sign, and the sign of T_c is reversed. The constant term in front of Eq. (2-85) must be changed to $(1 + T_c/K_m)/K_p$. This means simply that when the velocity becomes negative, the oscillation of $\theta_m(t)$ appears to take place about the axis $(1 + T_c/K_m)/K_p$. Since the sign of the Coulomb friction torque depends on the direction of the output velocity, in general, the Coulomb friction tends to reduce the overshoot or undershoot of the output displacement. If, under any condition, the output velocity reaches zero and the magnitude of the error $e(t)$ is less than T_c/K_m, the torque developed by the motor, which is proportional to $e(t)$, is insufficient to overcome the Coulomb friction torque, and thus the system has to stop. In other words, the output will stop at the moment any overshoot or undershoot first occurs inside the region bounded by $(1 - T_c/K_m)/K_p$ and $(1 + T_c/K_m)/K_p$. Figure 2-30 illustrates three typical displacement responses of the system with Coulomb friction when the input is a unit step function. Therefore, in general, the effect of the Coulomb friction is to reduce the overshoot and oscillation of a system response, but the system may have a steady-state error as determined by the Coulomb friction torque.

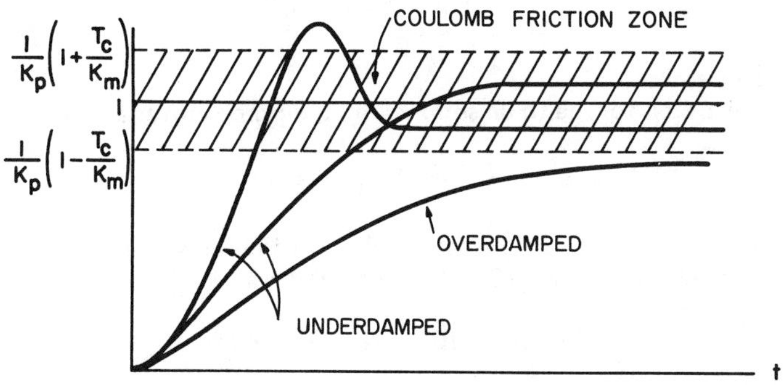

Fig. 2-30. Typical step responses of a system with Coulomb friction.

2.5 A TYPICAL INCREMENTAL MOTION CONTROL PROBLEM

In this section a typical incremental motion control problem will be considered for illustration purposes.

Consider a paper drive mechanism of a line printer. The operating parameters of the system are given below:

Print speed	1000 lines per minute
Pulley radius	$r_p = 0.4777$ inch
Load inertia	$J_L = 0.005$ oz-in-sec^2
Paper displacement	$d = 0.125$ inch (8 lines per inch)
Start-stop time	10 milliseconds
Dwell time	20 milliseconds
Load friction	$T_L = 10$ oz-in

Assuming that the system is ideal in the sense that the angular velocity of the load is as shown in Fig. 2-31, notice that it takes half the time to accelerate and half the time to decelerate the load. For the incremental paper displacement given, the corresponding incremental angular displacement is

$$\theta = d/r_p = 0.125/0.4777 = 0.261 \text{ rad}$$

Therefore, there are 24 incremental steps per revolution of the pulley.

$$\omega_{ave} = \theta/0.01 = 26.1 \text{ rad/sec} \tag{2-86}$$

From Fig. 2-31, the peak angular velocity is simply two times the average value. Thus,

$$\omega_{max} = 2 \times 26.1 = 52.2 \text{ rad/sec} \tag{2-87}$$

Or, the peak angular velocity is 8.31 revolution per second or 200 steps per second.

The angular acceleration, which is assumed to be constant, is

$$a = \omega_{max}/0.005 = 10440 \text{ rad/sec}^2 \tag{2-88}$$

or 40,000 steps/sec^2.

The acceleration torque to be supplied by the prime mover is given by

$$T = Ja + T_L \tag{2-89}$$

where J is the total inertia of the load and the prime mover. Let the inertia of the prime mover be 0.006 oz-in-sec^2. Then Eq. (2-89) gives

$$T = 0.006(10440) + 10 = 72.6 \text{ oz-in} \tag{2-90}$$

Normally, in selecting a prime mover it is necessary to include some safety factor so that a peak torque of 100 oz-in should be considered.

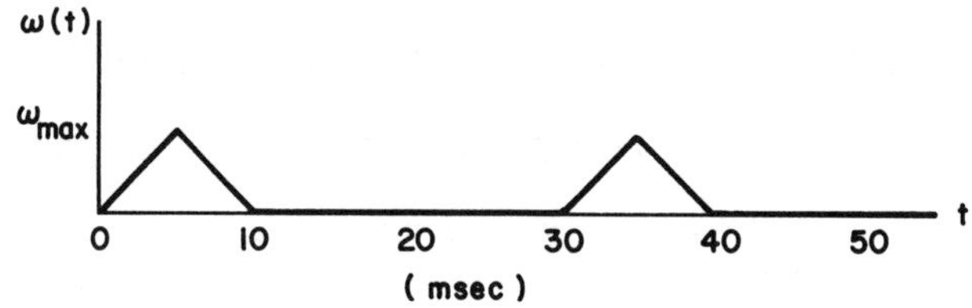

Fig. 2-31. Velocity response of a printer.

3 Sensors And Encoders

B. C. KUO

3.1 INTRODUCTION

Sensors and encoders are important system components in incremental motion control systems. Systems with dc motors are almost always closed-loop systems, and encoders are used to feedback the motor position and/or speed for comparison with the references. Closed-loop step motor systems require encoders to generate pulse command for continuous operation. Even under open-loop operation, step motors often have encoders coupled to their shafts for step confirmation and final position corrections.

In this chapter some of the sensors and encoders that are frequently used in incremental motion control systems and devices are described. Their principles of operation and applications are discussed. These devices include potentiometers, optical incremental encoder, magnetic pickup, electromagnetic transducers, and resolvers.

3.2 POTENTIOMETERS

Potentiometer is an electromechanical transducer which converts mechanical into electrical energy. The input to the device is in the form of a mechanical displacement, either linear or rotational. When a voltage is applied across the fixed terminals of the potentiometer, the output voltage, which is measured across the variable terminals, is proportional to the input displacement either linearly or according to some nonlinear relation.

Rotary potentiometers are available commercially in single-revolution or multi-revolution form. Some of the potentiometers have limited motion such as one or more revolutions, and some have unlimited mechanical motion. They commonly use wire-wound or conductive plastic resistance elements. Figure 3-1 shows the photo of a rotary potentiometer, and Fig. 3-2 shows a linear potentiometer, which also has a built-in operational amplifier. For precision control of incremental motion, the conductive plastic potentiometer is more preferable, since it has infinite resolution, long rotational life, good output smoothness, and low static noise.

Typical resistance tolerance of standard units is between ±5 percent and ±10 percent, and special units can have a tolerance of ±1 percent. The linearity of the resistance is in the range of ±0.25 to ±0.5 percent.

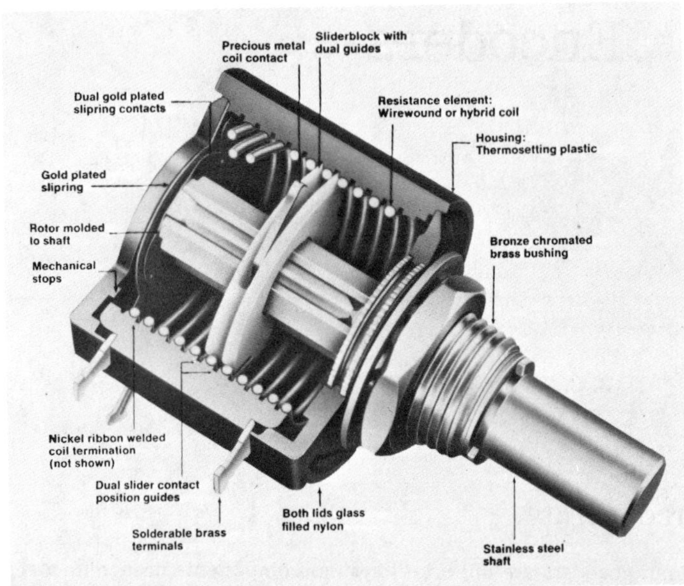

Fig. 3-1. A ten-turn rotary potentiometer (Courtesy of Helipot Division of Beckman Instruments, Inc.).

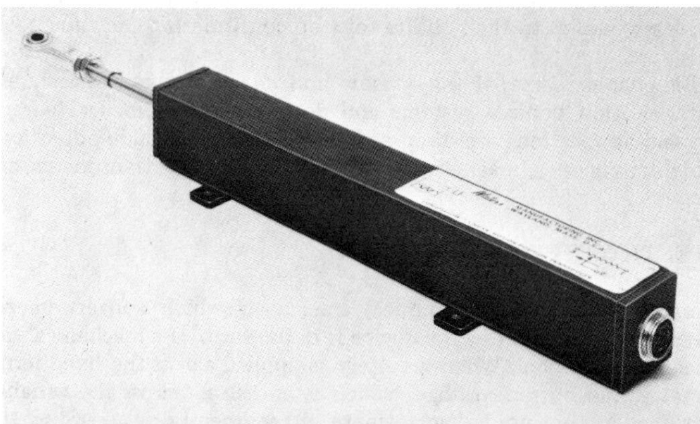

Fig. 3-2. A linear motion potentiometer with built-in operational amplifier (Courtesy of Waters Manufacturing, Inc.).

Figure 3-3 shows the equivalent circuit representation of a potentiometer, linear or rotary. Since the voltage across the variable terminal and reference is proportional to the shaft displacement of the potentiometer, when a voltage is applied across the fixed terminals, the device can be used to indicate the absolute position of a shaft position or the relative position of two shaft positions.

Figure 3-4(a) shows the arrangement when the housing of the potentiometer is fixed at reference, the output voltage e(t) will be proportional to the shaft position $\theta_c(t)$. For a rotary potentiometer $\theta_c(t)$ represents angular displacement in radians or degrees. Then,

$$e(t) = K_s\theta_c(t) \tag{3-1}$$

where K_s is a proportional constant. For an N-turn potentiometer, the total displacement of the variable arm is $2\pi N$ radians. The proportional constant K_s is given by

$$K_s = V/2\pi N \quad \text{volts/rad} \tag{3-2}$$

where V is the magnitude of the reference voltage applied to the fixed terminals. A more flexible arrangement is obtained by using two potentiometers connected in parallel, as shown in Fig. 3-4(b). This arrangement allows the comparison of two remotely located shaft positions. The output voltage is taken across the variable terminals of the two potentiometers, and is given by

$$e(t) = K_s[\theta_1(t) - \theta_2(t)] \tag{3-3}$$

Figure 3-5 illustrates the block diagram representations of the arrangements in Fig. 3-4.

In dc motor control systems potentiometers are often used for closed-loop position control. Figure 3-6 shows the schematic diagram of a typical dc motor

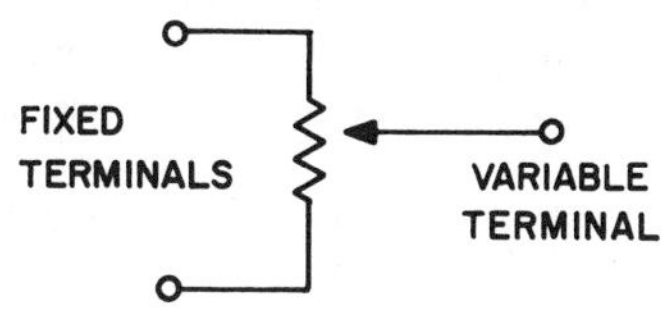

Fig. 3-3. A circuit representation of potentiometer.

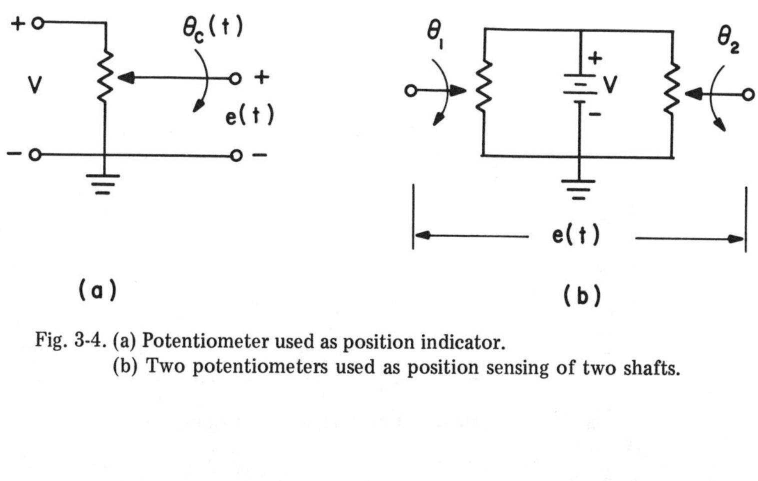

Fig. 3-4. (a) Potentiometer used as position indicator.
(b) Two potentiometers used as position sensing of two shafts.

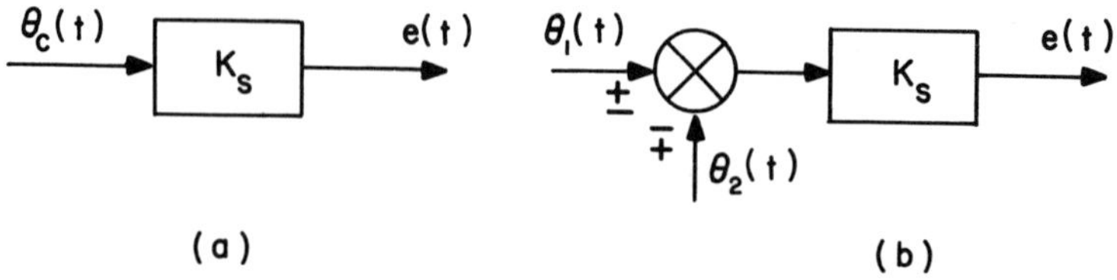

Fig. 3-5. Block diagram representations of potentiometer arrangements in Fig. 3-3.

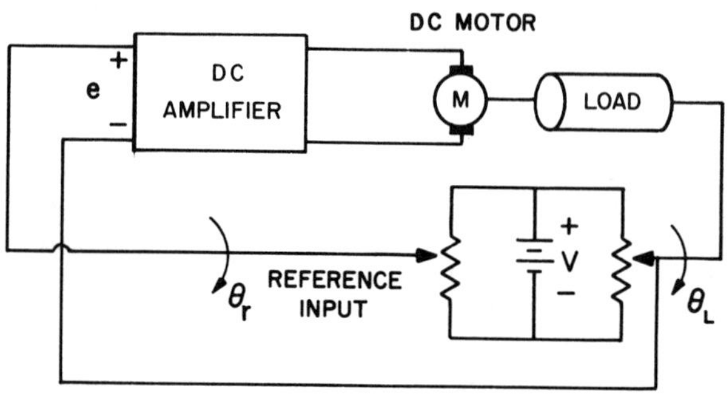

Fig. 3-6. A dc-motor positional control system with potentiometers as error sensor.

position control system. The potentiometers are used in the feedback path to compare the actual load position with the desired reference position. If there is a discrepancy between the load position and the reference, an error signal is generated by the potentiometers which will drive the motor in such a way that this error is minimized quickly.

Precision potentiometers are also commonly used during experimentation with dc motors and step motors for position control. For instance, a potentiometer can be coupled directly to the shaft of a step motor to indicate the step response of the motor. Figure 3-7 illustrates a typical step response of a step motor when the output voltage of the potentiometer is displayed on an oscilloscope.

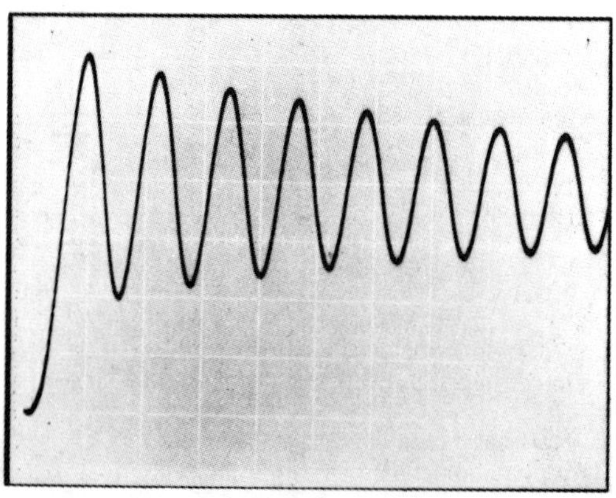

Fig. 3-7. A typical step motor single-step response.

3.3 THE INCREMENTAL ENCODER

Incremental encoders are needed in incremental motion control systems for position confirmation and for feedback signal generation. In general, encoders fall into two main categories, the *absolute* and the *incremental*. In simplest terms the *absolute encoder* provides a distinct digital code indicative of each particular least significant increment of resolution. The *incremental encoder*, on the other hand, provides a pulse for each increment of resolution but does not distinguish which increment. The choice of which type to use rests on economics and operational objectives. For the most part, the need for absolute encoders has much to do with the concern for data loss during power failure or the application involving periods of mechanical motion without the readout under power. With the improvement in reliability of solid-state devices and the expanded application of digital control systems, viability of operation and economics of application have given the incremental encoder increased recognition as a highly desirable element in incremental motion control. In contrast to the absolute device which loses its absolute characteristic if the code format is repeated and which cannot easily float a zero reference point, the incremental encoder's simplicity, economy, and ease of position zero reset provide for digital system design versatility. Therefore, most modern incremental motion control systems use incremental encoders for position confirmation and feedback purposes.

Encoders are available in rotary and linear forms (Figs. 3-8 and 3-9). Since primary actuators are predominantly rotary and since linear motion can be conveniently converted to rotary, and vice versa, by a variety of mechanical couplings, the linear encoder is used only in specific applications where the particular geometry and physics of motion are advantageous and where the code scale length and/or its inherent exposed vulnerability do not present a problem. The rotary encoder, being a totally enclosed device, is more readily applied by the mechanical design engineer, is generally of lower cost, endures severe environments, offers unlimited length of position sensing, and is available in great varieties of packaging and electronic capabilities.

The rotary incremental encoder is defined as a device which produces electrical pulses at equal angular increments of shaft displacement. The purpose of the shaft encoder is to provide a means for incremental position sensing or pulse-rate velocity sensing. Although opto-electronics is the major means for a pulse or code generation, magnetic pick-up and brush contact configurations have played a significant part in the history of encoder development. However, the increasing demand for higher resolution, higher pulse frequencies, greater accuracy and lower cost, coupled with the

Fig. 3-8. A rotary incremental encoder (Courtesy DISC Instruments, Inc.).

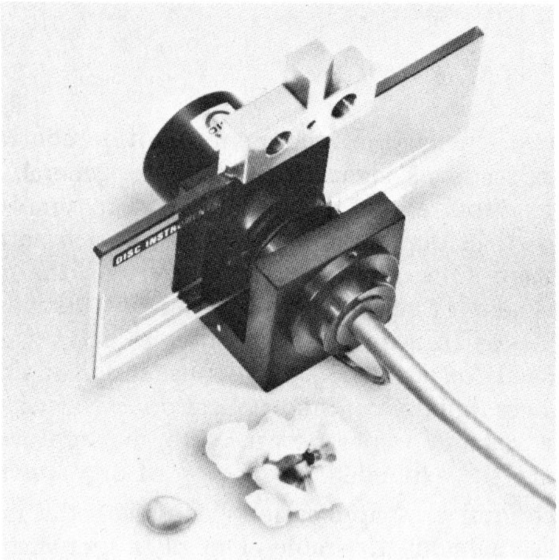

Fig. 3-9. A linear incremental encoder (Courtesy DISC Instruments, Inc.).

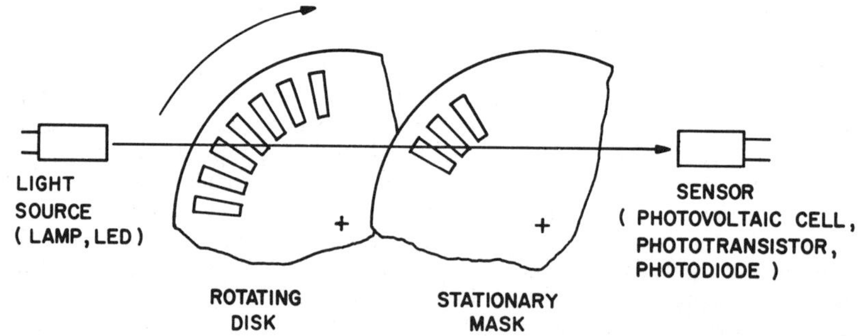

Fig. 3-10. Typical incremental opto-mechanics.

proliferation of solid-state electronics, has dictated the choice of optical disc (or drum) encoding at least for the foreseeable future.

An incremental encoder has four basic parts, a light source, a rotating disk (rotor or commutator), a stationary mask (stator) and a sensor, as shown in Fig. 3-10. The disk has alternate opaque and transparent sectors, any pair of which represents an incremental period. The mask is used to pass or block the light beam between the light source and the photosensor located behind the mask. The inclusion of the mask with a transparent slit corresponding to that of the rotor disk is determined by the size of the photosensitive device and its field of view in relation to the sector width. For encoders with a low resolution, the stator is generally not necessary, but for fine-resolution encoders, a multiple-slit stator is often used to maximize reception of the shutter light, as shown in Fig. 3-11.

The light source could be a light bulb, LED, or a fiber optic bundle.

In practice, the construction of the commutator and mask may be accomplished with glass, mylar, photoplast*, or metal. The glass material provides the best stability and the highest resolution potential. The metal, chemically etched, provides the

*Registered trademark of Eastman Kodak Company.

lowest cost, with mylar and photoplast falling in between glass and metal in both cost and performance. Recent developments in high-production techniques of chrome-on-glass deposition, printing, etching and cutting are diminishing the cost/performance trade-offs of chrome-on-glass versus the lower performance materials.

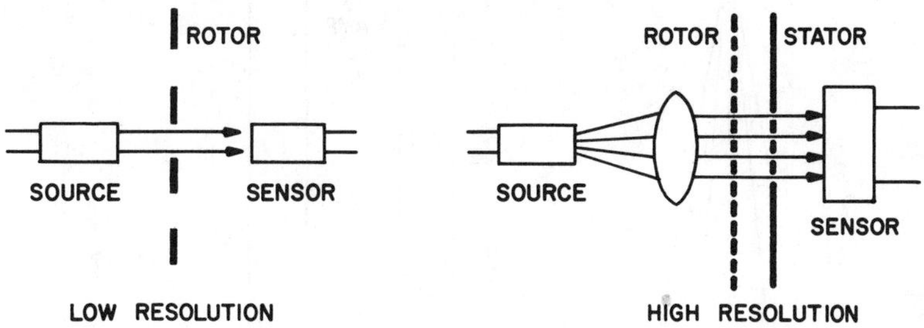

Fig. 3-11. Single versus multi-slit shuttering.

Resolution Of Incremental Encoders

The resolution of an encoder is the number of basic output signal periods presented during one complete shaft rotation. Resolution is specified in pulses per revolution, cycles per 360 mechanical degrees, cycles per degree, etc. Typical commercially available encoders have resolutions that range from 15 pulses per revolution to 10,000 pulses per revolution. The resolution of a linear encoder is expressed in pulses per inch (ppi).

The choice of the optical system, the quality of the code discs, and the mechanical precision of disc rotation all determine the resolution of the encoder. In general, the resolution is directly related to the ability of the encoder to generate the alternating opaque and transparent sectors on the disc and the capability of discerning these sectors. In Fig. 3-10 it is implied that the light passes from the source to the sensor in a straight-line beam, while in fact achieving this is no mean feat. A collimated light can be produced by use of lenses or parabolic reflectors. The need for collimation is illustrated in Fig. 3-12. In practice, there must be a gap between the disc and the mask and, as illustrated, angled light detracts from the shuttering efficiency. If the gap should vary during disc rotation, the shuttering will change and possibly be manifested in the output signals. Assuming a minimum practical gap of 0.001 inch between the disc and mask, it is clear that for resolutions and disc diameters which combine to produce sector widths of less than 0.001 inch the tolerability on non-collimation decreases rapidly.

The number of slots or windows (transparent) on the disk is equal to the resolution in pulses per revolution (ppr) of the encoder. In other words, the number of opaque and transparent sectors is equal to the incremental period of the encoder output. A majority of the incremental encoders available commercially have equal widths in the opaque and transparent sectors. Under this condition the width of the slot of the code pattern on the encoder disk is readily calculated from the formula:

$$W = \frac{\pi D}{2N} \tag{3-4}$$

where

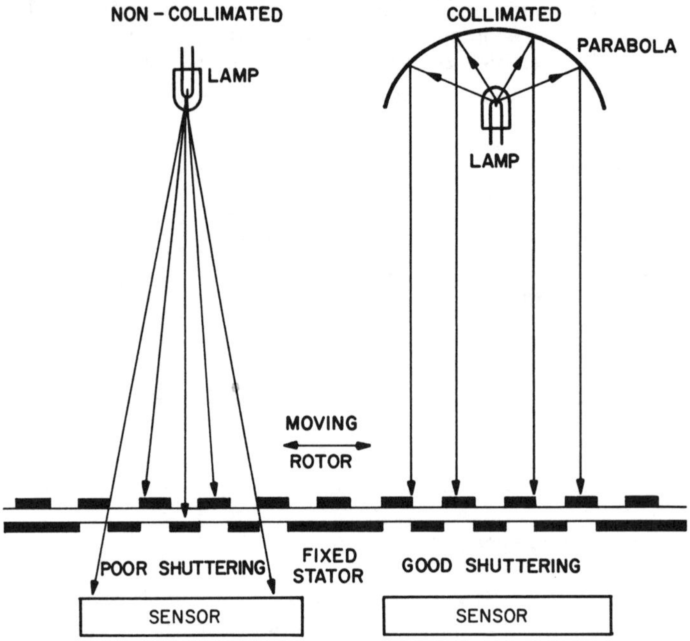

Fig. 3-12. Collimation/non-collimation effects.

D = mean diameter of code pattern

N = resolution of code disk in ppr

These parameters are illustrated in Fig. 3-13. A close approximation can be made by using D as the outside diameter of the encoder disk; then Eq. (3-4) is approximated by

$$W = \frac{0.75\pi D}{2N} \tag{3-5}$$

For example, a 2-inch disk with a resolution of 2000 ppr requires a slot width of 0.002356 inch. A practical limit on the slot width is 10^{-3} inches.

Accuracy And Frequency Modulation Of Incremental Encoders

The accuracy of an incremental encoder is totally unrelated to its resolution, and is a measure of its ability to define the location of any pulse with respect to any other pulse within the resolution chosen. In general, accuracy is specified in degrees, minutes or seconds of arc.

All the parameters mentioned in the last section on resolution also affect the accuracy of an encoder. However, the parameters which contribute to inaccuracy are not confined to one area. Starting with the disk pattern itself we may have accumulative and nonaccumulative sector-to-sector position errors and varying opaque-to-transparent sector ratios. Figure 3-14 illustrates these two imperfect situations.

In addition to these sources of error, the lack of concentricity of sector pattern

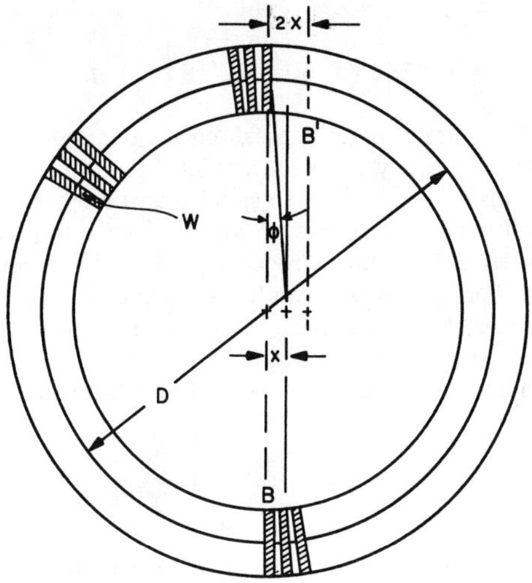

LINEARITY ERROR = 2X / W

WHERE W = πD / RESOLUTION = PERIOD WIDTH

X = ECCENTRICITY (PATTERN TO ROTATIONAL AXIS)

2X = TOTAL PERIOD ERROR IN 180° REVOLUTION

D = MEAN PATTERN DIAMETER

Fig. 3-13. Eccentricity/linearity relationship.

to the axis of rotation will also contribute to inaccuracy. Dimensional and assembly inaccuracies accumulate toward eccentricity, and result in a cyclic nonlinearity.

The generation of optical disk masters and the reproduction techniques employed at this state of the art are such that the code pattern itself is a minor error contributor to encoder inaccuracy. The total error contributed by the commutator disk generally would not exceed 15 arc seconds.

The primary source of encoder error is the eccentricity of the commutator code pattern with respect to the true center of rotation of the shaft being measured. With reference to Fig. 3-13, eccentricity is denoted by X (inches). The angular error due to eccentricity is given by

$$\sin\phi = \frac{X}{D12} = \frac{2X}{D} \tag{3-6}$$

Thus,

$$\phi = \sin^{-1}\frac{2X}{D} \tag{3-7}$$

For example, a code disk pattern with a 2-inch mean diameter and a 0.0005-inch pattern eccentricity, would have an angular error of

$$\phi = \sin^{-1}\frac{0.001}{2} = 0.02865 \text{ degrees} \tag{3-8}$$

or

$$\phi = 0.02865 \times 60 = 1.719 \text{ arc-min} \tag{3-9}$$

Angular accuracies of 2.5 arc-min are typical among commercially available encoders.

Another form of inaccuracy in an incremental encoder is *frequency modulation.* Frequency modulation is typically used to specify accuracy when the encoder is to be used as a tachometer for speed information or feedback. Frequency modulation is defined as the deviation from the center frequency when the encoded shaft is rotated at a constant velocity. When a disk is mounted off center it will have a moving rather than a stationary center, thus imparting a continuous variation in the speed with which its peripheral lines pass the sensor aperture.

It can be shown that the frequency modulation due to eccentricity is approximately given by

$$\text{percent FM} = \frac{2X}{D} \times 100\% \tag{3-10}$$

In practice, frequency modulation specifications of less than 1 percent peak to peak are typical.

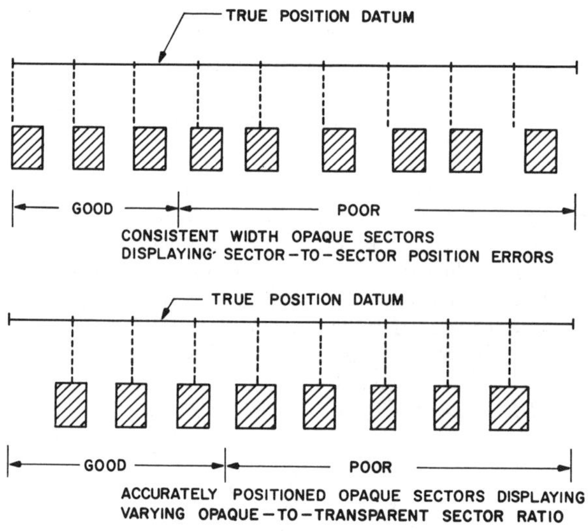

Fig. 3-14. Code pattern errors.

Stability Of The Encoder Output

Stability of the encoder refers to the ability of maintaining the specified accuracy under actual operating conditions.

Stability of the encoder output, while certainly affected by structural movements resulting from temperature excursions or from external deformation forces, is primarily the domain of light source and electronics. A shift in the light source, e.g., filament movement due to shock, can result in degradation of collimation, or a variation in emission pattern due to aging or power supply change can upset the signal level so as to alter the output relationships. The prime causes of electronic inability to preserve the output signal characteristics are temperature and power supply variation. The application of encoders in the industrial environment demands full consideration of these two aspects.

Encoder Outputs

For most purposes the low signal levels obtained directly from the encoder optoelectronic sensors are not adequate for control or signal processing. Therefore, amplifiers and waveform shapers are incorporated into the encoder packages.

The sensor waveforms are generally triangular or sinusoidal depending on the resolution required. Square-wave signals compatible with digital logic are derived by using a linear amplifier followed by a comparator.

Figure 3-15(a) shows a typical rectangular output waveform of a single-channel encoder device. In this case pulses are produced for both directions of shaft rotation. However, the signal can be used only for unidirectional control applications with no direction indication capability.

A dual-channel encoder with two sets of output pulses is necessary for direction sensing and other complex control functions such as the closed-loop acceleration and deceleration control of a step motor*. When the phasing of the two output pulse trains is 90 degrees apart electrically, the two signals are said to be in quadrature, as shown in Fig. 3-15(b). As shown by the signals in Fig. 3-15(b), the two signals uniquely define 0 to 1 and 1 to 0 logic transitions with respect to the direction of rotation of the encoder disk, so that a direction-sensing logic circuit can be easily constructed to decode the signals.

On some incremental encoders a marker pulse or index pulse which may be used as zero reference is provided. This marker provides a once-per-revolution pulse, usually with a specific relationship to the data channels, which is used to reference a mechanical position or clear an accumulated count.

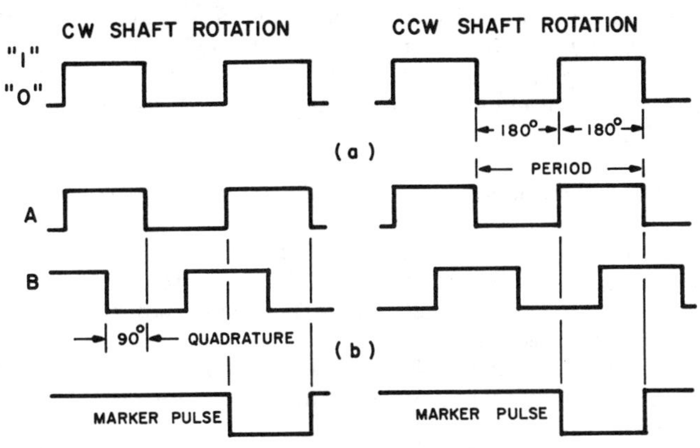

Fig. 3-15. (a) Typical rectangular output waveform of a single-channel encoder device (bidirectional). (b) Typical dual-channel encoder signals in quadrature (bidirectional).

*Refer to Chapter 9, Volume II, Step Motors And Controls.

Light Source And Sensors

The long standing source of illumination for optical encoders has been the incandescent lamp. Except for failures caused by abnormal shock and vibration and by excessive inadvertent overvoltage, the record of lamp reliability has been excellent. Selection of properly designed lamps and their intentional use at derated voltages will give many years of service prior to failure by filament evaporation.

The light emitting diode (LED) is a solid-state device which has been used broadly in the industry. The LEDs are extremely shock and vibration resistant, and under proper use their anticipated life extends far beyond the reasonable use expectancy of most systems being designed today. In addition, infrared LEDs operate in that portion of the spectrum where most photosensitive devices have their peak response.

Three types of photosensitive elements are commonly used in encoders, all are silicon solid-state devices, and the choice of which to use is not a simple matter of device parameters alone. Shuttering technique, electronic circuitry, performance objectives, package design, manufacturability and cost all combine to direct the selection.

The photodiode is operated in the reverse-voltage mode and functions as a conductive change device. The device is characterized by: very high speed of response, low signal level, small sensitive area (typically 20 × 20 mils), and limited package selection.

The phototransistor has become perhaps the most popular light sensing device as evidenced by its availability in low cost packaging. Fundamentally a transistor with a photodiode collector-to-base structure, the phototransistor is characterized by: medium-to-high speed of response, high output signal, wide variation in output level due to gain (h_{fe}) spread, small sensitive area (typically to 60 × 60 mils max.) and essentially standard transistor packaging.

The third device is the photovoltaic cell, commonly referred to as a solar cell. Unlike the two others, this device generates a voltage in proportion to the light energy incident upon it, and, therefore, requires no direct connection to a power source. This feature, in addition to the following characteristics, contributes toward making photovoltaic cell a pragmatic choice for incremental encoders: wide range of cell sensitive area (typically 50 × 50 mils to 1 × 1 inch), good sensitivity uniformity, wide spectral response, no packaging confinements, single chip matched pair availability, choice of voltage or current mode operation, low cost, and medium speed of response.

Practical Design Considerations

In addition to resolution and accuracy, other parameters and variables such as inertia and operating temperature should also be taken into consideration when selecting and designing an incremental encoder. These important design parameters and variables are summarized as follows:

1. *Resolution.* The practical limit of resolution is when the window size of the code pattern becomes less than 5×10^{-4} inches; a more realistic limit is approximately 10^{-3} inches.

2. *Accuracy/Frequency Modulation.* As mentioned previously, accuracy is theoretically unrelated to resolution. In practice, the encoders with high resolutions will be more accurate due to the accuracy to which the code pattern can be centered. Angular accuracies of 2.5 minutes of arc are typical. Frequency modulations of less than 1 percent peak to peak can be expected. Angular accuracies of 30 seconds of arc

or frequency modulation specifications of less than 0.05 percent peak to peak are readily achievable where the code pattern can be centered to an accurately rotating shaft as an integral assembly.

3. *Inertia.* When employing an incremental encoder one cannot entirely ignore the inertia of the rotor of the encoder. The inertia contributed to the system by the incremental encoder will range from 10^{-5} oz-in-sec^2 for an 1.5-inch outside diameter unit up to 1.4×10^{-4} oz-in-sec^2 for 2.5-inch models.

4. *Environmental Variables.* Among all the environmental variables which may affect the system performance, temperature is the most important. The operating temperature extremes dictate the amount of current that can be sent through the light source and the degree of matching required of the differential light sensor pairs. The storage temperature extremes dictate material selection, particularly the materials interface of the glass disk and metal hub. Where storage temperatures below −40°C are expected, the thermal coefficients of expansion between the glass disk and the metal hub must be as closely matched as possible. This is typically accomplished using soda lime glass and 416 stainless steel. The stainless steel hub obviously will increase the disk/hub inertia significantly compared to a standard aluminum or magnesium hub.

Operating temperatures of 0°C to 55°C and storage temperatures of −20°C to 85°C do not require special testing or materials selection and are, therefore, more economical.

As an illustration the specification sheet of an incremental encoder manufactured by Disc Instruments, Inc. is duplicated in Fig. 3-16.

Encoder Kits

Most incremental encoders made commercially come in the form of a sealed unit. The user simply selects the appropriate unit for his application and then makes proper installation by providing the mounting bracket and coupling the encoder shaft to the output shaft of the system. Usually there are no external adjustments necessary or available to the encoder.

There are conditions under which the standard encoders are inconvenient to use. For instance, in the closed-loop control of step motors, it is desirable to be able to adjust the relative positions of the light source and receiver with respect to the windows of the code pattern of the rotor disk so that various switching angle effects may be achieved*. Therefore, some encoder and motor manufacturers have incremental encoders available in kit form. In high-volume systems such as printers and electric typewriters, it is not uncommon to find the original equipment manufacturers (OEM) design and assemble their own encoder system in kit form.

The basic encoder kit includes a simple code disk on a hub and one or two optical encoder assemblies each comprised of a light source and sensor. Sometimes the package also includes an electronic circuit to condition the output signals of the sensors.

As shown in Fig. 3-17, a typical encoder kit makes use of the output shaft and the end bell of the motor it services. The encoder disk is mounted directly on the motor shaft, and the optical assembly is fastened to the end bell of the motor. Provisions can be made for adjusting the relative position of the optical assembly with respect to the window positions of the disk. Additional optical assemblies can be mounted in the same fashion if necessary.

Figure 3-18 illustrates an incremental encoder, TRim-step, manufactured by Trump-Ross Industrial Controls, which can be mounted directly on the end of a motor.

Figure 3-19 describes an optical assembly manufactured by Warner Electric

BULLETIN 860

ROTASWITCH® INCREMENTAL SHAFT ENCODERS

860 SERIES

SIZE 1.5″ MINIATURE

For applications where a modular type encoder is not practical and where optimum use of space is required, Disc Instruments has designed the inexpensive 860 Series ROTASWITCH Optical Shaft Encoder. Its 1.5″ diameter and 1.5″ length allow it to be easily installed in the limited space of high density peripheral equipment. Available in either unidirectional or bidirectional models, the 860 Series ROTASWITCH incorporates the latest in solid state design and will provide years of uninterrupted service. A GaAs Light Emitting Diode source is standard in this model. Size 15 Servo Mount standard.

Aimed at the peripheral market where high reliability and low cost are of primary concern, this particular ROTASWITCH Shaft Encoder is a natural for O.E.M. applications.
Available in various output configurations and a multiple selection of standard pulse rates, this miniature ROTASWITCH Shaft Encoder can be modified to fit virtually any application requirement.
For further information, write or call our Application Engineering Department. They will be more than happy to answer any encoder question you may have.

SPECIFICATIONS

Mechanical

Physical Characteristics . . See reverse side
Weight . . . . 3 ounces max.
Starting Torque at 25°C . . .08 inch-ounces max.
(.8 inch-ounces with sealed bearings)
Running Torque at 25°C . . .07 inch-ounces max.
(0.8 inch-ounces with sealed bearings)
Moment of Inertia . . . . . . 4.0 x 10⁻⁵ inch-ounces sec.² max.
Shaft Rotation . . . . Continuous and reversible
Slew Speed . . . . 5,000 rpm
Shaft . . . . ⅛" or ¼" dia. stainless steel, as specified.
Shaft Loading
⅛" Dia. . . . Axial . . . . 1 lb. max.
Radial . . . 1 lb. max.
¼" Dia. . . . Axial . . . 3 lb. max.
Radial . . 5 lb. max.
Bearings . . . . ABEC Class 5 or better
(Class 7 on special order)
Bearing Life . . . . $\frac{16 \times 10^6}{\text{rpm}}$ = hours min.

Electrical

Code . . . . Incremental
Index . . . . Optional, once per revolution
Pulses Per Revolution . . . . As specified; see Pulse Rate Availability Sheet
Accuracy . . . . ± 2.5 min. of arc standard
Illumination Source . . . . . LED
Sensor . . . . Phototransistor or solar cells
Output Signal
Sine Wave Models . . . . . Data . . 1V P-P min. with 10k load
Index . . 0.25V min. with 10k load

Square Wave Models
TTL Compatible
(vcc = 5 vdc ± 10%) Logic "1" VCC @ 1K ohm.
Logic "0" 0.5V max. @ 10 ma max. sink current.

HTL Compatible
(vcc = 12 vdc or
15 vdc + 10%) . . . . Logic "1" VCC @ 2.2K ohm; 3.3K ohm if 15 vdc. Logic "0" 0.7V max. @ 10 ma max. sink current.

Input Power Requirement
TTL Compatible and
Sine Wave Units . . . . + 5 vdc ± 10%
125 ma max.
HTL Compatible Units . . + 12 vdc or +15 vdc ± 10%
125 ma max.
Operating Speed . . . . Data . . 75kHz
Index . . 15kHz
Operating Temperature . . 0°C to + 55°C
Electrical Connections . . . 18″ min. shielded, jacketed cable

Fig. 3-16. A typical specification sheet of an incremental encoder (Courtesy DISC Instruments, Inc.).

PHYSICAL CHARACTERISTICS

ELECTRICAL CONNECTIONS

Models 861 and 865:
Red . . . . . . VCC*
Black . . . . . Common
White . . . . Output A
Green . . . . Index (Option)
Shield . . . . Floating

Models 862 and 866
Red . . . . . . VCC*
Black. . . . Common
White . . . . Output A
Green . . . . Output B
Orange. . . Index (Option)**
Shield . . . . Floating

* + 5V DC on TTL and Sine Wave Models
+ 12V or + 15V DC on HTL Models
** Index not available on Model 862

SHAFT	DIA X
A	.1247 +.0000 (3.167) −.0002
B	.2497 +.0000 (6.335) −.0002

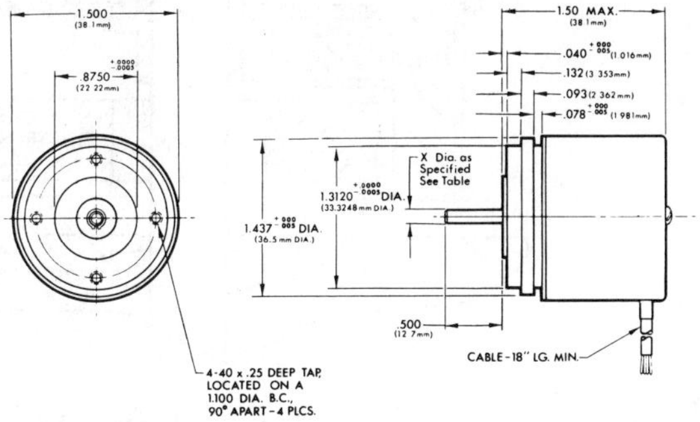

OUTPUT CONFIGURATIONS (Outputs shown CW rotation)

SQUARE WAVE

50/50 ± 10% DUTY CYCLE

861

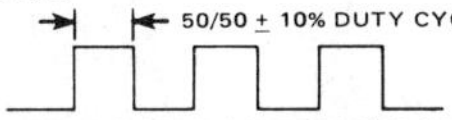

SINGLE OUTPUT, SQUARE WAVE, TTL COMPATIBLE

90 ± 45 ELECTRICAL PHASING

862

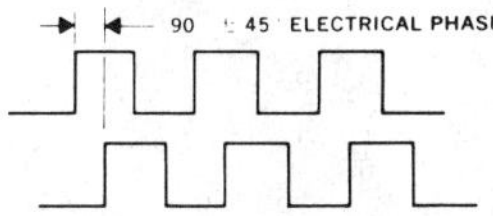

QUADRATURE OUTPUT, SQUARE WAVE, TTL COMPATIBLE

SINE WAVE

865

SINGLE OUTPUT, SINE WAVE

90° ± 45° ELECTRICAL PHASING

866

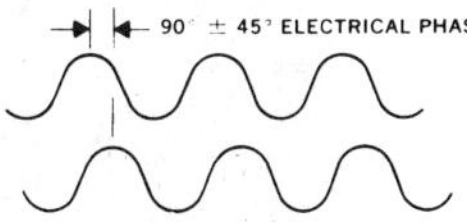

QUADRATURE OUTPUT, SINE WAVE

MODEL DESIGNATION

Model 862 S — 500 — O A L S — TTL

*Basic Model

Pulses Per. Rev. (As specified)

O — No Index
I — Index ** (optional)

Shaft Diameter
A — .1247 +.0000 −.0002
B — .2497 +.0000 −.0002

Source
L — LED

Sensor
P — Phototransistor
S — Solar Cell

Output (Sq. Wave Models)
TTL Compatible
HTL Compatible (optional, specify + 12 or + 15 vdc)

*861 . . . . Single square wave output
862 . . . . Dual square wave output
**Index pulse not available on Model 862

865 . . . . Single sine wave output
866 . . . . Dual sine wave, quadrature output

ORDERING INFORMATION

Specify:
1. Model number
2. "S" for Servo Mount (standard)
3. Pulses per revolution
4. Index or No Index (optional)
5. Shaft diameter ("A" or "B")
6. Source (L.E.D.)
7. Sensor (solar cell or phototransistor)
8. Output (TTL or HTL, if square wave model. If HTL, specify +12 or +15 vdc)

REPRESENTED IN YOUR AREA BY:
(REFERENCE ENCLOSED REPRESENTATIVE LIST.)

DISC INSTRUMENTS, INC. • 102 East Baker Street • Costa Mesa, California 92626 • Phone (714) 979-5300 • TWX 910-595-1987 DISC CSMA

377-9-7.5OCL PRINTED IN U.S.A.

Fig. 3-16. (continued)

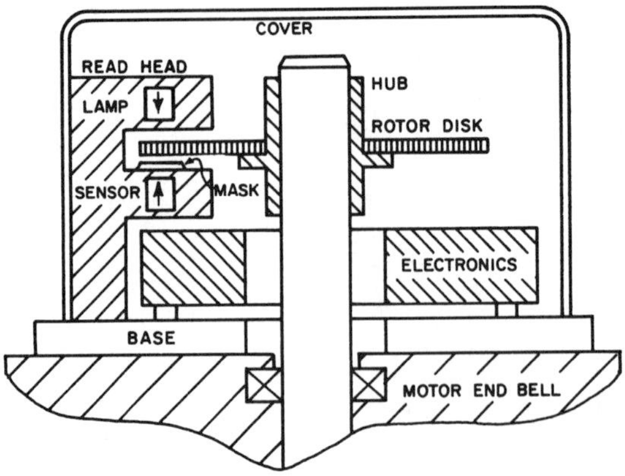

Fig. 3-17. A typical encoder kit.

Fig. 3-18. Standard TRim-step encoder mounted on a step motor (Courtesy Trump-Ross Industrial Controls).

Brake and Clutch Company. Two of these units can be used in conjunction with an encoder disk for bidirectional position confirmation and/or closed-loop acceleration and deceleration control of step motors.

3.4 RESOLVERS AS INCREMENTAL ENCODERS

Resolvers have been widely used in control systems. Typically, resolvers may be used for transformation of coordinates, error detection, shifting of time-phase, and data transmission.

The conventional resolver is an iron-core device. The schematic diagram of a four-winding type resolver is shown in Fig. 3-20. The stator has a two-pole arrangement; the windings on these two poles are electrically independent and are separated by 90 degrees in space phase. The two rotor windings are spaced 90 mechanical degrees apart, as shown in Fig 3-20. A three-winding type resolver would have only

MOTION CONTROL SYSTEMS

TECHNICAL DATA BULLETIN

OPTICAL SWITCH

MCS–1600-1

MCS–1600-2

DESCRIPTION

The MCS-1600 series optical switches are completely solid state sensing devices with essentially unlimited life. The light source is a light emitting diode which produces a narrow, high intensity infrared beam across the sensing slot. The beam is detected by a phototransistor which activates the Schmitt Trigger. Internal hysterisis prevents output jitter (switch bounce) even if the switch is activated by a very slowly moving object. The output is compatible with DTL/TTL logic systems. Output is short circuit proof to ground.

MCS-1600-1:

Designed to be used in systems where electrical symmetry is important such as dual shaft position encoders for step motors, etc.

MCS-1600-2:

Designed for uses where electrical symmetry is not a critical factor such as single pickup step motor shaft encoders, zero position detectors, etc.

TYPICAL APPLICATIONS

Precision Switch	(Output switches high when opaque material is passed across sensing slot. Repeatability ±.0025")
Tachometers or Rotary Transducer	(High speed, high accuracy angular transducer. Constant pulse amplitude down to 0 RPM. Transducer disc to be supplied by user. Suggested dimensions per Fig. 1)
Level Sensing	
Counters	
Position Sensing	

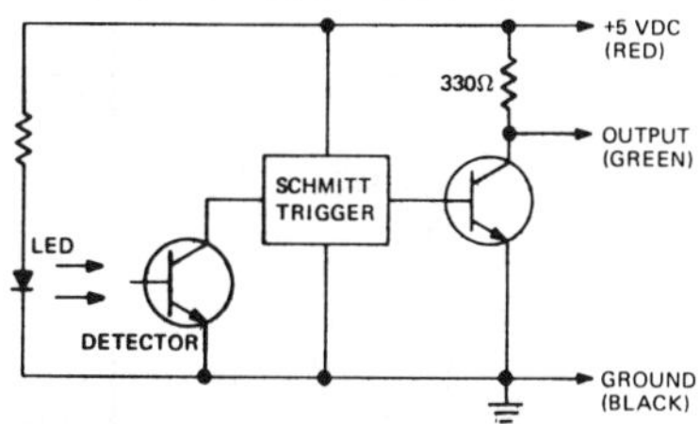

SIMPLIFIED CIRCUIT – OPTICAL SWITCH

SPECIFICATIONS

ELECTRICAL:

Input Power: +5V DC ±5% @ 100 ma
Output: Detector light - 0.5V DC max. with 12 ma sink current
Detector dark - 3.4V DC min. with 4 ma load
Pulse Rise and Fall Times: 0.2 μsec typical
Pulse Rate: up to 25,000 pulses per second
Temperature Range:
Operating: -20 °C to +55°C
Storage: -45°C to +100°C
Electrical Symmetry (when used with transducer disc as shown in Fig. 1):
MCS-1600-1: ±7%
MCS-1600-2: ±12%

MECHANICAL:

Optical Window	- .035" X .110"
Dimensions	- See Fig. 3
Supplied Cable Length	- 13"
Mounting	- 2 holes spaced .578" for a # 2-32 thread cutting screw

WARNER ELECTRIC BRAKE & CLUTCH COMPANY Beloit, Wisconsin 53511

Fig. 3-19. Specifications of optical encoder (Courtesy Warner Electric Brake and Clutch Company).

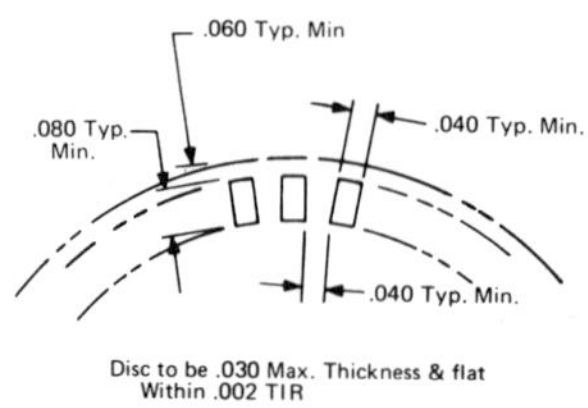

FIG. 1 – RECOMMENDED TRANSDUCER DISC DIMENSIONS

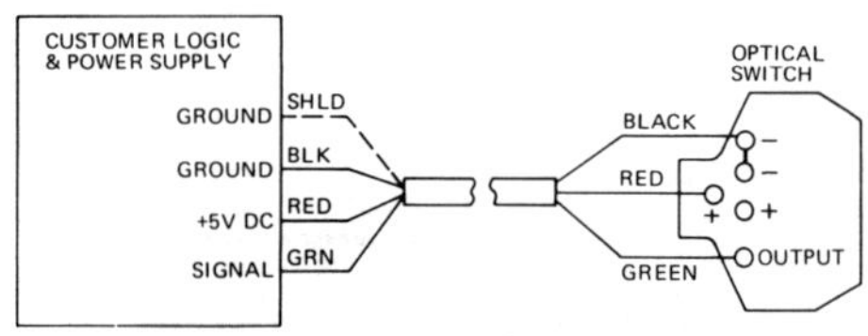

FIG. 2 – TYPICAL CONNECTIONS

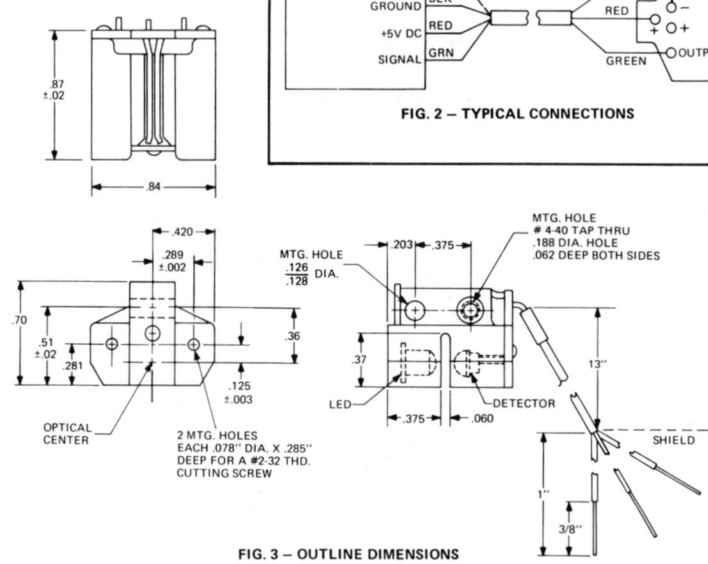

FIG. 3 – OUTLINE DIMENSIONS

P-525

Printed in U.S.A.

Fig. 3-19. (continued)

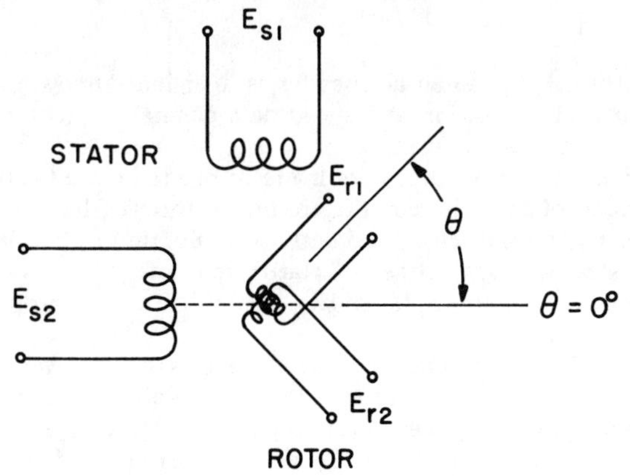

Fig. 3-20. Schematic diagram of a four-winding resolver.

one winding on the rotor.

When constant voltages E_{s1} and E_{s2} are applied to the two stator windings, the voltages induced on the rotor windings are:

$$E_{r1} = E_{s1}\cos\theta + E_{s2}\sin\theta \tag{3-11}$$

$$E_{r2} = E_{s1}\cos\theta - E_{s2}\sin\theta \tag{3-12}$$

Figure 3-21 shows that the voltages across the rotor windings are sum and difference of cosine and sine of the rotor displacement and are periodic with respect to the rotor displacement θ. Furthermore, the two rotor voltages are 90 degrees apart and are said to be in *quadrature*. It can be seen that the set of two rotor voltages completely defines the rotor position. The two quadrature voltages can also be used to determine the direction of rotation. However, for position control purposes, each rotor voltage produces on two null points every revolution or 360 degrees of the rotor. Thus the conventional resolver, having a small number of poles, can be used only for position nulling in control systems.

For incremental motion control and sensing, a large number of poles are needed on the stator, which is very difficult to accomplish with iron-core devices.

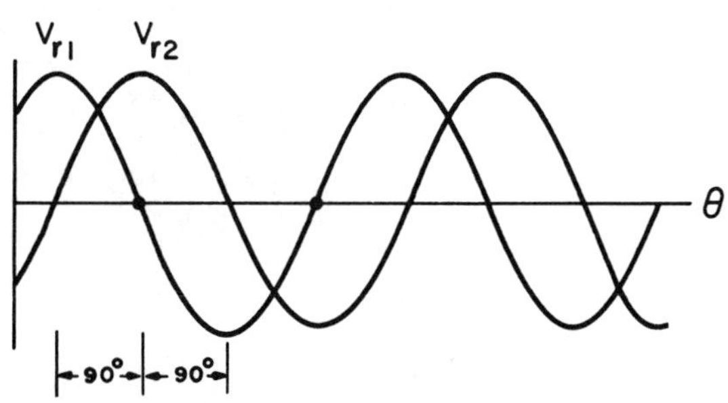

Fig. 3-21. Rotor voltages of a four-winding resolver.

The Inductosyn [4]

The Inductosyn* incremental encoder is a high-accuracy multipole transducer which may be used for position and speed data generation. The Inductosyn is available in both linear and rotary form.

The stator and rotor windings are made of printed circuits on an insulated layer bonded to metallic plates. The circuits are in the form of hairpin turns and may be printed on base materials such as stainless steel, aluminum, beryllium and titanium. Figure 3-22 illustrates a typical set of stator and rotor disk assemblies of a linear Inductosyn encoder. Figure 3-23 shows the rotor and stator disks of a rotary Inductosyn.

The rotor pattern of single-coil disk consists of an even number of series-connected conductors with hairpin turns. Each conductor of the rotor pattern defines one electrical pole, and two poles define one electrical cycle. For example, a pattern containing 256 poles corresponds to 128 cycles. This means that in association with the appropriate stator as the rotor rotates 360 degrees the device provides 128 cycles. In other words, each cycle corresponds to 2.8125 degrees.

Because of the printed-circuit construction, it is possible to provide a large number of poles. It is common to find commercial units with 256, 360, 512, 720, and 2048 poles.

Figure 3-24 shows the layout of the stator and rotor disks of an Inductosyn. The stator has an eight-segment winding, with alternate segments connected together to form two groups of four. The segments of one group are displaced from the other group by a distance of one-half a winding width or pole width. This displacement is

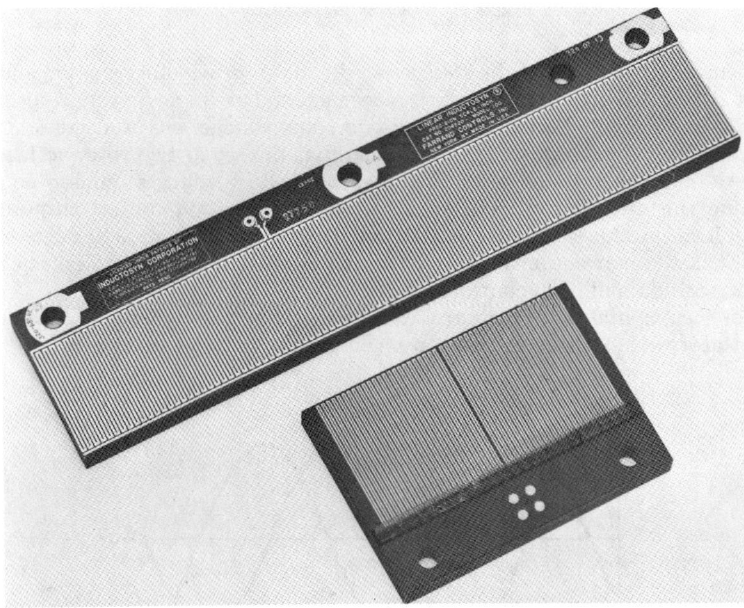

Fig. 3-22. Stator and rotor disk assemblies of a linear Inductosyn encoder (Courtesy Farrand Controls).

*Registered trademark of Inductosyn Corporation.

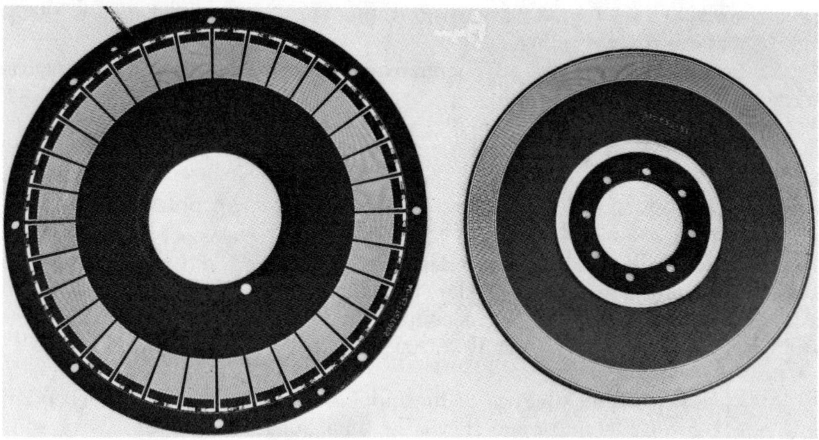

Fig. 3-23. Stator and rotor disk assemblies of a rotary Inductosyn encoder (Courtesy Farrand Controls).

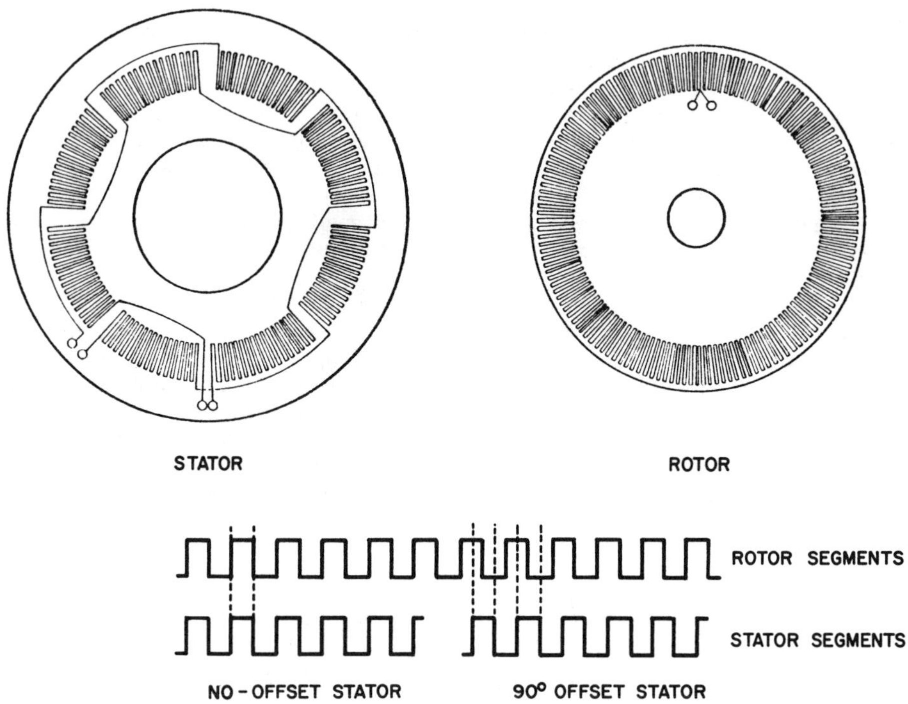

Fig. 3-24. Typical layout of the stator and rotor disks of an Inductosyn.

equivalent to a 90-degree phase difference. The rotor has one symmetrical winding. The bottom part of Fig. 3-24 illustrates the phase relations between the stator windings and the rotor winding.

When the stator windings are energized with constant voltages, the voltage across the rotor is of the form:

$$e_r = K(\cos N\theta/2 + \sin N\theta/2) \tag{3-13}$$

where K is a proportional constant and N is the number of poles of the stator. The periodic signal represented by Eq. (3-13) will have N/2 cycles per revolution, and two null points per cycle. Thus, the encoder output will have N detectable null points per revolution. As an incremental encoder, the null points can be used for the definition of distinct angular positions. When coupled to a null-detection circuit, the encoder output can also be used to compute the average speed of a rotating shaft over a given time period.

Another way of making use of the Inductosyn as an incremental encoder is to apply equal-amplitude sinusoidal signals in time quadrature to the stator windings. The rotor output is then a constant-amplitude sinewave signal having a time phase angle that changes linearly as a function of the rotor position.

3.5 MAGNETIC PICKUPS AS ENCODERS

Magnetic pickups provide a practical and economical way of motion sensing in numerous industrial applications.

The magnetic-pickup encoder consists of essentially a pickup transducer mounted in close proximity to a metal discontinuity, such as sprockets, keyways, cams, bolt or rivet heads or a toothed wheel. A toothed wheel is perhaps the most common metal discontinuity used in a magnetic-pickup transducer, since it is readily available, and it is known to give predictable performance.

Figure 3-25 shows the physical layout of a magnetic-pickup encoder. The air gap between the pickup and the gear teeth should be as small as possible, typically 0.004 to 0.005 inches.

Figure 3-26 shows the internal construction of a typical magnetic pickup. The basic components of the pickup include a permanent magnet, a pole piece, and a coil, all enclosed in a housing. The permanent magnet sets up a magnetic flux path through the pole piece and the outside air space. When an external metal discontinuity approaches the tip of the pole piece, the magnetic field contracts which induces a voltage in one direction in the coil. When the metal piece moves away from the tip of the pole piece, the magnetic field expands, thus inducing a voltage in the opposite direction in the coil. Figure 3-27 illustrates a typical output voltage waveform when a

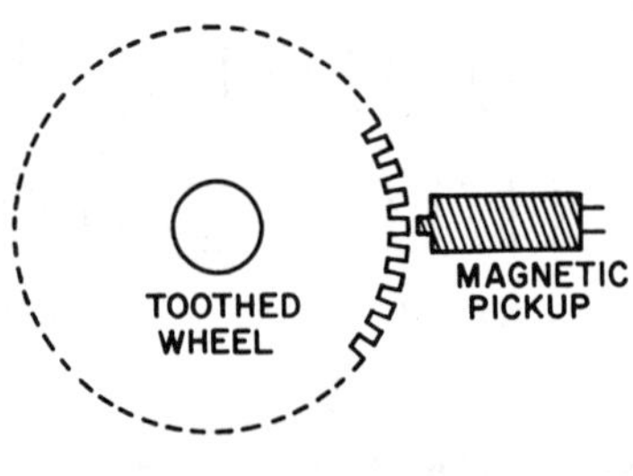

Fig. 3-25. Physical layout of a magnetic-pickup encoder.

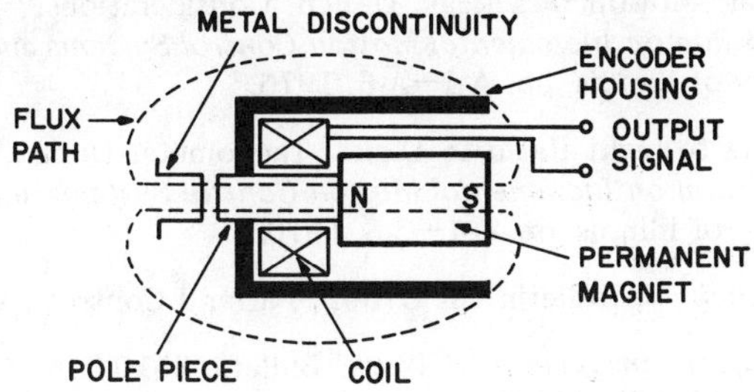

Fig. 3-26. Internal construction of a typical magnetic pickup.

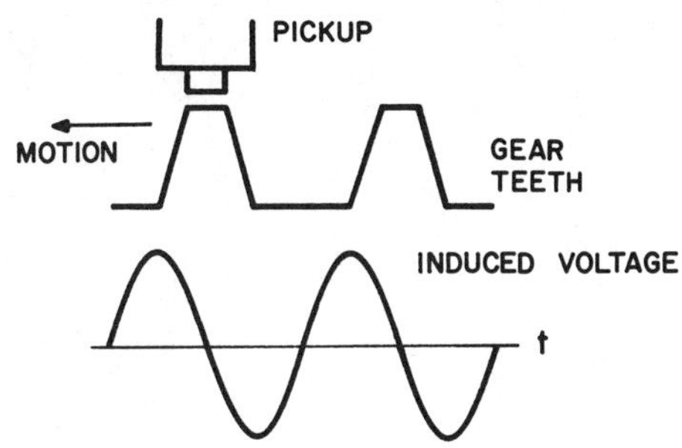

Fig. 3-27. Typical output voltage waveform of a magnetic pickup sensor.

sequence of gear teeth passes under the pole. The passage of one tooth pitch induces one cycle of the ac voltage. Therefore, if 100 gear teeth pass the pole piece in one second, a voltage signal of 100 Hz is produced at the output terminals of the encoder. Since the induced voltage of the encoder has positive and negative peaks and zero crossings which correspond to distinct positions of a rotary shaft, the encoder can be used for incremental motion detection and other motion control and sensing purposes. Most of the anti-skid control systems for trailer and truck applications utilize magnetic-pickup encoders for wheel speed computation.

It should be pointed out that the induced voltage waveform of a magnetic pickup does not always resemble a sinusoidal signal. In general, the waveform depends greatly on the shape of the metal discontinuity. Depending on the particular application the shape of the induced output voltage waveform may not be important. For more details on the selections and applications of magnetic-pickup transducers, the reader may refer to the literature [5].

REFERENCES

1. P. A. Lajoie, "The Incremental Encoder—An Optoelectronic Commutator," *Proceedings, Second Annual Symposium on Incremental Motion Control Systems and Devices*, ed., B. C. Kuo, University of Illinois, pp. C-1—C-39, 1973.

2. M. D. Borman, "Optical Shaft Angle Encoders For The Measurement And Control

of Incremental Motion—Practical Design Considerations," *Proceedings, Fifth Annual Symposium on Incremental Motion Control Systems and Devices*, ed., B. C. Kuo, University of Illinois, pp. A-1—A-6, 1976.

3. P. Y. Hu, "Dark-to-Light Ratio In Digital Tachometer Design," *Proceedings, Fifth Annual Symposium on Incremental Motion Control Systems and Devices*, ed., B. C. Kuo, University of Illinois, pp. C-1—C-6, 1976.

4. *Inductosyn Principles*, Bulletin ER 67301, Farrand Controls, Valhalla, New York.

5. *Magnetic Pickups (transducers) At Work*, Bulletin 8510-674, Airpax Controls Division, Ft. Lauderdale, Florida.

4 Feedback Control System Theory

B. C. KUO

4.1 INTRODUCTION

Most incremental motion control systems have feedback effects in their operations in one way or another. When a dc motor system is devised for the control of position or velocity, feedback is almost always required to insure accuracy. Control systems with feedback are referred to as *closed-loop control systems;* their analysis and design are governed by control systems theory.

Control systems theory directly related to the subject of incremental motion control may be found in many standard textbooks [1]. As a convenience to the reader, an introductory treatment of control systems theory is given in this chapter. Due to the limitation of space, only the subjects directly related to the topics discussed in these volumes will be covered.

Control systems can generally be classified as *open-loop control systems* and *closed-loop control systems.* Figure 4-1 illustrates the basic building block of an open-loop control system. The system consists essentially of a controller and a controlled process. The controlled process generally includes such elements as the primemover, the load and the power amplifier, and the controller acts as a brain which executes the input signal in some prescribed manner. For example, Fig. 4-2 shows the schematic diagram of a dc-motor control system. In this case, the power amplifier, the motor, the gear train, and the load are considered to be the controlled process. The pulsewidth modulator which controls the time duration and polarity of the current in the motor windings is considered as the controller. The output of the system may be the load position or velocity depending on which is the controlled variable.

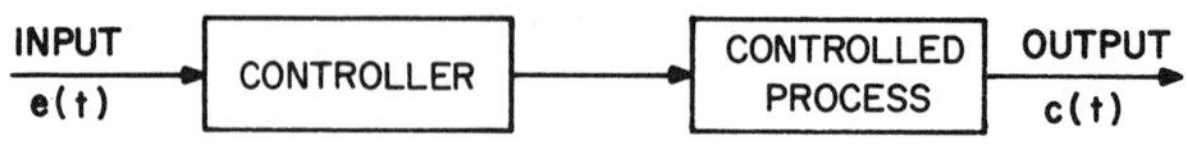

Fig. 4-1. Block diagram of an open-loop control system.

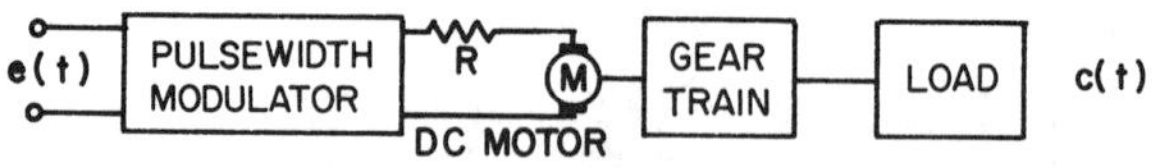

Fig. 4-2. An open-loop dc-motor control system.

The open-loop system shown in Fig. 4-2 suffers from the characteristic that the output c(t) depends strictly on the input e(t). The only way to arrive at the desired output is to *estimate* the proper input e(t). Thus, an open-loop system cannot furnish accurate and reliable performance that is usually called for in commercial and industrial applications.

Figure 4-3 shows the block diagram of a closed-loop control system which utilizes a sensor to detect errors between the reference input and the output. This way, the system is designed so that the input to the controller, e(t), is regarded as an error which must be reduced to zero as quickly as possible. Analytically, the error sensor performs the operation,

$$e(t) = r(t) - c(t) \tag{4-1}$$

With reference to the dc-motor control system shown in Fig. 4-2, if the controlled variable is to be the speed of the load, we can feedback the speed through a speed measuring device, such as a tachometer or an encoder with a speed computing circuit, and compare the measured speed with the reference speed. At any given instant of time, the error in speed is sent to the pulsewidth modulator to control the dc motor, as shown in Fig. 4-4. Thus, when the measured speed is less than the desired speed, the error signal will be of such magnitude to produce a proportional current pulse to the motor to increase its speed. If the output speed is too high, the pulsewidth will be proportionally reduced.

Figure 4-5 illustrates the block diagram of a closed-loop step-motor control system. The reference input signal is in the form of a voltage. The error signal, e(t), is a voltage which is converted by the voltage-controlled oscillator into a pulse train to drive the step motor. The step motor would respond to the magnitude and the sign of the error and rotate in the proper direction until the error is nulled.

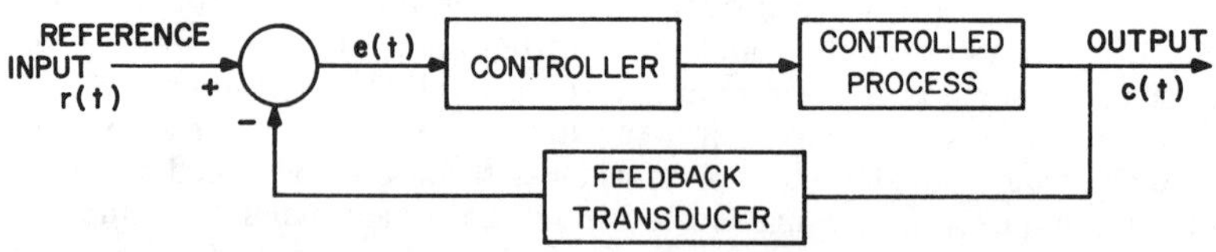

Fig. 4-3. Block diagram of a closed-loop control system.

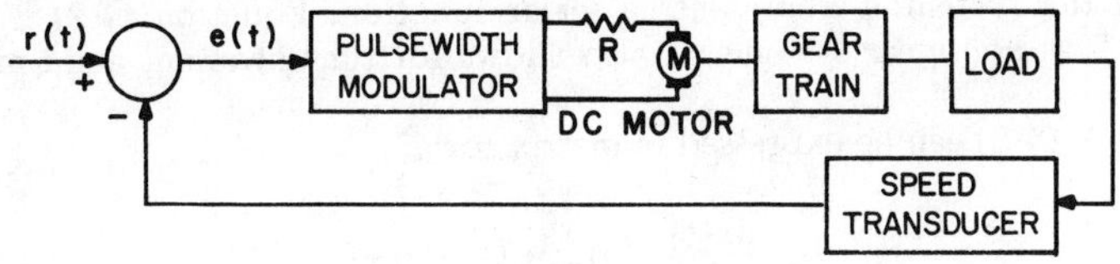

Fig. 4-4. A closed-loop dc-motor speed control system.

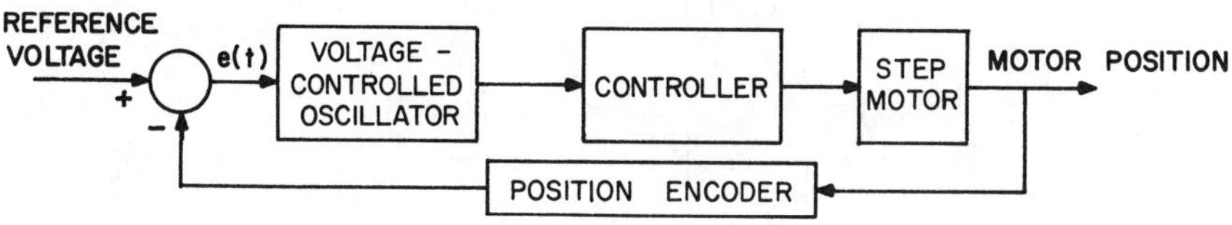

Fig. 4-5. A closed-loop step-motor control system.

4.2 MATHEMATICAL MODELING OF CONTROL SYSTEMS

In order to understand how a control system works and to carry out rigorous analysis and design of control systems, it is important to establish the mathematical model of the system. For instance, in order to understand how a dc motor may work in a control system, it is important to first establish the mathematical relations between the input voltage, current, magnetic flux in the air gap, torque, velocity and position, of the motor. Experience has shown that the design of a control system can seldom be accomplished without any analytical and/or computer simulation studies.

The equations that govern the dynamics of a control system are usually algebraic equations, such as Eq. (4-1), and integral-differential equations. These equations form the basis of the mathematical modeling of a dynamical system. Since most physical systems consist of nonlinear effects of one form or another (saturation, hysteresis, backlash, dead zone, etc.), the modeling equations are generally nonlinear. In Chapter 2 the integral-differential equations of mechanical elements used in control systems have been discussed.

For analysis and design purposes, integral-differential equations are difficult to handle. The control systems engineer has found that a system of first-order differential equations are more convenient for analysis and design purposes. These first-order differential equations are also called the *state equations*. The state equations can generally be applied to linear systems as well as nonlinear systems. When the system is considered to be linear, or is approximated by a linear model, the *transfer function* concept can be used.

State Equations And Output Equations

In general, the state equations of a nth-order system can be written as

$$\frac{dx_i(t)}{dt} = f_i(x_1, x_2, \ldots, x_n, u_1, u_2, \ldots, u_p, t) \tag{4-2}$$

$i = 1, 2, \ldots, n$, where $x_i(t)$ denotes the ith *state variable* and $u_1, u_2, \ldots, u_p$ are the p inputs of the system; f_i represents a scalar function. Equation (4-2) is valid for linear as well as nonlinear systems. It also allows for time-invariant and time-varying elements.

Equation (4-2) can be expressed in matrix form,

$$\frac{d\mathbf{x}(t)}{dt} = \mathbf{f}(\mathbf{x},\mathbf{u},t) \tag{4-3}$$

where

$$\mathbf{x}(t) = \begin{bmatrix} x_1(t) \\ x_2(t) \\ \vdots \\ x_n(t) \end{bmatrix} = n \times 1 \text{ state vector} \tag{4-4}$$

$$\mathbf{u}(t) = \begin{bmatrix} u_1(t) \\ u_2(t) \\ \vdots \\ u_p(t) \end{bmatrix} = p \times 1 \text{ input vector} \tag{4-5}$$

As an illustrative example, consider the armature circuit of a dc motor shown in Fig. 4-6. The differential equation of the circuit is

$$v(t) - v_b(t) = R_a i(t) + L_a \frac{di(t)}{dt} \tag{4-6}$$

where v(t) is the applied voltage and $v_b(t)$ is the back emf.

Defining i(t) as the state variable x(t), Eq. (4-6) is rearranged to give

$$\frac{dx(t)}{dt} = -\frac{R_a}{L_a} x(t) + \frac{1}{L_a}[v(t) - v_b(t)] \tag{4-7}$$

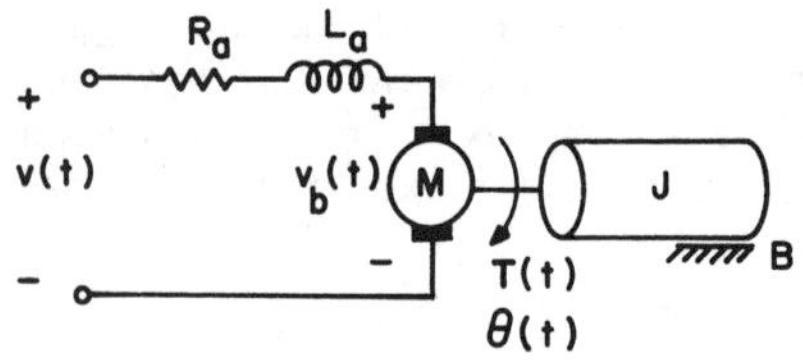

Fig. 4-6. A dc-motor system.

Thus we can identify the last equation with the state equation form in Eq. (4-3). In the present case, the system is linear and the state equation of Eq. (4-7) is linear. In general, for linear time-invariant systems, the state equations are written as

$$\frac{dx(t)}{dt} = Ax(t) + Bu(t) \tag{4-8}$$

where

$$A = n \times n \text{ coefficient matrix}$$

$$B = n \times p \text{ coefficient matrix}$$

For Eq. (4-7) it is apparent that A and B are scalars, and

$$A = -R_a/L_a \tag{4-9}$$

$$B = 1/L_a \tag{4-10}$$

and $u(t) = v(t) - v_b(t)$.

The differential equation that describes the mechanical load of the motor system in Fig. 4-6 is

$$T(t) = J\frac{d^2\theta}{dt^2} + B\frac{d\theta}{dt} \tag{4-11}$$

where T(t) is the torque developed by the motor, J is the total moment of inertia of the motor and load, and B is the viscous frictional coefficient.

Let the state variables be defined as $x_1(t) = \theta(t)$ and $x_2(t) = d\theta(t)/dt$; Eq. (4-11) is decomposed into two state equations,

$$\frac{dx_1(t)}{dt} = x_2(t) \tag{4-12}$$

$$\frac{dx_2(t)}{dt} = -\frac{B}{J}x_2(t) + \frac{1}{J}T(t) \tag{4-13}$$

In terms of the matrix form of Eq. (4-8), we recognize that

$$A = \begin{bmatrix} 0 & 1 \\ 0 & -\frac{B}{J} \end{bmatrix} \tag{4-14}$$

$$B = \begin{bmatrix} 0 \\ \frac{1}{J} \end{bmatrix} \tag{4-15}$$

In this case, the motor torque, T(t), is identified as the input to the mechanical system, u(t).

The controlled variables of a control system are also called the outputs. The output variables are generally related to the state variables and the input variables through algebraic equations of the form:

$$\mathbf{c}(t) = \mathbf{Dx}(t) + \mathbf{Eu}(t) \tag{4-16}$$

where $\mathbf{c}(t)$ denotes the $q \times 1$ output vector; D is a $q \times n$ coefficient matrix, and E is a $q \times p$ coefficient matrix. For the mechanical system shown in Fig. 4-6, if the motor angular displacement θ is considered as the output, we have

$$c(t) = \theta(t) = x_1(t) \tag{4-17}$$

Thus,

$$D = [1 \quad 0] \tag{4-18}$$

and $E = 0$.

In general there are many ways of determining the state variables and the state equations [1]. State equations are used in this volume to model dc motor systems (Chapter 15). They are also used extensively in Volume II, Step Motors and Controls (Chapters 2, 3, 12).

Transfer Functions

Transfer functions are used extensively in the analysis and design of linear systems. The starting point in the derivation of the transfer function is the integral-differential equation or the state equation and the output equation. The important assumption in the transfer function analysis is that *all the initial conditions of the system are assumed to be zero.*

The transfer function of a linear system is defined as the ratio of the Laplace transform of the output variable to the Laplace transform of the input variable. For example, taking the Laplace transform on both sides of Eq. (4-11), and assuming zero initial condition, i.e., the initial displacement and velocity of the motor are zero, we have

$$T(s) = Js^2\Theta(s) + Bs\Theta(s) \tag{4-19}$$

Thus, the transfer function with $T(t)$ as input and $\theta(t)$ as output is

$$\frac{\Theta(s)}{T(s)} = \frac{1}{Js^2 + Bs} \tag{4-20}$$

Most of the models related to dc motors in this volume are represented by transfer functions, since accurate results are generally obtained with linear dc motor equations. On the other hand, step motors are extremely nonlinear and their dynamics cannot be modeled by linear equations, and thus the transfer function method is not practical in this case.

Characteristic Equation

For linear time-invariant systems the *characteristic equation* completely defines the stability of the system. The characteristic equation is defined as

$$|sI - A| = 0 \tag{4-21}$$

which denotes the equation by setting the determinant of sI − A to zero. The matrix A is defined in the state equation of Eq. (4-8), and s is a complex variable which is also synonymous with the Laplace transform variable.

Another way of defining the characteristic equation is simply setting the denominator of the transfer function to zero. Thus, the characteristic equation of the system described by the transfer function of Eq. (4-20) is simply

$$Js^2 + Bs = 0 \tag{4-22}$$

Or, as an alternative, using the A matrix in Eq. (4-14), we have

$$|sI - A| = \begin{vmatrix} s & -1 \\ 0 & s + \frac{B}{J} \end{vmatrix} = s(s + \frac{B}{J}) = 0 \tag{4-23}$$

which is the same result as in Eq. (4-22).

The roots of the characteristic equation are also referred to as the *eigenvalues of A*. These roots completely define the stability of a linear system. A linear system is *asymptotically stable* (or simply stable) if any finite initial conditions of the system outputs eventually decay to zero as time progresses from the initial time. This also corresponds to the condition that any bounded input will produce bounded outputs.

For a linear system to be stable, all the roots of its characteristic equation must be found in the left-half of the s-plane. Roots that are on the imaginary axis and are simple give rise to sinusoidal sustained oscillations. Roots that are in the right-half of the s-plane correspond to responses that will increase in magnitude as time progresses.

Given the characteristic equation, the Routh-Hurwitz criterion [1] is a method of determining the location of the roots with respect to the right-half and the left-half s-plane without actually solving for the roots.

Relative Stability

The stability referred to in the preceding section is also termed as *absolute stability*; i.e., stable or unstable. However, if a system is stable, it is necessary to indicate how stable it is. This is called *relative stability*. The location of the roots of the characteristic equation also gives indication on relative stability.

In Chapter 2, the unit step responses of a second-order system with the standard-form characteristic equation

$$s^2 + 2\zeta\omega_n s + \omega_n^2 = 0 \tag{4-24}$$

are presented for various values of ζ, the *damping ratio*. The two roots of Eq. (4-24) are written

$$s_{1,2} = -\zeta\omega_n \pm j\omega_n\sqrt{1-\zeta^2} \tag{4-25}$$

These roots are plotted in Fig. 4-7 for a fixed value of ω_n, the *natural undamped frequency*, and variable ζ. The characteristics of the system responses that correspond to the various root locations are illustrated in Fig. 4-7. Also as shown in Fig. 2-21, as ζ increases from zero the damping improves, and the response becomes less oscillatory. When $\zeta = 1$, the damping is critical (critical damping), and the response is not oscillatory, and there is no overshoot in the step response.

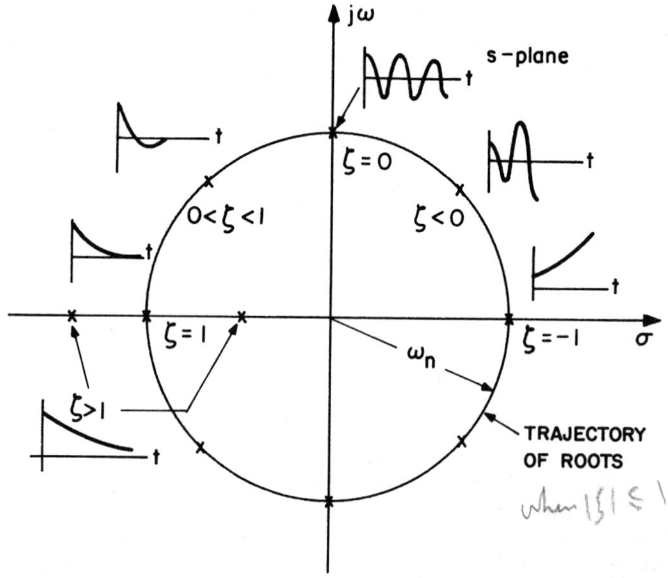

Fig. 4-7. Trajectory of roots of a second-order equation.

The natural undamped frequency, ω_n, is the frequency of the undamped oscillation when the damping is zero ($\zeta = 0$). As seen from Eq. (4-25), when $\zeta = 0$, the two roots are at $\pm j\omega_n$, on the imaginary axis.

4.3 ROOT LOCUS DIAGRAM

The trajectory of roots of the characteristic equation shown in Fig. 4-7 is formally known as a *root locus diagram*. When designing a control system it is often desirable to investigate the effects of varying one system parameter, such as the gain of the amplifier, the torque constant of the motor, motor resistance or inductance. The root locus diagram, thus, is a valuable design tool which indicates the variation of the characteristic equation roots when a given system parameter varies (from $-\infty$ to $+\infty$).

As an illustrative example, consider the closed-loop control system shown in Fig. 4-8. The characteristic equation of the system is

$$s(s+1)(s+2) + K = s^3 + 3s^2 + 2s + K = 0 \tag{4-26}$$

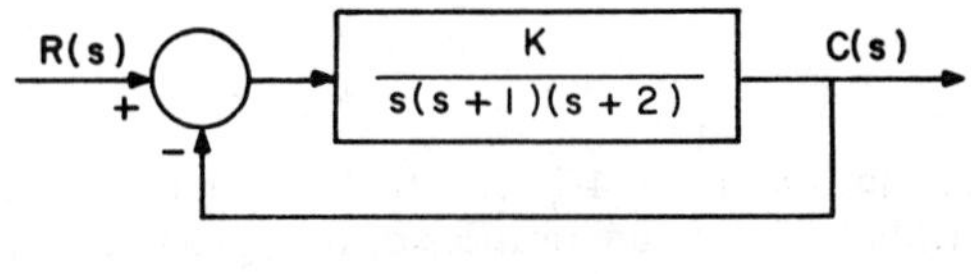

Fig. 4-8. A closed-loop control system.

In this case K may represent the gain of an amplifier that is located in the forward path of the system. The complete root locus diagram of Eq. (4-26) as K varies from $-\infty$ to ∞ is shown in Fig. 4-9.

Several simple facts become apparent from this root locus diagram:

1. The system is unstable for $K < 0$ (one root in the right-half s-plane), and for $K > 6$ (two roots in the right-half plane).

2. When $K = 6$, the characteristic equation has two roots on the $j\omega$ axis at $s = \pm j\sqrt{2}$. Thus, the natural undamped frequency is $\sqrt{2}$ rad/sec.

3. When $K = 0.38$ the damping is critical.

In general, the construction of the root locus diagram does not rely on the actual solution of the characteristic equation for every value of a large number of values of K. A set of rules of construction [1] is available which allows the designer to sketch the root locus diagram based on the pole-zero configuration of the open-loop transfer function.

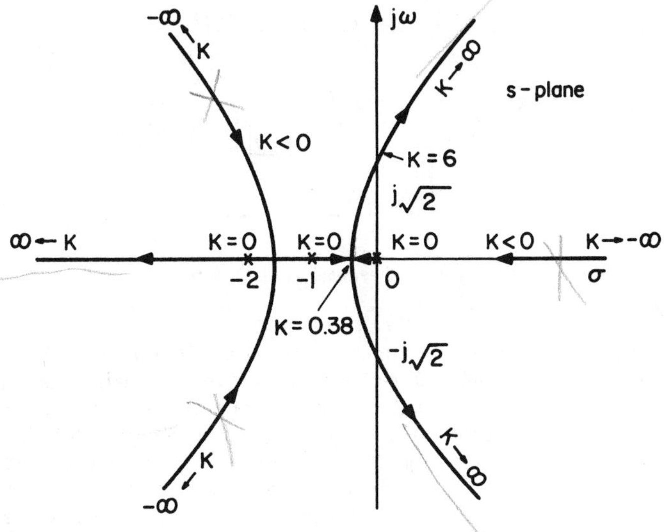

Fig. 4-9. Root locus diagram of Eq. (4-26).

4.4 FREQUENCY-DOMAIN ANALYSIS

The frequency response of a system is defined as the output when the input is a sinusoid. Although the actual input of a control system is rarely a pure sinusoid, frequency-domain analysis in reality offers a number of well-established tools which may be used for design and analysis purposes.

For linear systems it is generally possible to find a direct correlation between the time response and the frequency response so that when information on the latter is known one can predict the time-domain performance of the system.

Consider that a feedback control system is represented by the block diagram shown in Fig. 4-10. The closed-loop transfer function is written as

$$M(s) = \frac{C(s)}{R(s)} = \frac{G(s)}{1 + G(s)} \tag{4-27}$$

For sinusoidal steady state, we let s = jω, and Eq. (4-27) becomes

$$M(j\omega) = \frac{C(j\omega)}{R(j\omega)} = \frac{G(j\omega)}{1 + G(j\omega)}$$

The last expression can be written in terms of magnitude and phase,

$$M(j\omega) = |M(j\omega)| \underline{/\phi(j\omega)} \tag{4-28}$$

where

$$|M(j\omega)| = \left| \frac{G(j\omega)}{1 + G(j\omega)} \right| \tag{4-29}$$

$$\underline{/\phi(j\omega)} = \underline{/G(j\omega)} - \underline{/1 + G(j\omega)} \tag{4-30}$$

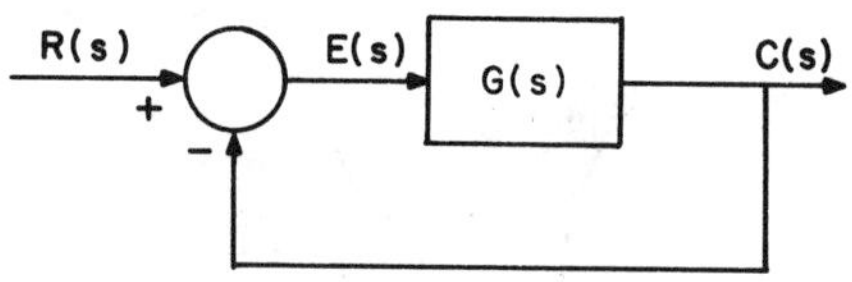

Fig. 4-10. Feedback control system.

In general, when the transfer function M(s) is of the minimum-phase type, that is, no poles or zeros in the right-half s-plane and no pure time delays, the magnitude and phase of M(jω) are uniquely related. Thus, frequently the analysis and design of a control system may be carried out by using only the magnitude plot, |M(jω)| versus ω. Furthermore, one of the most significant characteristics of control systems theory is that the behavior of the closed-loop system may be determined by investigating the properties of the open-loop transfer function G(s). This is because the roots of the characteristic equation are also the zeros of the function 1 + G(s). In other words, the roots of the characteristic equation must satisfy

$$1 + G(s) = 0 \tag{4-31}$$

or

$$G(s) = -1 \tag{4-32}$$

The root locus diagram is constructed systematically by finding the conditions under which Eq. (4-32) is satisfied. Therefore, in general, the absolute and relative stability of a closed-loop control system can be studied in the frequency domain by investigating the properties of the magnitude and phase properties of G(jω).

Bandwidth And Resonance

Two of the most important criteria in the frequency-domain study are the *bandwidth* and parameters related to resonance. Figure 4-11 shows a typical magnitude-versus-frequency plot of either $M(j\omega)$ or $G(j\omega)$ of a control system. In general, a system may have several resonance points, and a curve such as in Fig. 4-11 is determined experimentally by using a network analyzer.

The bandwidth of a system (open loop or closed loop as the case may be) is defined as the frequency at which the magnitude has dropped to 70.7 percent of the zero-frequency magnitude (Fig. 4-11).

Intuitively, when the bandwidth of a system is high, the system will be able to pass higher-frequency components, and thus the time response of the system will be relatively faster. The reverse is true for a low-bandwidth system. Theoretically, a system with an infinite bandwidth is an all-pass system, whereas a system with a zero bandwidth will transmit nothing. Therefore, in practice, the control engineer often uses bandwidth to describe the quality of his system.

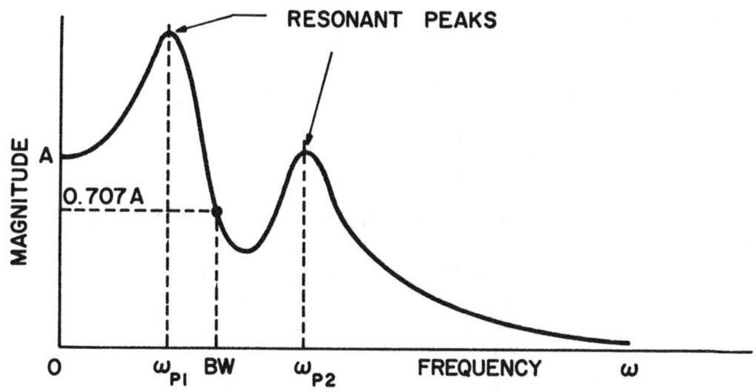

Fig. 4-11. A typical magnitude-versus-frequency plot of a control system.

The resonance phenomenon represents the characteristics of a system to possess high gains at certain narrow bands of frequencies. Figure 4-11 illustrates two such resonances in the magnitude curve. Resonances are typically described by the frequencies (resonant frequencies) at which they occur as well as the magnitudes (resonant peaks) at the peak of the resonances.

If any of these resonances should fall in the significant frequency range in which the system operates, they must be suppressed or eliminated by proper design, or the system may be subject to instability or poor relative stability.

Consider the second-order system transfer function,

$$M(j\omega) = \frac{\omega_n^2}{(j\omega)^2 + 2\zeta\omega_n(j\omega) + \omega_n^2} \tag{4-33}$$

The magnitude of $M(j\omega)$ is

$$|M(j\omega)| = \frac{1}{[(1-u^2)^2 + (2\zeta u)^2]^{1/2}} \tag{4-34}$$

where

$$u = \omega/\omega_n \tag{4-35}$$

Figure 4-12 shows the $|M(j\omega)|$ versus u plots for various values of ζ. As has been stated earlier, the frequency response is directly related to the time response since there is an established relation between the latter and ζ (Fig. 2-21). Thus, we see that in general small values of ζ will correspond to high resonance peaks. When $\zeta \leqslant 0$ the resonance peak is infinity. In addition, there is also a well-defined relation between ζ and bandwidth.

For the transfer function given in Eq. (4-33) we can show that the resonant frequency is

$$\omega_p = \omega_n\sqrt{1 - 2\zeta^2} \tag{4-36}$$

and the resonant peak is

$$\text{Resonant Peak} = \frac{1}{2\zeta\sqrt{1 - \zeta^2}} \tag{4-37}$$

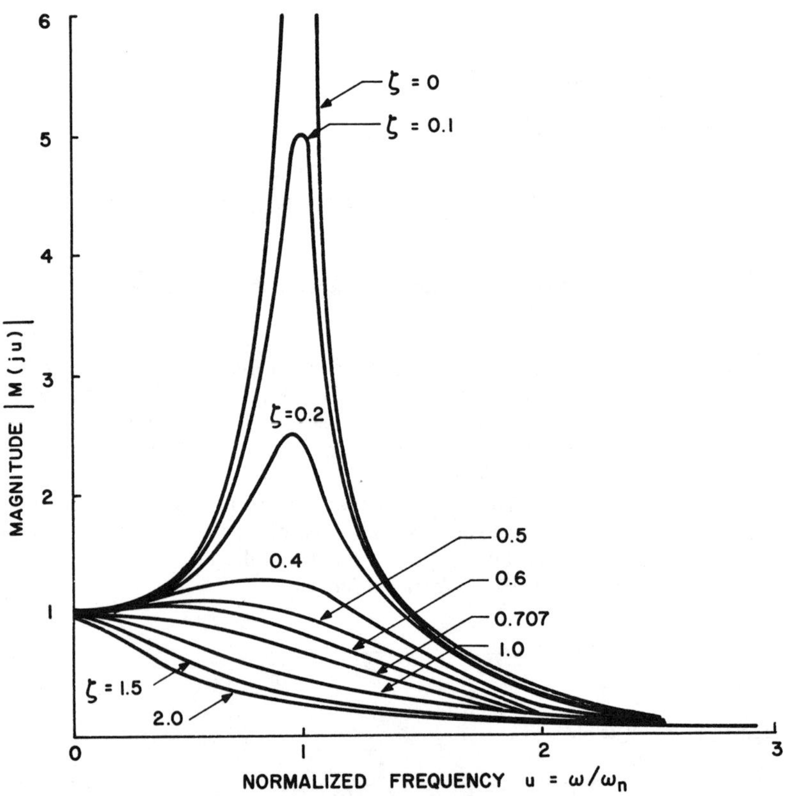

Fig. 4-12. Magnitude-versus-frequency plots.

The bandwidth is given by

$$BW = \omega_n[(1 - 2\zeta^2) + \sqrt{4\zeta^4 - 4\zeta^2 + 2}]^{1/2} \tag{4-38}$$

For higher-order systems, there would be more parameters than just ζ and ω_n, and it is impractical to attempt to derive the resonant frequency, resonant peak and bandwidth analytically. In practice, a graphical technique using the Bode plot [1] is preferred for high-order systems.

Bode Plots Of Transfer Functions

The Bode plot is a straight-line approximation of the frequency response in magnitude (decibels) and phase versus frequency (radians/second).

We shall describe the Bode plot technique by means of an example. Let the transfer function of a linear control system be

$$G(s) = \frac{K(1 + T_1 s)}{s(1 + T_2 s)} \tag{4-39}$$

where K, T_1 and T_2 are real constants.

Letting $s = j\omega$, Eq. (4-39) becomes

$$G(j\omega) = \frac{K(1 + j\omega T_1)}{j\omega(1 + j\omega T_2)} \tag{4-40}$$

The magnitude of $G(j\omega)$ is written

$$|G(j\omega)| = \left| \frac{K(1 + j\omega T_1)}{\omega(1 + j\omega T_2)} \right| \tag{4-41}$$

The magnitude of $G(j\omega)$ in decibels (db) is obtained by multiplying the logarithm to the base 10 of $|G(j\omega)|$ by 20; we have

$$20\log_{10}|G(j\omega)| = 20\log_{10}|K| + 20\log_{10}|1 + j\omega T_1| - 20\log_{10}\omega - 20\log_{10}|1 + j\omega T_2| \tag{4-42}$$

The phase of $G(j\omega)$ is

$$\underline{/G(j\omega)} = \underline{/K} + \underline{/1 + j\omega T_1} - \underline{/j\omega} - \underline{/1 + j\omega T_2} \tag{4-43}$$

Equations (4-42) and (4-43) show that for real poles and zeros of the transfer function, the magnitude in db and phase plots of $G(j\omega)$ are composed essentially of the algebraic sum of the following three terms:

1. Constant K
2. Poles or zeros at the origin $(j\omega)^{\pm n}$ $(n = 1,2,\ldots)$
3. Simple pole or zero $(1 + j\omega T)^{\pm 1}$

The logarithmic plots of the amplitudes of these functions and the phase plots are discussed as follows.

For the constant terms, K, in db is simply

$$20\log_{10}|K| = \text{constant} \quad (\text{db}) \tag{4-44}$$

For the pole or zero at $\omega = 0$,

$$20\log_{10}|\omega^{\pm n}| = \pm 20n\log_{10}\omega \tag{4-45}$$

which is the equation of a straight line in either semi-logarithmic coordinates (db versus ω) or rectangular coordinates (db versus $\log\omega$). The slopes of these straight lines are $\pm 20n$ db.

For the simple zero at $\omega = -1/T$, we let

$$G(j\omega) = 1 + j\omega T \tag{4-46}$$

Then,

$$20\log_{10}|G(j\omega)| = 20\log_{10}\sqrt{1 + \omega^2T^2} \tag{4-47}$$

$$\underline{/G(j\omega)} = \tan^{-1}\omega T \tag{4-48}$$

A linear asymptotic approximation is used in the construction of the magnitude curve. At very low frequencies, $\omega T \ll 1$, Eq. (4-47) can be approximated by

$$20\log_{10}|G(j\omega)| \cong 20\log_{10}0 = 0 \ \text{db} \tag{4-49}$$

which is a horizontal line, and is referred to as the low-frequency asymptote. At very high frequencies, $\omega T \gg 1$, and Eq. (4-47) becomes

$$20\log_{10}|G(j\omega)| \cong 20\log_{10}\sqrt{\omega^2T^2} = 20\log_{10}\omega T \tag{4-50}$$

which is a straight line with a slope of 20 db per decade of frequency, and is called the high-frequency asymptote. The intersect of the low-frequency and the high-frequency asymptote is at $\omega = 1/T$. This frequency of intersect is also known as the *corner frequency* of the Bode plot. The actual magnitude plot of Eq. (4-47) is a smooth curve. The error is approximately 3 db at the corner frequency and approximately 1 db at one octave above and below the corner frequency. Figure 4-13 shows the asymptotic approximation and the actual magnitude plot of the Bode diagram of Eq. (4-46).

The phase plot of Eq. (4-46) is also shown in Fig. 4-13 as a function of ω. In this case the phase plot can also be approximated by a straight line that extends from 0 degrees at $\omega = 0.1/T$ to 45 degrees at $\omega = 10/T$. The actual phase plot deviates from the straight line by not more than 6 degrees.

For the transfer function

$$G(j\omega) = \frac{1}{1 + j\omega T} \tag{4-51}$$

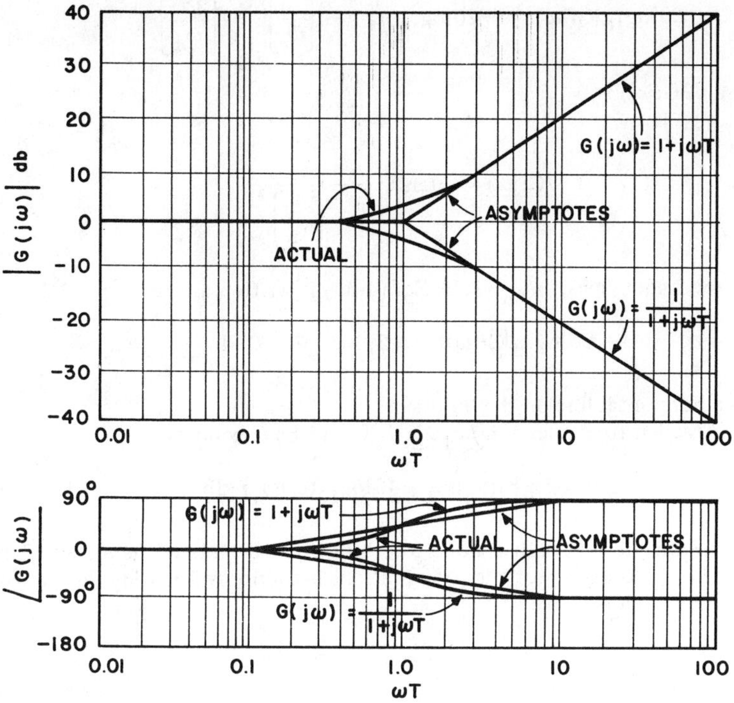

Fig. 4-13. Bode plots of G(s) = 1 + Ts and G(s) = 1/(1 + Ts).

if we follow through the same analysis as given above, we would obtain a magnitude curve which is the mirror image of that of Eq. (4-46). Similarly, the phase plot of Eq. (4-51) is also the mirror image of that of Eq. (4-46). These plots are also shown in Fig. 4-13.

Now we can make use of the component curves of the simple real poles to construct the Bode plot of the transfer function of Eq. (4-39). The resultant magnitude and phase plots are obtained by adding the component curves point by point at the corresponding frequencies.

When a transfer function has complex poles or zeros, we can represent the transfer function of a quadric function as

$$G(s) = \frac{1}{\dfrac{s^2}{\omega_n^2} + \dfrac{2\zeta}{\omega_n} s + 1} \tag{4-52}$$

Setting s = jω, the last equation becomes

$$G(j\omega) = \frac{1}{1 - \left[\dfrac{\omega}{\omega_n}\right]^2 + j\omega \dfrac{2\zeta}{\omega_n}} \tag{4-53}$$

The magnitude of G(jω) in decibels is

$$20\log_{10}|G(j\omega)| = -20\log_{10}\sqrt{\left[1-\left[\frac{\omega}{\omega_n}\right]^2\right]^2+\left[\frac{2\zeta\omega}{\omega_n}\right]^2} \tag{4-54}$$

The phase of $G(j\omega)$ is

$$\underline{/G(j\omega)} = -\tan^{-1}\left[\frac{\frac{2\zeta\omega}{\omega_n}}{1-\left[\frac{\omega}{\omega_n}\right]^2}\right] \tag{4-55}$$

At very low frequencies, $\omega/\omega_n \ll 1$; Eq. (4-54) is written as

$$20\log_{10}|G(j\omega)| \cong -20\log_{10}1 = 0 \text{ db} \tag{4-56}$$

which is a horizontal line on the 0 db axis.

At very high frequencies, $\omega/\omega_n \gg 1$; Eq. (4-54) becomes

$$20\log_{10}|G(j\omega)| \cong -40\log_{10}(\omega/\omega_n) \text{ db} \tag{4-57}$$

which is a straight line with a slope of -40 db per decade of frequency. The intersection of the two asymptotic approximations is found by equating Eq. (4-56) to Eq. (4-57), from which we have

$$\omega = \omega_n \tag{4-58}$$

Thus, $\omega = \omega_n$ is considered to be corner frequency of the Bode plot of the second-order transfer function in Eq. (4-52).

The error between the actual and the asymptotic straight-line approximation of the magnitude plot of Eq. (4-52) depends on ζ. Figure 4-14 shows the magnitude and phase plots of the second-order transfer function.

Due to the additive characteristics of the Bode plot, the plot is valuable for analysis as well as design problems.

Relative And Absolute Stability From The Bode Diagram—Gain And Phase Margins

The Bode diagram can be used for the studies of relative and absolute stability of a feedback control system that has a minimum-phase transfer function. The stability criterion of a closed-loop system in the frequency domain is the Nyquist criterion which states that the closed-loop system with the transfer function in Eq. (4-26) is stable if the plot of $G(j\omega)$ in the polar coordinates does not enclose the $(-1,j0)$ point. This is translated to the following requirements on the Bode plot of $G(j\omega)$:

1. At $|G(j\omega)| = 0$ db $\qquad \underline{/G(j\omega)} > -180°$

and

2. At $\underline{/G(j\omega)} = -180°$ $\qquad |G(j\omega)| < 0$ db

For relative stability, we use the *gain margin* and the *phase margin* to indicate how stable the system is.

The gain margin is a measure of how much the loop gain can be increased before reaching instability. Similarly, the phase margin indicates how much the phase of $G(j\omega)$ can be increased before the closed-loop system becomes unstable.

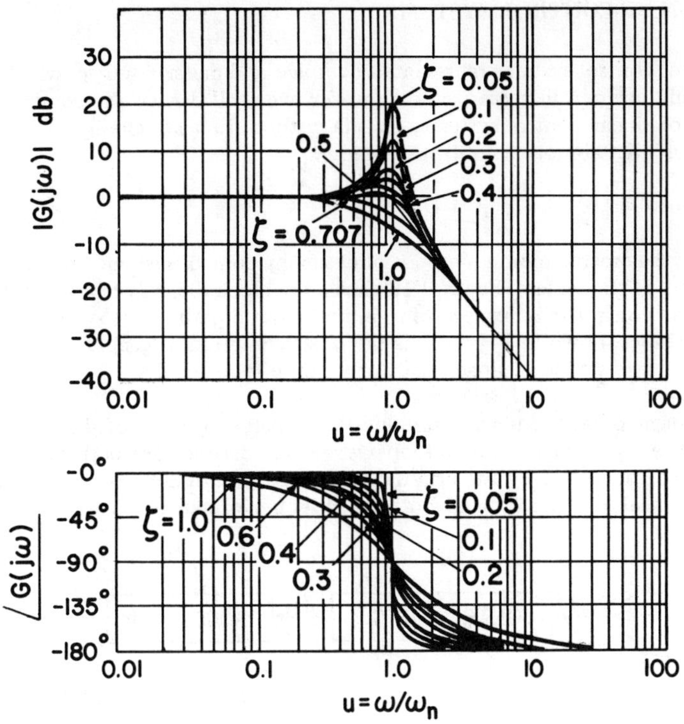

Fig. 4-14. Magnitude and phase plots of the second-order transfer function in Eq. (4-52).

Figure 4-15 illustrates the interpretation of the gain margin and the phase margin on a Bode diagram of G(jω).

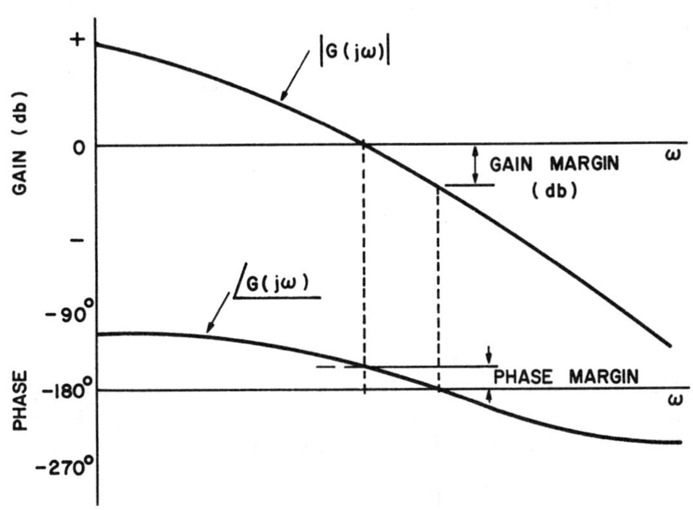

Fig. 4-15. Interpretation of gain margin and phase margin on Bode diagram.

4.5 DESIGN OF CONTROL SYSTEMS

The design of control systems represents a complex and massive subject which cannot be thoroughly covered in these pages. However, we shall discuss the common design techniques of linear control systems, especially those that are closely related to the design of incremental-motion control systems.

The PID Controller

One of the most common methods of designing control systems is to first fix the configuration of the system. Figure 4-16 shows the block diagram of a control system for which the controller is fixed a priori in the forward path of the system. Although the configuration shown in Fig. 4-16 is the most common, in general, the controller can also be placed in the feedback path or a minor-loop feedback path, or combinations of these.

The design objective is to determine the transfer function of the controller so that the performance of the closed-loop system is as desired. The performance of the system may be described by the criteria such as the gain and phase margins defined above, or the peak overshoot, bandwidth, or other performance indices.

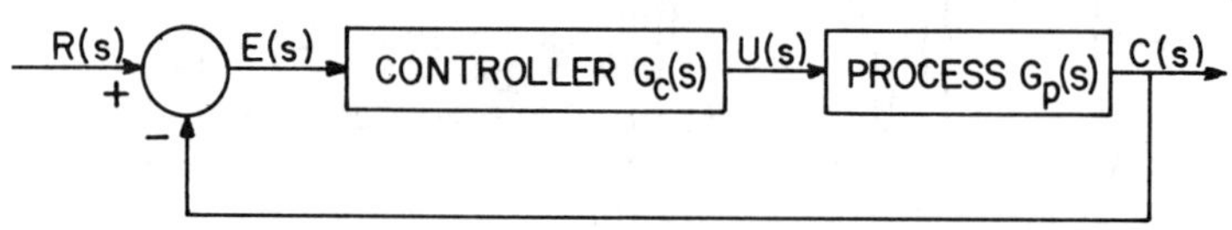

Fig. 4-16. Control system with controller.

In the classical design approach, the controller $G_c(s)$ is determined more or less by trial and error. The most common controller used in practice consists of controlled signals in the form of proportional, integral and derivative of the error e(t); thus, the term, PID controller. In the time domain the actuating signal that is applied to the controlled process $G_p(s)$ is described by

$$u(t) = K_p e(t) + \int K_i e(t)dt + K_d \frac{de(t)}{dt} \tag{4-59}$$

where K_p, K_i and K_d are constants of proportional, integral, and derivative controls, respectively.

In terms of the Laplace transform, the PID controller is represented by the transfer function

$$G_c(s) = K_p + \frac{K_i}{s} + K_d s \tag{4-60}$$

In practice, depending on the nature of the controlled process and the desired performance, only one of two components of the PID control may be needed. For instance, a PD controller is represented by the transfer function

$$G_c(s) = K_p + K_d s \tag{4-61}$$

When $G_c(s)$ is cascaded with the controlled process $G_p(s)$, it has the effect of adding a zero at $s = -K_p/K_d$ to the open-loop transfer function. The result is that the stability condition of the closed-loop system is improved by the addition of the zero to its open-loop transfer function. As an illustrative example, let the controlled process of the system in Fig. 4-16 be

$$G_p(s) = \frac{K}{s(s+1)(s+2)}$$

The root locus diagram of the closed-loop system with $G_c(s) = 1$ is already shown in Fig. 4-9. The system is unstable for all values of K greater than or equal to 6. Let the controller be of the PD form as given in Eq. (4-61). Then, the open-loop transfer function is

$$G(s) = G_c(s)G_p(s) = \frac{K(K_p + K_d s)}{s(s+1)(s+2)} \tag{4-62}$$

Figure 4-17(a) illustrates the root locus diagram of the system when K_p/K_d is arbitrarily chosen to be 3. In this case, the roots of the closed-loop characteristic equation are always in the left-half s-plane so that the system is stable for all positive K. In Fig. 4-17(b) the root locus diagram for a very small value of K_p/K_d is shown. In this case, the loci of the complex roots are drawn further into the left-half plane, and the system will have a still more improved relative stability than the system that corresponds to the diagram in Fig. 4-17(a). Thus, the conclusion is that by increasing the derivative constant K_d the relative and absolute stability of the system are improved.

In practice the derivative control can also be implemented by feeding back the derivative of the output variable, such as the common use of a tachometer to feedback the output speed of a dc motor for stabilizing effects.

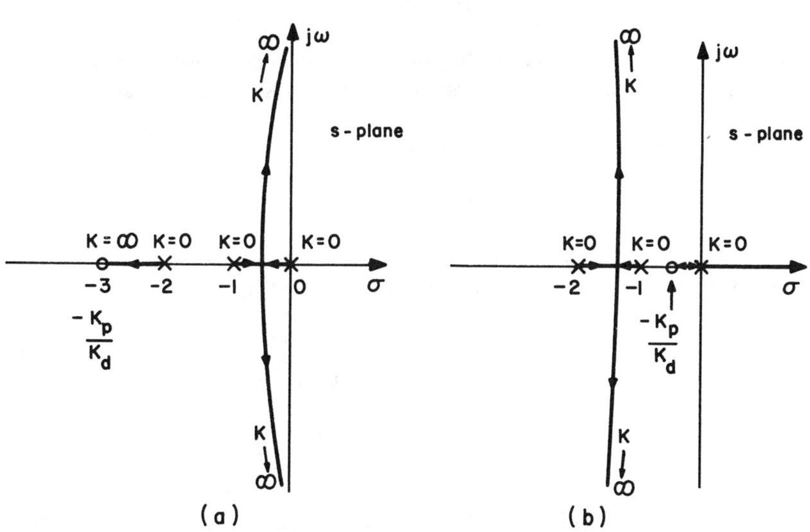

Fig. 4-17. Root locus diagram of system with PD controller.

The integral control is generally used for the purpose of improving the steady-state error of the closed-loop control system. However, when properly applied the integral control can also improve the relative stability of the system.

Consider that the system shown in Fig. 4-16 has the controlled process

$$G_p(s) = \frac{K}{s(1 + 0.1s)} \tag{4-63}$$

Since the system is of the second-order, it is always stable for $0 \leqslant K < \infty$. The steady-state error of the system for a given input is given by

$$e_{ss} = \lim_{t \to \infty} e(t) = \lim_{s \to 0} sE(s) \tag{4-64}$$

if the function sE(s) does not have any poles on or to the right of the $j\omega$ axis of the s-plane. Evaluating E(s) from Fig. 4-16 and substituting into Eq. (4-64), we have

$$e_{ss} = \lim_{s \to 0} \frac{sR(s)}{1 + G_p(s)G_c(s)} \tag{4-65}$$

For $G_c(s) = 1$ and $G_p(s)$ as given in Eq. (4-63), and a unit-step input, the steady-state error becomes

$$e_{ss} = \lim_{s \to 0} \frac{1}{1 + \dfrac{K}{s(1 + 0.1s)}} = 0 \tag{4-66}$$

When the input is a unit ramp function, $R(s) = 1/s^2$,

$$e_{ss} = \frac{1}{K} \tag{4-67}$$

Let us attempt to find a controller that will reduce the steady-state error to zero for a ramp input, and at the same time the relative stability of the system should not be reduced.

Let the transfer function of the controller be

$$G_c(s) = K_p + K_i/s \tag{4-68}$$

which is known as the PI controller.

Thus, the open-loop transfer function of the system becomes

$$G(s) = G_c(s)G_p(s) = \frac{K(K_p + K_i/s)}{s(1 + 0.1s)}$$

$$= \frac{KK_i(1 + K_p s/K_i)}{s^2(1 + 0.1s)} \tag{4-69}$$

Now since G(s) has a double pole at $s = 0$, we can show that the steady-state error due to a ramp input is zero.

If we choose the ratio K_i/K_p to be very small, from the transient response viewpoint Eq. (4-69) can be approximated by

$$G(s) \cong \frac{KK_p}{s(1 + 0.1s)} \tag{4-70}$$

Thus, the characteristic equation of the system is

$$s^2 + 10s + 10KK_p = 0 \tag{4-71}$$

whereas that of the original system is

$$s^2 + 10s + 10K = 0 \tag{4-72}$$

If we choose K_p to be greater than one, the damping ratio of the compensated system is improved to

$$\zeta = \frac{10}{2\sqrt{KK_p}} \tag{4-73}$$

as compared with the value of

$$\zeta = \frac{10}{2\sqrt{K}} \tag{4-74}$$

for the original system.

Pole-Zero Cancellation Compensation

The controlled process of a practical control system often has poles and zeros that are complex and very close to the imaginary axis, and sometimes these may even be in the right half of the s-plane. To compensate this type of system one immediately would think of the possibility of introducing a controller with poles and zeros to cancel the undesirable zeros and poles of the process.

Unfortunately, the poles and zeros of a controlled process are not stationary during the operation of the system. Furthermore, there usually are many assumptions that must be made in arriving at the transfer function of the controller so that the locations of the poles and zeros are not certain. Therefore, in reality, the pole-zero cancellation is almost never exact. However, even with inexact cancellation the transient contribution due to the uncancelled poles or zeros is still small so that the compensation scheme can still be very effective. It should be pointed out that if the undesirable poles or zeros are in the right-half s-plane the pole-zero cancellation compensation will not work, since the inexact cancellation in poles and zeros will result in characteristic equation roots that are in the right-half s-plane, and the resultant closed-loop system would be unstable.

Lead-Lag And Notch Compensations

It is useful to consider the design of the controller in terms of filter characteristics. For instance, the PD controller used earlier can be considered as a high-pass filter, whereas the PI controller is referred to as a low-pass filter. The PD controller is also a lead controller since the phase characteristic of the controller is positive for all frequencies. Similarly, the PI controller always has a negative phase for all frequencies and is referred to as lag compensation.

In general other lead and lag as well as combinations of lead and lag compensations can be used for improving the performance of control systems. A simple lead or lag controller is represented by the transfer function

$$G_c(s) = \frac{1 + T_1 s}{1 + T_2 s} \tag{4-75}$$

where $T_1 > T_2$ for lead and $T_1 < T_2$ for lag.

A lag-lead or notch controller is obtained by cascading a lag with a lead controller; e.g.,

$$G_c(s) = \left[\frac{1 + T_1 s}{1 + T_2 s}\right]\left[\frac{1 + T_3 s}{1 + T_4 s}\right] \tag{4-76}$$

There are no exact procedures one can follow in making use of these lead and lag or lag-lead controllers. Usually the designer has to combine his own experience with trial and error to arrive at the best design configuration for a particular system. For more details on the subject of control systems design, the reader may refer to [1].

4.6 SAMPLED-DATA AND DIGITAL CONTROL SYSTEMS [2,3]

The control systems discussed thus far in this chapter all have one feature in common; i.e., the signals found in the systems are all functions of the continuous time variable t. There are many control systems in which the signals at various parts of the system are in the form of sampled-data or digitally-coded data. For example, the phase-locked loop (Chapter 14) and the step motor systems (Volume II) are all modelled as sampled-data controlled systems. A control system with a controller implemented by a digital computer, such as a microcomputer, is modelled as a digital control system.

Sampled-data and digital control systems are characterized by a system component known as the *sampler*. The sampler is used to model the sampling and the digital operations. Figure 4-18 illustrates the block diagram of a typical sampled-data control system. The zero-order hold (z.o.h.) is a device which performs the filtering on the sampled signal. The mathematical descriptions of the sampler and the z.o.h. will be given in the following. The block diagram of a typical digital control system is shown in Fig. 4-19. The digital controller is modelled with a sampler at the output and the input of the controller.

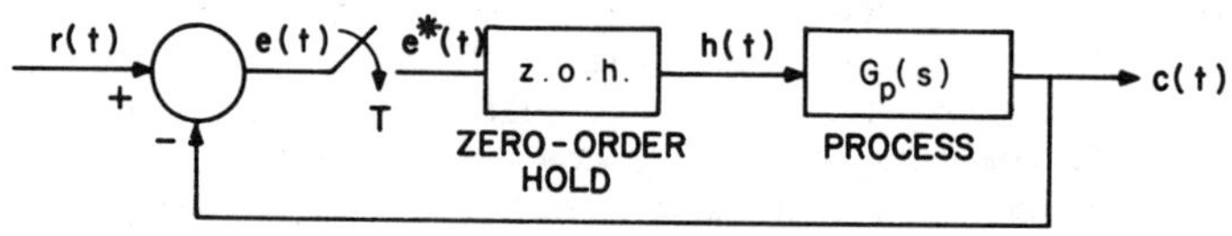

Fig. 4-18. Block diagram of a sampled-data control system.

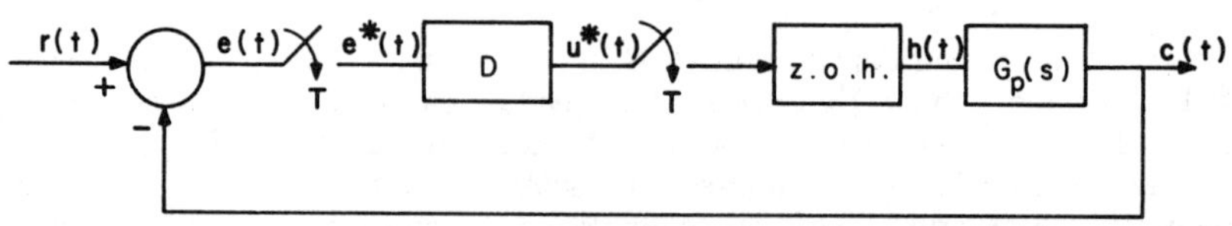

Fig. 4-19. Block diagram of a digital control system.

Mathematically, if the sampling duration of the sampler is small, the output of the sampler can be approximated as a sequence of impulses [1,2], and the sampler is called the *ideal sampler*. Thus, if the input to the ideal sampler is denoted as e(t), the output of the sampler is

$$e^*(t) = \sum_{k=0}^{\infty} e(kT)\delta(t - kT) \tag{4-77}$$

where $\delta(t)$ is the unit impulse function, and T is the *sampling period* in seconds.

Taking the Laplace transform on both sides of Eq. (4-77), we have

$$E^*(s) = \sum_{k=0}^{\infty} e(kT)e^{-kTs} \tag{4-78}$$

The input-output relation of the ideal sampler is shown in Fig. 4-20.

The zero-order hold is a device which holds the value of an impulse at a given sampling instant for the entire sampling period until the next input arrives. Thus, if the sampled signal e*(t) shown in Fig. 4-20 is sent through the zero-order hold, the output of the zero-order hold is as shown in Fig. 4-21.

The transfer function of the zero-order hold is given by

$$G_{ho}(s) = \frac{1 - e^{-Ts}}{s} \tag{4-79}$$

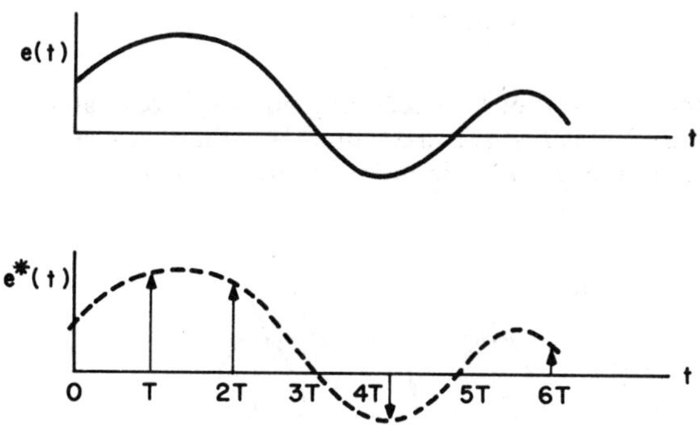

Fig. 4-20. Input-output relation of the ideal sampler.

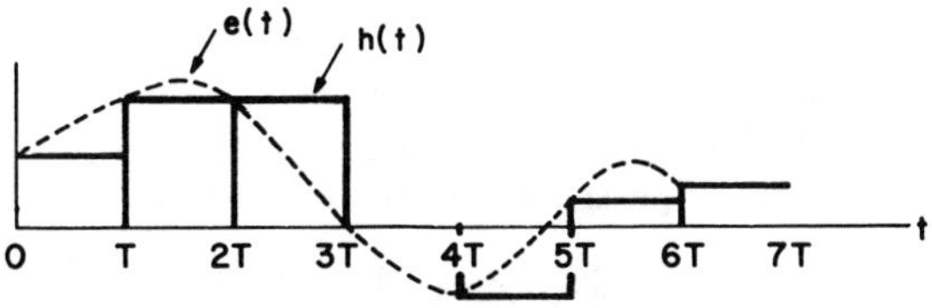

Fig. 4-21. Output of zero-order hold.

It should be noted that the ideal sampler is a linear device since it satisfies the principle of superposition. However, unlike all linear systems it does not have a transfer function. An approximate transfer relation for the ideal sampler can be obtained by expanding e*(t) into a Fourier series, since it is a periodic function. This results in

$$e^*(t) = e(t) \sum_{n=-\infty}^{\infty} \frac{1}{T} e^{-jn\omega_s t} \tag{4-80}$$

where ω_s is the sampling frequency in radians per second, and is related to the sampling period T through

$$\omega_s = 2\pi/T \tag{4-81}$$

Taking the Fourier transform on both sides of Eq. (4-80) yields

$$E^*(j\omega) = \frac{1}{T} \sum_{n=-\infty}^{\infty} E(j\omega - jn\omega_s) \tag{4-82}$$

This equation represents the input-output relation of the ideal sampler in the frequency domain. If the highest frequency contained in $E(j\omega)$ is less than $\omega_s/2$, the amplitude spectrum of $E^*(j\omega)$ would be as shown in Fig. 4-22. Now if the sampled signal $E^*(j\omega)$ is sent through a low-pass filter with the cut-off frequency at $\omega_s/2$, only the fundamental component of $E^*(j\omega)$ will be passed. Then,

$$E^*(j\omega) \cong \frac{1}{T} E(j\omega) \tag{4-83}$$

that is, the transfer function of the ideal sampler is approximately 1/T. However, it should be emphasized that this approximation is of limited use, and, in general, the input-output relation in Eq. (4-77) should be used.

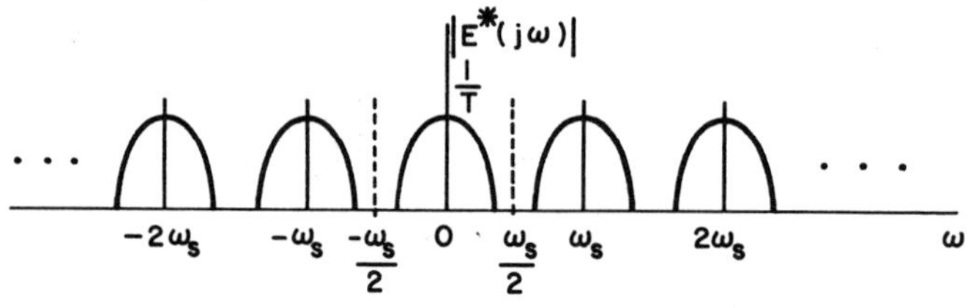

Fig. 4-22. Amplitude spectrum of $E^*(j\omega)$.

The z-Transform

Because the transfer function relations in sampled-data systems include exponential functions of the form, e^{Ts}, it is convenient to introduce the relation

$$z = e^{Ts} \tag{4-84}$$

where z is a complex variable. Equation (4-84) defines the *z-transform*.

To illustrate the z-transform analysis of a sampled-data control system, consider that the controlled process of the system of Fig. 4-18 is

$$G_p(s) = \frac{K}{s} \tag{4-85}$$

a simple integrator.

If we sample the output of the system, c(t), by a fictitious sampler with a sampling period T, the result is c*(t). The transfer relation between e*(t) and c*(t) can be written

$$G^*(s) = \frac{C^*(s)}{E^*(s)} \tag{4-86}$$

The z-transform of G*(s) is defined as

$$G(z) = \sum_{k=0}^{\infty} g(kT)z^{-k} \tag{4-87}$$

or

$$G(z) = \mathfrak{z}[g(t)] \tag{4-88}$$

The same definition can be applied to e(t) and c(t) whose z-transforms are E(z) and C(z), respectively.

To find G(z), we write

$$G(z) = \mathfrak{z}\left[\frac{1 - e^{-Ts}}{s}\frac{K}{s}\right] = (1 - z^{-1})\mathfrak{z}\left[\frac{K}{s^2}\right]$$

$$= (1 - z^{-1})\mathfrak{z}[Kt] \tag{4-89}$$

Thus,

$$G(z) = (1 - z^{-1}) \sum_{k=0}^{\infty} K(kT)z^{-k} \tag{4-90}$$

The infinite series in the last equation is written in a closed form,

$$G(z) = (1 - z^{-1})\frac{Tz}{(z-1)^2} = \frac{T}{z-1} \tag{4-91}$$

Thus, the open-loop transfer function of the system in Fig. 4-18 is given by Eq. (4-91), and the closed-loop transfer function is

$$\frac{C(z)}{R(z)} = \frac{G(z)}{1 + G(z)} = \frac{T}{z - 1 + T} \tag{4-92}$$

The characteristic equation is still defined as the denominator of the closed-loop transfer function set to zero. In this case it is

$$z - 1 + T = 0 \tag{4-93}$$

However, the stability boundary in the z-domain is the unit circle $|z| = 1$, since the z-transformation in Eq. (4-84) transforms the imaginary axis of the s-plane onto the unit circle in the z-plane. Thus, the system described by the closed-loop transfer function of Eq. (4-92) is $T < 1$.

4.7 CONCLUSIONS

In this chapter we introduced the fundamentals of the analysis and design of analog and sampled-data control systems. Due to the limited space and the objectives of this book, the coverage on these subjects is necessarily brief. The reader should refer to the references listed at the end of the chapter for more details.

REFERENCES

1. B. C. Kuo, *Automatic Control Systems*, 3rd ed., Prentice-Hall, Inc., 1975.
2. B. C. Kuo, *Analysis And Synthesis of Sampled-Data Control Systems*, SRL Publishing Company, 1963.
3. B. C. Kuo, *Digital Control Systems*, SRL Publishing Company, 1977.

5 DC Motors In Incremental Motion Systems

E. K. PERSSON

5.1 INTRODUCTION

In this chapter we will discuss the suitability of different types of dc motors for use in incremental motion servo systems. The function of the motor in serving as the actuating component for the system is to provide the torque needed to accomplish the system's incremental motion demands. There are other devices available which are designed specifically to accomplish incremental motion functions, such as the mechanical Geneva movement and electric step motors. Both have intrinsic "preprogrammed" motion steps built in; i.e., a given motion command will result in a predetermined motion step. The dc motor, on the other hand, has no such features, and, in order to function in an incremental motion servo system, it must be included in a control loop consisting of shaft connected transducers and amplifiers to control the motion. We may well ask, then, why dc motors should be considered over, for instance, step motors in incremental motion systems since they do not naturally "step." The answer is that dc motors offer superior response and improved flexibility in step rates and step sizes compared to the step motor. It turns out that the dc motor is used where the inherent limitations of step motors begin to be dominant.

Some classes of dc motors are more suitable for incremental motion purposes than others. In addition, members of the subclasses of these motors have specific design advantages which allow them to fulfill certain motion requirements more efficiently than others. In the following discussion we will examine the basic operating principles of dc motors; with these principles in mind, we will then compare the relative advantages and disadvantages of each of the several dc motor types. Finally, we will suggest general selection criteria for evaluating the effectiveness of particular motors in incremental motion applications.

5.2 OPERATIONAL PRINCIPLES

The dc motor is basically a torque transducer whose most important design characteristic is expressed in a linear relationship: the shaft torque of a dc motor is directly proportional to its armature current. A dc motor's shaft torque is produced by the interaction between a magnetic field and a set of conductors; this principle is illustrated in Fig. 5-1. Here, a current-carrying conductor is established in a field containing a magnetic flux ϕ , and the conductor is located at a distance r from the center of rotation. The relationship between the shaft torque and the current is:

$$T = K\phi I \tag{5-1}$$

where T is the shaft torque in Newton-meters; ϕ is the magnetic flux in webers; I is the current in amperes; and K is a proportionality constant. Thus, the shaft torque is directly proportional to the product of the magnetic flux and the current.

When the conductor moves in the magnetic field, a voltage is generated across it. This voltage, which is proportional to the shaft velocity, tends to oppose the current flow. The relationship between this back-emf voltage and the shaft velocity is:

$$E = K\phi\omega \tag{5-2}$$

where E denotes the back emf in volts; ϕ is the magnetic flux in webers; and ω is the velocity expressed in radians/second. Equations (5-1) and (5-2) form the fundamental basis for dc motor operation.

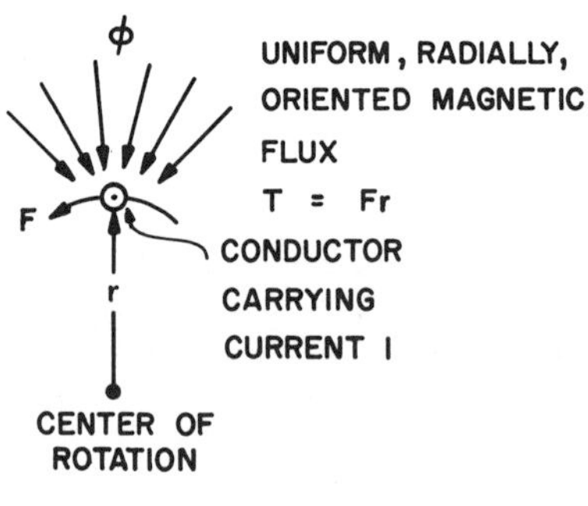

Fig. 5-1. Illustration of torque production in a dc motor.

5.3 BASIC CLASSES OF DC MOTORS

DC motors can be divided into several broad categories based on the method they employ to create the magnetic field, and on the basic design and structure of the armature. We will first review the two basic classes of magnetic flux supply: the variable-magnetic-flux motors and the constant-magnetic-flux motors. We will then examine the three basic classes of armature design: the iron-core dc, the surface-wound dc and the moving-coil motor. We will also consider the brushless dc motor as a special class which uses the same technology as the brush-type dc motors with the exception that the commutation is performed by electronic rather than mechanical means.

Variable-Magnetic-Flux Motors

There are two basic kinds of variable-magnetic-flux motors: those in which the field winding is connected in series with the armature winding and those in which the field winding is separately excited.

An example of the series-field-winding type is illustrated in Fig. 5-2. Since the magnetic flux in this motor is proportional to the current, and since the field flux is thus variable, we have a relationship between speed and torque which is nonlinear as shown in Fig. 5-3. Such motors are useful for specific conditions where high torque at low speeds and low torque at high speeds are called for. Examples of such uses are in the traction-drive systems.

An example of the separately-excited field winding type of motor is shown in Fig. 5-4. This type of dc motor is commonly called the shunt motor. In this motor, the variable magnetic flux is independent of the armature current; as a result, the machine constants can be externally controlled over a wide range. However, since the

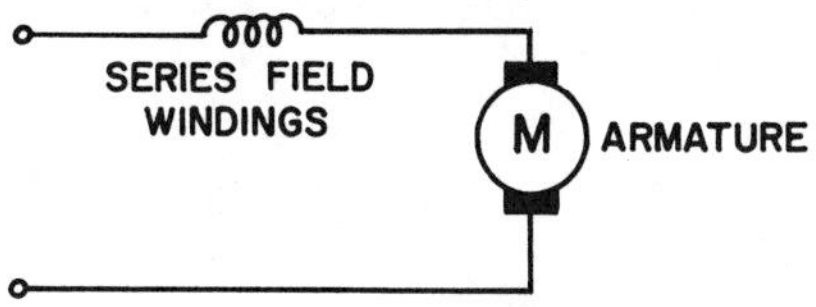

Fig. 5-2. A series dc motor armature and field connection.

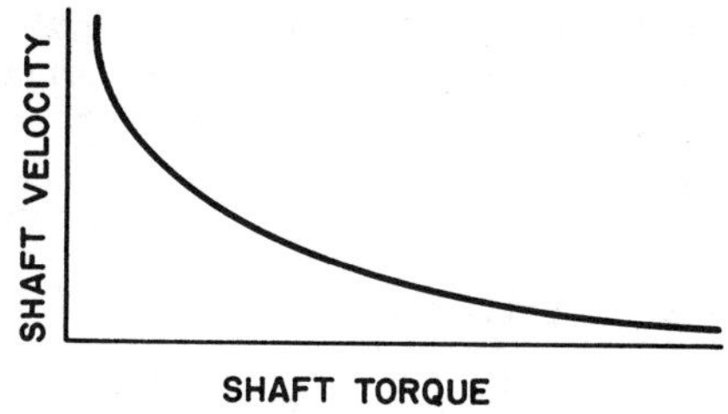

Fig. 5-3. Speed-torque characteristics of a series dc motor under constant voltage conditions.

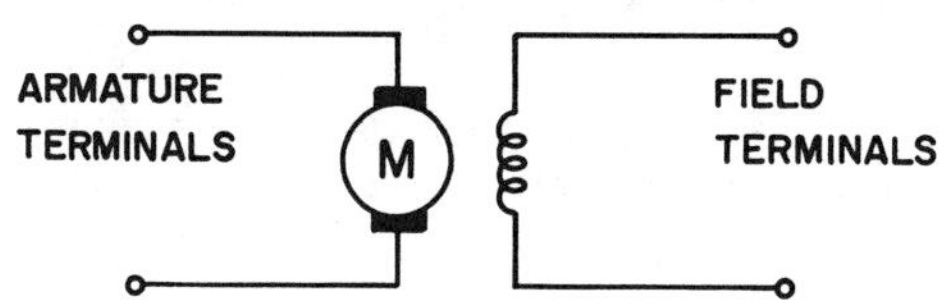

Fig. 5-4. DC motor with separately variable field.

field winding generally has a large electrical time constant, it is not always appropriate to use a variable-field motor for incremental motion applications unless the servo system can accommodate such a time constant. Such a motor may be useful in cases where high-torque servo-controlled motion is demanded, requiring a high torque constant and, yet, requiring that the motor should be able to perform rapid "slew" motion under light torque demand. In such a system the high-torque activities are performed under high-field excitation conditions and the "slew" motion is performed under low-current "field-weakened" conditions. This lessens the back-emf voltage of the motor and allows high velocities to be achieved with the same power supply as was used for the high-torque mode. In Fig. 5-5, we see a torque-speed characteristic of the shunt motor under constant-field excitation and constant-armature voltage. Under high-torque conditions it is usual to observe the effects of magnetic saturation causing a significant "droop" of the speed-torque characteristics, as seen in the picture.

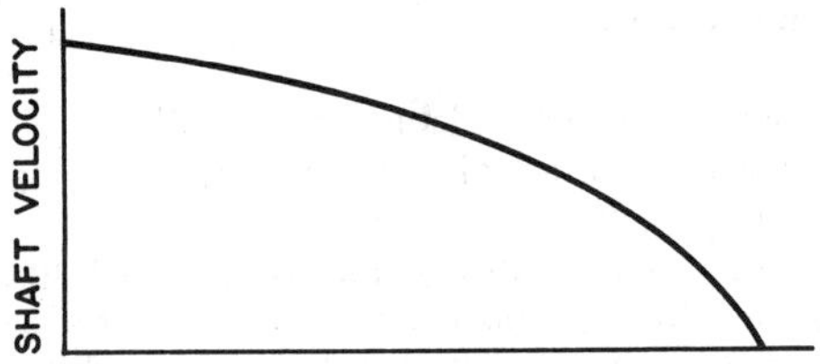

Fig. 5-5. Speed-torque characteristics of a shunt motor under constant armature voltage conditions and constant-field excitation conditions.

Constant-Magnetic-Flux Motors

The most common field excitation system in use today is the permanent-magnet dc motor shown in Fig. 5-6. In this system, because the field flux is essentially constant, the ratio between armature current and torque is fixed. Thus, Eq. (5-1) becomes simply:

$$T = K_t I \tag{5-3}$$

and Eq. (5-2) becomes

$$E = K_e \omega \tag{5-4}$$

As can be seen, this system allows for a very simple linear relationship between armature current, torque, and velocity. The electrical expression for such a motor is as follows:

$$V = K_e \omega + L\frac{di}{dt} + Ri \tag{5-5}$$

where V is the applied voltage, K_e is the back-emf voltage constant, L is the armature inductance, and R is the motor terminal resistance. The dynamic equation of the motor is

$$T_g = J\frac{d\omega}{dt} + B\omega + T_f + T_L \tag{5-6}$$

where T_g is the generated torque, J is the sum of the moments of inertia of the motor and the load, B is the viscous damping coefficient, T_f is the internal friction torque, and T_L is the load torque.

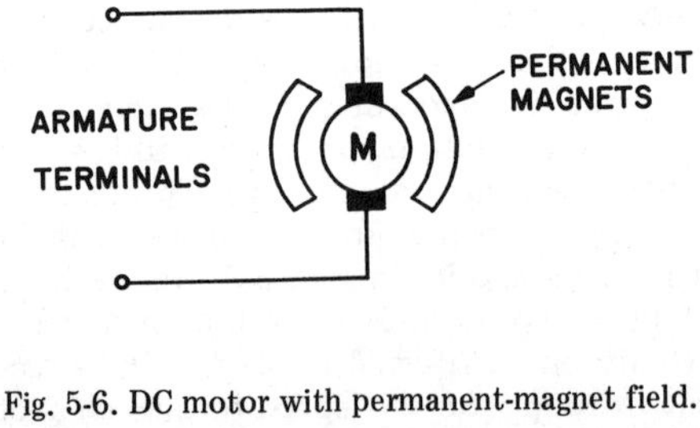

Fig. 5-6. DC motor with permanent-magnet field.

The same equations could be written for a separately-excited (i.e., wound-) field motor with a constant-field excitation. However, the advantage of the permanent-magnet field over the wound-field structure is that no power is dissipated in the field. The permanent-magnet dc motor is, therefore, more efficient and requires less space than an equivalent motor with a wound field. Also, due to the availability of permanent magnets with high coercive force characteristics, the linear relationship of Eq. (5-1) holds for higher armature currents in permanent-magnet dc motors than in their wound-field dc motor counterparts.

The simple mathematical relationship which exists between electrical input parameters and mechanical output characteristics in the permanent-magnet dc motor makes it very popular for use in incremental motion servo systems; use of this type of motor can simplify both the servo calculations (because the motor can be treated linearly) and the system design.

Several basic configurations of permanent-magnet dc motors exist. The various designs aim at satisfying a wide range of performance demands, most of which deal with the torque-to-inertia ratio. This ratio is especially important when motor inertia is the critical parameter, such as in digital tape transport capstan drive motors. In these systems, the only load is the moment of inertia of the tape capstan, which is approximately equal to the moment of inertia of the motor In applications where the load inertia is dominant, a motor with a higher moment of inertia would be the logical choice.

Permanent-magnet dc motors can be classified according to both commutation structure and armature design. Commutation of the armature current in dc motors is either electrical or mechanical; mechanical commutation structures use commutators made of copper and brushes generally made of carbon or graphite compounds containing metal particles, while electrical commutation of the armature current is achieved without the use of mechanical means. (Thus, the term "brushless" is used to describe this type of dc motor.)

The mechanical commutating motors can further be broken down into three types of armature design: iron core, surface wound, and moving-coil motors. We will now examine each of these three types of mechanically-commutated dc motors, as well as the above-mentioned brushless dc motor.

Iron-Core Motors

The rotor and stator configuration of an iron-core motor is shown in Fig. 5-7. The magnet structure in this example is shown as a segment of magnet material—in this case, a Barium Ferrite magnet (other materials, such as ALNICO or Rare Earth compound, could also have been used). The magnetic flux passes through a laminated rotor structure. The rotor contains slots, into which wire is placed. (The wire is mechanically well-supported in such a rotor structure since it can be impregnated with varnish or resins and mechanically bonded to the iron core.) The magnetic flux, therefore, must pass through a small air gap and over a low-reluctance path to reach the opposite pole. The iron-core structure has the highest moment of inertia and the largest rotor inductance of all the permanent-magnet motors under consideration. In addition, the rotor may have preferred "detent" positions due to unequal reluctance paths in the magnetic structure. (This is not normally a problem, however, because proper design—skewing the rotor slots, for example—can minimize or even eliminate this tendency.) The iron-core motor is by far the type of dc motor most widely used in incremental motion applications with low and moderate performance demands, especially since the iron-core rotor structure has high thermal capacity and can take overloads for extended periods of time without damage.

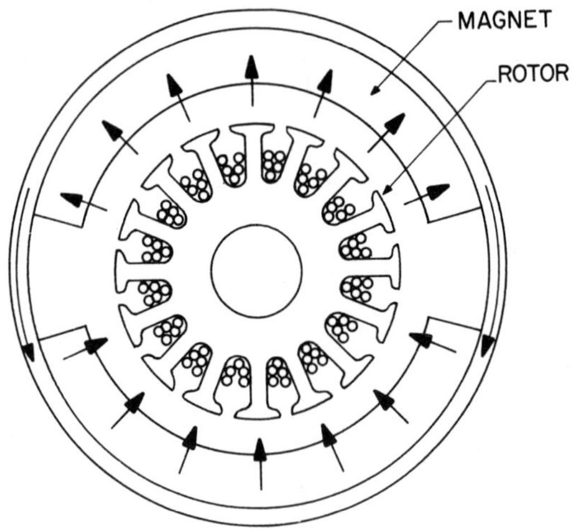

Fig. 5-7. Cross-sectional view of a permanent-magnet iron-core dc motor.

Surface-Wound Motors

Figure 5-8 shows the rotor design of a surface-wound motor. The conductors are bonded to the surface of a cylindrical rotor structure, which is made of laminated discs fastened to the motor shaft. Since no slots are used in this design, the armature has no tendency to assume preferred "detent" positions. And, since the conductors exist in the air gap between the rotor discs and the permanent-magnetic field, this design has lower inductance than the iron-core structure. However, because the air

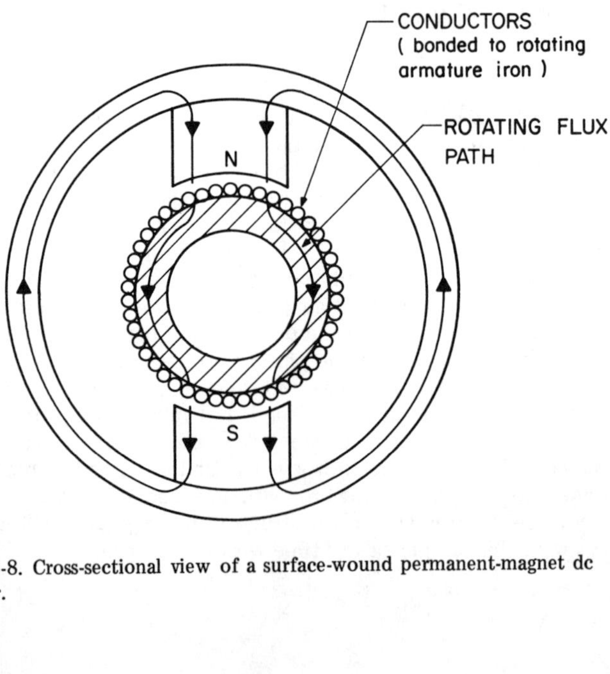

Fig. 5-8. Cross-sectional view of a surface-wound permanent-magnet dc motor.

gap between the magnet and the low-inductance rotor is larger than in the iron-core motor, a longer magnet is required in order to provide a magnetic flux equivalent to that of the iron-core motor. Surface-wound motors are, therefore, more expensive to produce and have larger outside diameters than equivalent iron-core motors.

Moving-Coil Motors

Moving-coil motors are designed to have very small moments of inertia. This is achieved through rotor design: the rotors consist of conductors placed in the air gap between a stationary flux return path and the magnet structure, as shown in Fig. 5-9. In this example, the conductor structure is supported by nonmagnetic materials—usually epoxy resins and fiberglass—to form a hollow cylinder. One end of the cylinder forms a hub, which is attached to the motor shaft. A cross-section view of such a motor is shown in Fig. 5-10. As a further illustration, Fig. 5-11 and Fig. 5-12, respectively, show photographs of a moving-coil armature and its magnet and housing counterpart.

Since all unnecessary elements have been removed from the armature of a moving-coil motor, its moment of inertia is very low. However, it has a larger magnetic air gap than the two models discussed earlier, and, therefore, requires an even larger magnetic structure than the previous examples to produce an equivalent air gap magnetic flux. Hence, the price one pays for the high torque-to-inertia ratio in such a motor is a rather large magnet structure. In addition, the rotor structure has a very low thermal capacity; if overloaded, it can easily be damaged. When moving-coil motors are used in high-performance incremental motion applications, it is common to cool the rotor structure by means of forced-air cooling. The moving-coil armature circuit, because the conductors are not in direct contact with iron, has a very low inductance: values of less than 100 microhenries are common.

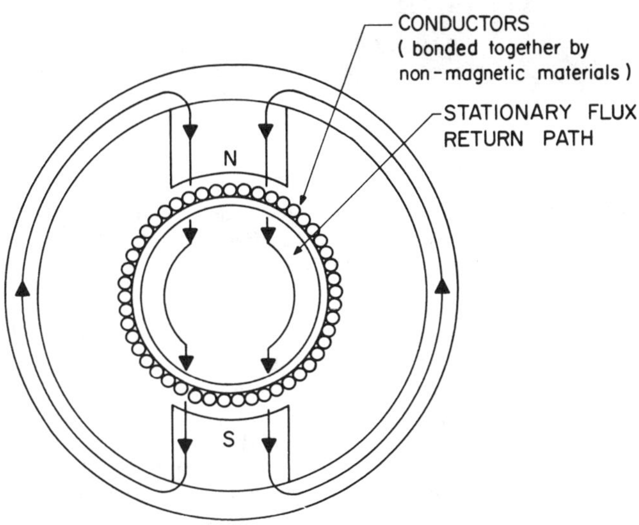

Fig. 5-9. Cross-sectional view of a moving-coil permanent-magnet dc motor.

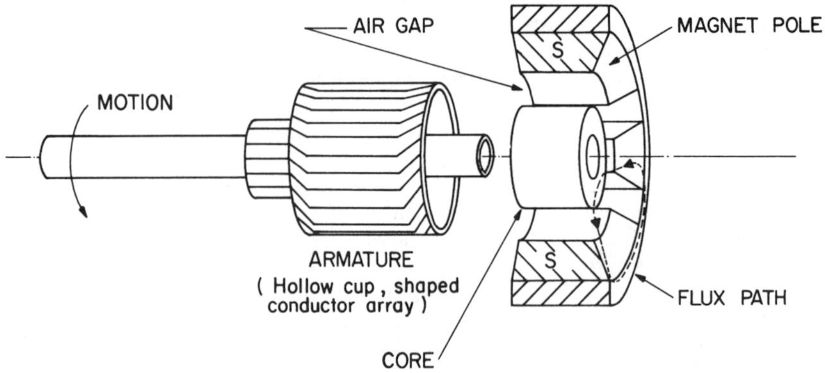

Fig. 5-10. Cross-sectional side view of a moving-coil dc motor.

Fig. 5-11. Moving-coil armature of a hollow cup-shaped cylinder of bonded aluminum wires. The output shaft is made from Alumina ceramic materials for high tensional stiffness. (Courtesy, Electro-Craft Corporation)

Another version of the moving-coil motor is shown in Fig. 5-13. This type of motor has a disc-shaped armature structure which is made up of a multitude of conductors—sometimes in several layers—forming a conductor array. The conductor array is utilized as a commutation surface. This motor type has, in the past, been called a "printed motor," in reference to early models in which so-called "printed circuits" were first used to form the conductor arrays; today this motor is made from stamped and welded copper conductor paths. It has found many uses, especially in applications where short axial length is a critical factor. The disc armature in this type of motor generally has a higher moment of inertia than the cylindrical moving-coil armature for two reasons: the end turns are the dominant inertia-producing element, and the moment of inertia of a disc increases by the fourth power of its diameter.

Another example of a moving-coil motor is shown in Fig. 5-14. Here we see a 7 horsepower (continuous rated) moving-coil motor used in a machine tool applica-

Fig. 5-12. Housing and magnet structure of a moving-coil motor, showing the magnet and back iron flux return path. (Courtesy, Electro-Craft Corporation)

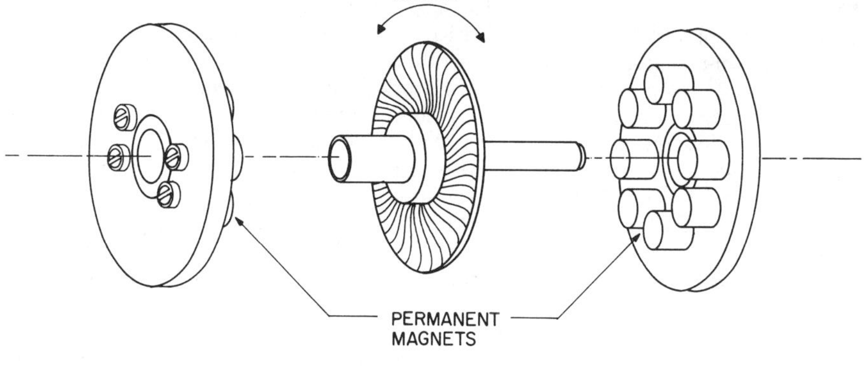

Fig. 5-13. "Exploded" view of a disc-type moving-coil dc motor.

tion. This motor can attain peak torque levels of 100 ft-lb and can take current pulses of up to 800 amperes without demagnetization.

At present, the moving-coil motor is the best actuator choice for high-performance incremental motion systems; it has the highest torque-to-inertia ratio and the lowest inductance of all the motors so far discussed. As an example of the acceleration capabilities of such motors, consider the fact that, in high-performance digital tape transports, capstan drive acceleration levels of 10^6 rad/sec^2 are commonly achieved. (In some instances these motors use shafts made of alumina ceramic materials in order to minimize torsional resonance between the motor armature and the capstan.)

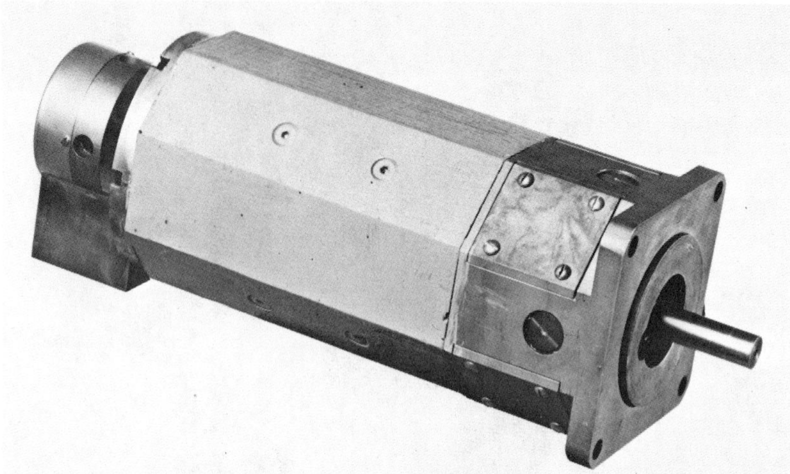

Fig. 5-14. Industrial-type moving-coil dc motor, designed for incremental motion. (Courtesy, Electro-Craft Corporation)

Brushless DC Motors

Brushless dc motors differ from the above-mentioned dc motors in that they employ electrical (rather than mechanical) commutation of the armature current. The configuration of the brushless dc motor most commonly used—especially for incremental motion applications—is one in which the rotor consists of magnets and "back iron" support, and whose commutated windings are located external to the rotating parts (Fig. 5-15). (Compared to the dc motors previously considered, this is an inverted configuration.) Depending on the specific application, brushless dc motors can be used when a low moment of inertia is called for. In order to achieve low moment of inertia in this type of motor, the rotor structure must be manufactured from a magnetic material which offers a high ratio of magnetic flux and a high coercive force in relation to its mass. ALNICO 8 and Barium Ferrite are such materials; Rare Earth magnets are even more suited for this purpose. The brushless dc motor winding commutation requires special attention since economic considerations (the number of switches required) generally dictate that fewer commutations be made in brushless motors than in conventional dc motors.

Because the commutation arrangement in brushless motors exists external to the motor, we must examine the switching arrangement in conjunction with the motor in order to understand the electronic control requirements. In Fig. 5-16, we see a block diagram of the essential parts of a brushless dc motor. The shaft encoder shown in the picture is necessary to synchronize the commutation switching with the angular position of the rotor. Figures 5-17 and 5-18 show some alternative controller (also called inverter) configurations commonly used in brushless dc motors. Many variations of the circuits shown are possible, but it is beyond the scope of this brief overview to fully discuss all of the electronic commutation schemes. Reference [3] gives a more detailed treatment of the variety of control circuits which can be used.

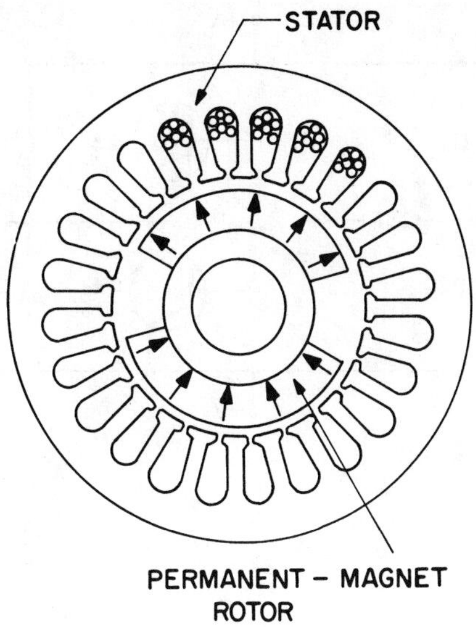

Fig. 5-15. Cross-sectional view of a permanent-magnet brushless dc motor.

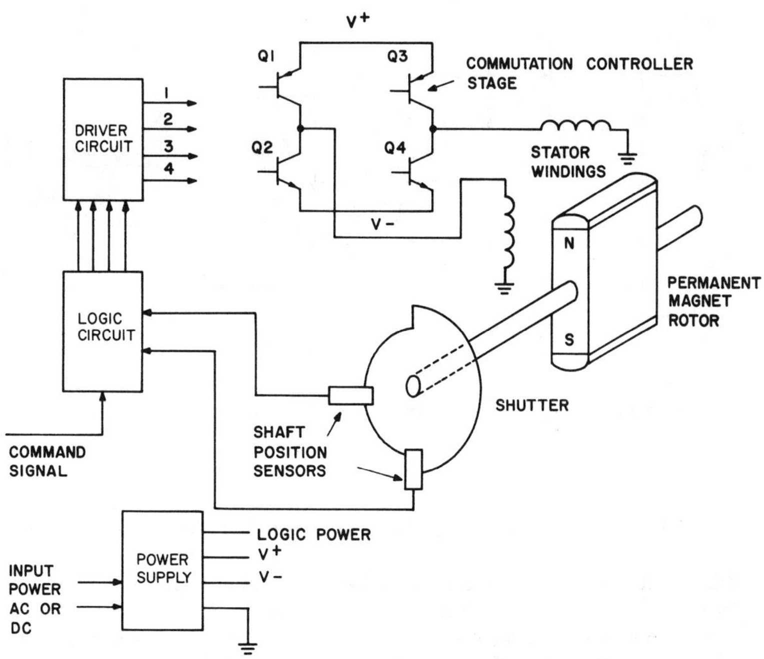

Fig. 5-16. The essential parts of a brushless dc motor.

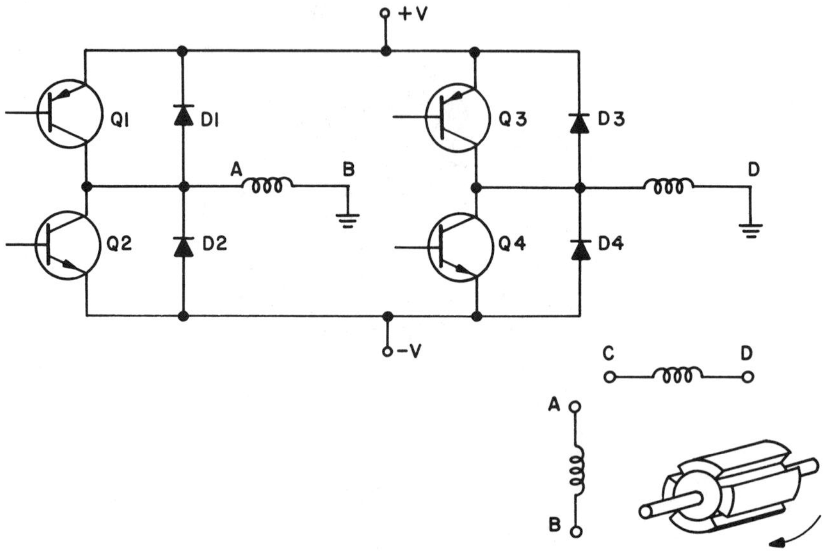

Fig. 5-17. An example of a two-phase brushless dc motor inverter configuration using two power supplies.

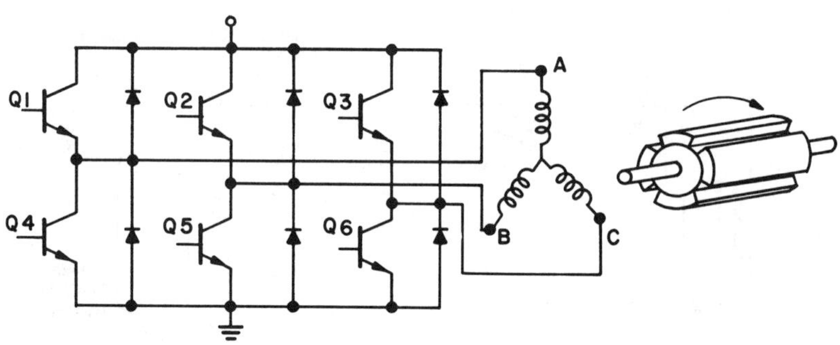

Fig. 5-18. A typical three-phase brushless dc motor inverter, using a single power supply.

5.4 SELECTION CRITERIA FOR INCREMENTAL MOTION APPLICATIONS

Up to this point we have discussed some basic operating principles of dc motors as well as typical characteristics of basic dc motor models. As we have seen, each of the various motor designs has its own unique advantages and disadvantages. When a designer is faced with the problem of selecting the best motor for a given application, he must first define his specific needs in terms of system requirements. With this knowledge, he can then examine the various options of available motors with a view toward the necessary controller, power supply, and feedback transducer requirements. We will now discuss some of the primary performance characteristics required of the dc motor in order to meet the demands of incremental motion applications.

Motor Efficiency

Systems incorporating incremental motion often involve the production of high peak torques with little useful work done; that is, they involve acceleration and deceleration of inertial loads with small frictional demands. The motor, therefore, absorbs energy during both acceleration and deceleration, and the primary heat-producing element in the motor is armature resistance, and, if high shaft velocities are involved, we may have significant contributions to motor heating by viscous damping effects as well.

It is important to note that high motor efficiency per se is not the only criterion by which a dc motor for incremental motion use should be judged: it must be considered together with motor moment of inertia, motor inductance and peak current capability (to mention a few), in order to arrive at optimum *systems* efficiencies and performance capabilities.

Armature Inertia

Since the objective in incremental motion systems is the rapid starting and stopping of a rotating load, the motor's contribution to the system's inertia should be optimized as nearly as practicable, in order to fulfill the system specifications and requirements. (See Chapter 15 for a discussion on this subject.) The moving-coil motor discussed above offers the lowest inertia of all dc motors. If adequate cooling means are available, this motor design may be a good choice. However, the low thermal capacity of the rotor structure demands a thorough knowledge of the worst-case load conditions since the moving-coil armature (which can have a thermal time constant as low as 30 seconds) can be damaged by even short-term overload conditions. The iron-core structures, on the other hand, can in many cases sustain prolonged overload without damage; they can have thermal time constants as high as 30 minutes. Iron-core motor designs using high energy-product magnets such as ALNICO or Rare Earth types can achieve a drastically lower moment of inertia than motors using Ceramic magnets. With a variety of moving-coil and iron-core designs available today, we can find areas of inertia/torque capability overlap between the groups.

Armature Inductance

In order to achieve a rapid motor response, the armature inductance of the motor must be low. The electrical time constant of the motor can then permit the required current rise time without interfering with the servo operational requirements. However, sometimes a low armature inductance is needed to fulfill servo response demands at the same time that a relatively high-inductance value is needed in order to use pulsewidth-modulation (PWM) techniques for power control. In such cases, external inductors must be added on to low-inductance motors (such as the moving-coil motor) in order to keep the PWM ripple current at an adequate level.

Linear Current-To-Torque Relationships

As discussed earlier, the linear relationship which exists between armature current and output torque is a very important characteristic of dc motors when considering their suitability for use in incremental motion servo systems. Another way of stating this is that the torque constant K_t should not vary with the current level or the angle of rotation of the rotor. Conditions which can cause the torque constant to vary are magnet circuit saturation and/or a poor armature design. The former results in low torque constant at high torque levels (high peak torque dissipation) and the latter causes a cyclically varying torque with rotation due to an insufficient number of commutation points (causing cyclic torque disturbances in the servo loop).

Reluctance Torque Disturbances

The magnetic design of dc motors for incremental motion applications should be such that the armature does not have preferred shaft angle positions (i.e., that the rotor does not "cog" when rotated by external means while unenergized). The tendency to cog is the result of a design in which the rotor structure provides a lower reluctance to the magnetic field in some positions than in others. Thus, if it is necessary to stop and hold a rotor shaft in a given angular position, a rotor which "cogs" might require a constant current to hold the shaft in that specific position—if that position also happens to be located at a point where reluctance torque is present.

The Motor Torque Constant Versus Temperature

If the motor is to be used in an application where significant motor heating is likely to occur, the temperature coefficient of the magnetic field must be examined. This is necessary to determine whether the torque constant could be significantly altered at extreme temperature conditions (possibly causing thermal "runaway" at high operating temperatures). As it turns out, the lowest-cost permanent-magnet materials have a higher temperature coefficient of magnetic flux than others: Barium-Ferrite-type ceramic magnets, for example, have a temperature coefficient of −0.2%/°C, whereas ALNICO 5 has a temperature coefficient of −0.02%/°C, and ALNICO 8 has a temperature coefficient of less than −0.01%/°C. The newest of the magnet materials, the Rare Earth compounds, have a tendency to experience long-term irreversible demagnetizing effects if operated at high temperatures. Depending on specific compounds, they may show "aging" effects if operated above temperatures from 150°C to 200°C. The motor manufacturer's specifications will indicate the safe temperature regions.

Motor Peak Current Capability

Depending on the type of magnet used and on the air gap characteristics of the motor, motor magnets may demagnetize if the current exceeds a certain value. This is generally due to the intrinsic coercive force characteristics of the magnet materials. (The "strongest" magnets in this category are the Samarium Cobalt magnets, which are practically impossible to demagnetize without destroying other parts of the motor.)

A common trend is that all permanent magnets have a lower resistance to demagnetization at high operating temperatures. Since individual motor designers have varying degrees of sensitivity to the combination of peak currents and temperature, it is important to examine the motor specifications in order to determine whether current-caused demagnetization could create a potential problem in the motor under consideration.

Viscous Damping Coefficient

The viscous damping coefficient consists of several components, some of which are mechanical in nature and others are caused by electromagnetic actions in the structure. The presence of viscous damping can be inferred from the observation of increased no-load current at increased speeds.

The mechanical component of the viscous damping is partly due to friction caused by bearing lubrication components and bearing seals. The contribution to viscous friction by these elements usually is one that increases linearly with velocity. Another mechanical effect which may manifest itself at high shaft velocity is the so-called "windage" element due to the rapidly rotating parts in the motor.

The electromagnetic components of the viscous damping effect are caused by three principal elements: the circulating current, the eddy current and the hysteresis effects. The circulating-current effect is due to currents flowing in commutated coils in the presence of stray fields in the stator structure. This effect, which increases linearly with the velocity, results in heat losses similar to those generated by viscous friction. The eddy-current effect is usually noticed at high motor speeds and is caused by eddy currents generated in the armature conductors moving in the magnetic field. This effect tends to increase with the square of the motor velocity. Motors with large, or flat conductors tend to have a significant high-speed effect of eddy-current damping.

The third effect mentioned, the hysteresis effect, is caused by the change in magnetic flux in the rotor due to the rotation in the permanent-magnet field; the iron experiences a rapidly changing alternating magnetic field. The magnitude of the hysteresis effect depends on the armature lamination material quality. For a constant-flux-density field, the hysteresis effect will be directly proportional to the velocity.

The comments about the eddy-current and hysteresis effects were made on the assumption of a relatively constant armature current. The heating effects due to eddy-current and hysteresis effects may be greatly accentuated if the motor is operated in a pulsewidth modulation or pulse-frequency modulation mode with a significant ac armature current component in addition to the dc current. If magnet materials of low resistivity (such as $SmCo_5$) are used in the presence of significant ac current components, eddy-current heating may also occur in the magnet material causing heat.

It is, therefore, apparent that we must pay attention to operational conditions such as these if the motor is used in high-performance activities. An examination and awareness of all possible heat loss sources is important in the choice of the best motor for a given application.

Torsional Resonance Conditions

Since incremental motion systems are usually subject to step torque conditions, and since their motors are usually connected to feedback transducers as well as to the load, it is important to examine the system for the presence of torsional resonance frequencies. Torsional resonance causes oscillation in the closed-loop servo system. In order to reduce or eliminate its effects, the following should be attended to: (1) the motor shaft must have high torsional stiffness; (2) the feedback transducers must be properly connected to the motor shaft (preferably with a direct connection); and (3) the load must be properly coupled, either directly or by means of a coupling which does not offer significant torsional resonance problems.

A direct-coupled dc motor-tachometer is shown in Fig. 5-19. The motor and tachometer armatures are of the iron-core type, both are bonded on the same shaft in order to maximize torsional stiffness.

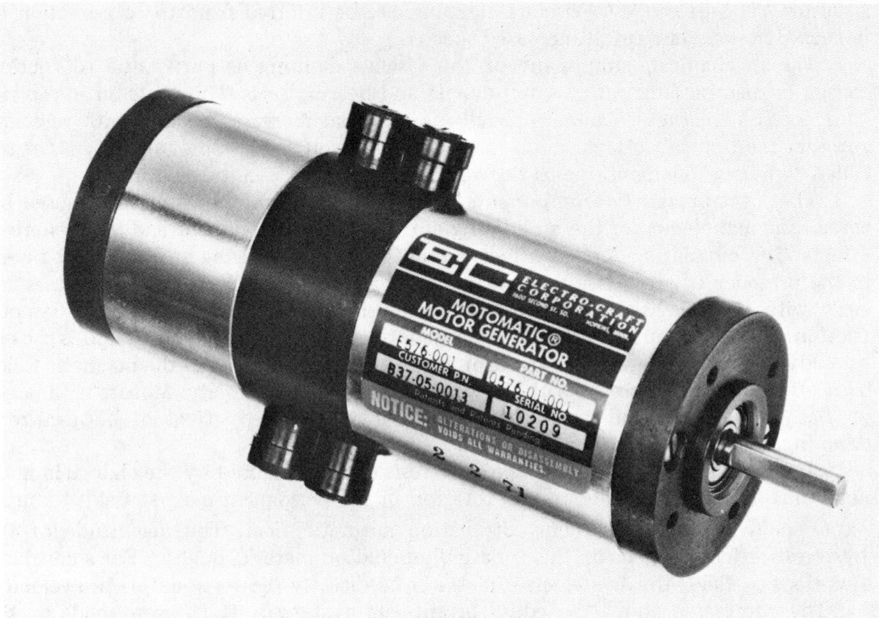

Fig. 5-19. A dc motor-tachometer for incremental motion control. (Courtesy, Electro-Craft Corporation)

5.5 CONCLUSION

The correct choice of a system actuator is the single most important step toward achieving the most highly efficient design in incremental motion servo systems. In this chapter we have discussed the merits of the dc motor in fulfilling this function. We have considered the dc motor's basic operating principles as well as the selection criteria for choosing an actuator for incremental motion applications. With a good understanding of the capabilities and shortcomings of the various dc motor types presented in this chapter, the system designer can more effectively meet the design requirements of the incremental motion application under examination.

REFERENCES

1. A. E. Fitzgerald, C. Kingsley, Jr., and A. Kusko, *Electric Machinery*, 3rd ed., McGraw-Hill Book Company, New York, 1971.
2. *DC Motors, Speed Controls, Servo Systems*, 3rd ed., Electro-Craft Corporation, Oct. 1975.
3. E. K. Persson, "Brushless DC Motors in High-Performance Servo Systems," *Proceedings, Fourth Annual Symposium on Incremental Motion Control Systems and Devices*, ed., B. C. Kuo, University of Illinois, pp. T-l—T-16, 1975.

6 Mathematical Modeling of DC Motors

D. K. GEHMLICH

6.1 INTRODUCTION

The basic direct-current dynamo, consisting of an armature, brushes, and field windings, either series, parallel, or combination, has been used for many years as a basic energy converter. Such motors have been used as prime movers in such applications as rolling mills, electric hoists, cranes, electric vehicles, and some pumps where variable speeds are necessary. The dc motor, along with an appropriate dc amplifier system and either position or velocity feedback, provides the power element in many guidance and control applications. Recent advances in the art of magnetic materials and solid-state amplification have made permanent-magnet dc motors even more popular as the drive elements of many closed-loop control systems. The proper design and compensation of such systems requires an adequate mathematical model for all of the components of the system. In this chapter, we develop a linear model for a separately-excited dc shunt motor and the permanent-magnet-type dc motor. The electromechanical model for the motor treats the input armature voltage and the load torque as input variables and the shaft velocity or position as output variables. The thermal model of the motor treats the input power or power dissipation as an input variable and the temperature rise in the device as the output variable.

6.2 THE ELECTROMECHANICAL MODEL

The significant components of a separately-excited dc motor are modeled as shown in Fig. 6-1. The armature is modeled as a constant resistance R_a in series with a constant

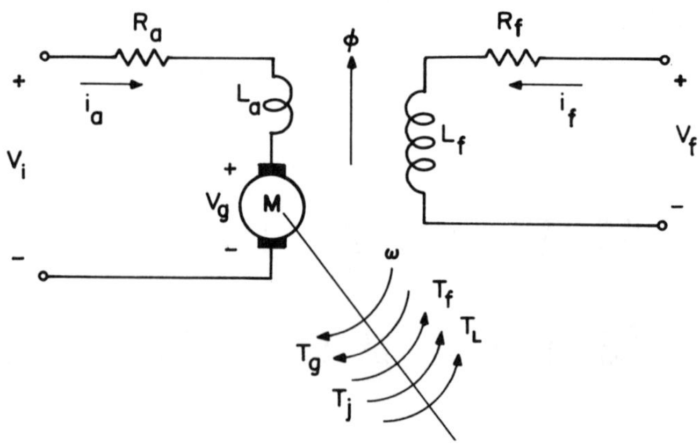

Fig. 6-1. Model of a separately excited dc motor.

inductance L_a, representing the armature winding inductance, and a voltage source v_g representing the generated voltage in the armature. The wound field is represented by a constant resistor R_f and a constant inductance L_f, with the net air gap field flux being designated by ϕ. The velocity of the armature shaft is represented by ω and the generated torque, motor friction torque, motor inertial torque, and the load torque (on the motor shaft) are indicated by T_g, T_f, T_J, and T_L, respectively.

The first of a series of basic equations modeling the dc motor is obtained by writing a voltage equation around the armature loop.

$$v_i(t) = R_a i_a(t) + L_a \frac{di_a(t)}{dt} + v_g(t) \tag{6-1}$$

The voltage term $v_g(t)$ in Eq. (6-1) is a generated voltage, which results when the conductors of the armature move through the field flux established by the field current i_f. According to Faraday's law, a conducting loop rotating in a constant magnetic field has induced in it a voltage

$$v(t) = \frac{d\lambda(t)}{dt} \tag{6-2}$$

where $\lambda(t)$ represents the magnetic flux linkages to the coil, and t is time. In a rotating commutator machine, each closed conducting path in the rotor has in it a voltage given by Eq. (6-2) above, where $d\lambda(t)/dt$ is proportional to the magnetic flux in the air gap and the angular velocity $\omega(t)$, or

$$v_g(t) = K\phi(t)\omega(t) \tag{6-3}$$

Finally, assuming the field current to be constant and ignoring the changes in the field flux due to armature reaction and other second-order effects, the field flux becomes a constant, and Eq. (6-3) becomes

$$v_g(t) = K_e\omega(t) \tag{6-4}$$

A force is exerted on the current-carrying conductors in the rotor in the presence of perpendicular field flux. The magnitude of this force is proportional to the magnitude of the current in the conductors, strength of the magnetic field, and the length of the conductors. Each of the conductors in the magnetic field contributes a component of the total force, and that force vector operating at a radius arm determined by the construction of the rotor produces the net developed electromagnetic torque, which, assuming the field flux to remain a constant, is proportional to the armature current,

$$T_g(t) = K_t i_a(t) \tag{6-5}$$

where K_t is the torque constant of the motor.

The mechanical power developed in the rotor is the product of developed torque and angular velocity

$$P_g(t) = T_g(t)\omega(t) \tag{6-6}$$

Of course, not all of this developed mechanical power in the rotor is delivered to the mechanical load connected to the shaft. Some of the developed power is lost in the motor through rotor wind resistance and friction and by hysteresis and eddy-current losses in the rotor iron, hereafter called friction losses, and some of the developed energy is stored as kinetic energy in the rotating mass of the rotor. The torque necessary to overcome the friction losses is designated as $T_f(t)$ and the torque necessary to accelerate the inertia is $T_J(t)$. At any instant of time, the developed torque must be equal and opposite to the sum of the torques necessary to overcome friction, inertia, and the load torque. Thus,

$$T_g(t) = T_f(t) + T_L(t) + J\frac{d\omega(t)}{dt} \tag{6-7}$$

where the inertial torque is represented by the product of the moment of inertia J and the angular acceleration $d\omega/dt$, and J is total moment of inertia of the rotor and the load referred to the motor shaft. That component of the total friction torque at the motor shaft that is a linear function of the rotor angular velocity is called the viscous friction component or the viscous damping component and is often separated from all other friction terms, as shown in the following equation:

$$T_g(t) = T_f(t) + T_L(t) + J\frac{d\omega(t)}{dt} + B\omega(t) \tag{6-8}$$

where B is the viscous frictional coefficient representing all motor and load viscous friction referred to the motor shaft, and $T_f(t)$ is the sum of all motor and load friction, windage, and iron-loss terms referred to the motor shaft, excluding the viscous friction. Equations (6-1), (6-4), (6-5), and (6-8) constitute a basic set of equations that model the dc motor and from which transfer functions for the dc motor operated in various modes may be obtained. Taking the Laplace transform of both sides of the basic set of equations and rearranging leads to

$$V_i(s) - V_g(s) = (R_a + sL_a)I_a(s) \tag{6-9}$$

$$V_g(s) = K_e\omega(s) \tag{6-10}$$

$$T_g(s) = K_t I_a(s) \tag{6-11}$$

$$T_g(s) - T_f(s) - T_L(s) = (B + sJ)\Omega(s) \tag{6-12}$$

The block diagram representation of these basic equations is shown in Fig. 6-2.

Before continuing on to the transfer function for the dc motor, it may be appropriate to pause here for several observations. First, let us assume that the voltage applied to the armature circuit of the motor is *constant* and that the motor has reached a *steady-state* operating point with a *constant load* applied. The mechanical power developed by the rotor is given by the steady-state constant version of Eq. (6-6), and when combined with that of Eq. (6-5) gives

$$P_g = T_g\omega = K_t I_a \omega \tag{6-13}$$

where all elements of the last equation are constants due to the steady-state condition. The electrical power absorbed by the armature must be the product of the voltage across it and the current through it. Hence,

$$P = V_g I_a = K_e \omega I_a \tag{6-14}$$

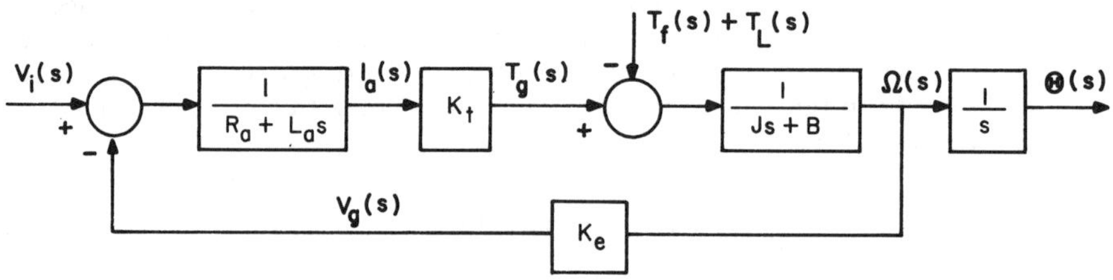

Fig. 6-2. Block diagram of a dc motor model.

Thus, because the mechanically developed power must be equal to the electrical power absorbed in the rotor, in the MKS system, $K_t = K_e$.

By now it has become apparent that the basic set of equations contains no references to the field current nor the field coil parameters of resistance and inductance. The assumption of constant field flux is a direct consequence of the assumption of constant field current; thus, the model to be derived from this basic set also represents a dc permanent-magnet motor in which the field flux is constant by virtue of a permanent-magnet field structure.

6.3 DC MOTOR TRANSFER FUNCTIONS

The block diagram of Fig. 6-2 represents a two-input system with the output being either the angular velocity ω or the angular position θ, or both. From Fig. 6-2 the output velocity of the system is written

$$\Omega(s) = G_1(s)V_i(s) + G_2(s)[T_f(s) + T_L(s)] \tag{6-15}$$

in which

$$G_1(s) = \left.\frac{\Omega(s)}{V_i(s)}\right|_{T_f(s) + T_L(s) = 0} \tag{6-16}$$

$$G_2(s) = \left.\frac{\Omega(s)}{T_f(s) + T_L(s)}\right|_{V_i(s) = 0} \tag{6-17}$$

The voltage-velocity transfer function $G_1(s)$ is

$$G_1(s) = \frac{\Omega(s)}{V_i(s)} = \frac{K_t}{(L_a s + R_a)(Js + B) + K_t K_e}$$

$$= \frac{K_m}{\alpha s^2 + \beta s + 1} \tag{6-18}$$

where

$$K_m = \frac{K_t}{R_a B + K_t K_e}$$

$$\alpha = \frac{L_a J}{R_a B + K_t K_e}$$

$$\beta = \frac{R_a J + L_a B}{R_a B + K_t K_e}$$

Equation (6-18) defines the second-order voltage transfer function for the dc motor under the assumption that T_f and T_L are zero. The equation can also be expressed as

$$G_1(s) = \frac{K_t}{R_a B(1 + \tau_e s)(1 + \tau_m s) + K_t K_e}$$

where

$\tau_e = L_a/R_a =$ electrical time constant

$\tau_m = J/B =$ mechanical time constant

If the armature inductance is very small, the electrical time constant can be neglected, and Eq. (6-18) becomes

$$G_V(s) = \frac{\Omega(s)}{V_i(s)} = \frac{K_t}{R_a(Js + B) + K_t K_e}$$

$$= \frac{K_m}{\tau s + 1} \tag{6-19}$$

where

$$\tau = \frac{R_a J}{R_a B + K_t K_e}$$

In Eq. (6-19), the constant K_m is typically called the motor constant.

The load torque transfer function $G_2(s)$ is given by

$$G_2(s) = \frac{\Omega(s)}{T_f(s) + T_L(s)} = \frac{-\dfrac{1}{Js + B}}{1 + \dfrac{K_t K_e}{(Js + B)(L_a s + R_a)}}$$

$$= \frac{-\dfrac{R_a}{K_t} K_m \left[\dfrac{L_a}{R_a} s + 1\right]}{\alpha s^2 + \beta s + 1} \tag{6-20}$$

which, if the armature inductance is negligible, reduces to

$$G_L(s) = \frac{\Omega(s)}{T_f(s) + T_L(s)} = \frac{-\dfrac{R_a}{K_t} K_m}{\tau s + 1} \tag{6-21}$$

Transfer Function Examples

A small dc motor (typical example, Electro-Craft Model E-576 Servo Motor) has the following specifications:

$V_i(\text{max}) = 30$ volts

$I_a(\text{max}) = 2$ amperes

$R_a = 3$ ohms

$L_a = 6$ millihenries

$K_t = 50 \times 10^{-3}$ Newton-meter/ampere

$T(\text{rated}) = 0.1$ Newton-meter

$J_m = 40 \times 10^{-6}$ kilogram-meter^2

$B_m = 35 \times 10^{-6}$ kilogram-meter^2/sec

$\omega(\text{rated}) = 300$ radians/sec

If we assume that the load inertia and friction referred to the motor shaft are

$J_L = 60 \times 10^{-6}$ kilogram-meter^2

$B_L = 65 \times 10^{-6}$ kilogram-meter^2/sec

then the voltage-velocity transfer function becomes [from Eq. (6-18)]

$$G_1(s) = \frac{17.9}{2.1 \times 10^{-4}s^2 + 0.107s + 1}$$

If the armature inductance is ignored, the transfer function becomes [from Eq. (6-19)]

$$G_V(s) = \frac{17.9}{0.107s + 1}$$

If a step voltage of 10 volts is applied to motor at t = 0, the angular velocity as a function of time is easily calculated.

$$\Omega(s) = \frac{10}{s}\left[\frac{17.9}{0.107s + 1}\right]$$

$$= \frac{179}{s(0.107s + 1)}$$

Taking the inverse Laplace transform of the last equation leads to

$$\omega(t) = 179(1 - e^{-9.35t})$$

and as $t \to \infty$ (steady state), $\omega \to 179$ rad/sec.

The load torque-speed transfer function becomes

$$G_2(s) = \frac{-1074(2 \times 10^{-3}s + 1)}{2.1 \times 10^{-4}s^2 + 0.107s + 1}$$

which for small L_a becomes

$$G_L(s) = \frac{-1074}{0.107s + 1}$$

Thus, for a step increase in load torque of 0.05 N-m, the speed change is

$$\Omega(s) = \frac{0.05}{s}\left[\frac{-1074}{0.107s + 1}\right]$$

$$= \frac{-53.7}{s(0.107s + 1)}$$

The time domain change in speed is

$$\omega(t) = -53.7(1 - e^{-9.35t})$$

In the steady state, the speed drop is 53.7 rad/sec with a load torque increase of 0.05 N-m.

6.4 POWER DISSIPATION

In this section we treat the dc motor as an energy conversion device in which electrical input energy is converted to mechanical output energy. Figure 6-3 shows the actual flow of power through such an energy conversion device. Some fraction of the electrical input power is immediately converted to heat in the windings of the armature. The rest of the electrical power is converted to mechanical power, some of which is lost through friction and windage losses, leaving the rest as mechanical rotational power available at the shaft.

The total electrical input power is the instantaneous product of the applied armature voltage and the armature current. Thus, the total instantaneous input power is

$$p_i(t) = v_i(t)i_a(t) \tag{6-22}$$

Combining Eqs. (6-1) and (6-4), we obtain the armature circuit voltage equation which relates the applied voltage to the armature current.

$$v_i(t) = R_a i_a(t) + L_a \frac{di_a(t)}{dt} + K_e\omega(t) \tag{6-23}$$

Substitution of Eq. (6-23) into Eq. (6-22) results in the expression for the total input power in terms of armature current and shaft angular velocity.

$$p_i(t) = R_a i_a^2(t) + L_a i_a(t)\frac{di_a(t)}{dt} + K_e i_a(t)\omega(t) \tag{6-24}$$

The armature current can be written in terms of the mechanical motor parameters and shaft torques by combining Eqs. (6-5) and (6-8)

$$i_a(t) = \frac{1}{K_t}\left[J\frac{d\omega(t)}{dt} + B\omega(t) + T_f(t) + T_L(t)\right] \tag{6-25}$$

which when substituted into Eq. (6-24) results in the following power expression:

$$p_i(t) = R_a i_a^2(t) + \frac{K_e B\omega^2(t)}{K_t} + \frac{K_e\omega(t)T_f(t)}{K_t} + \frac{K_e\omega(t)T_L(t)}{K_t}$$

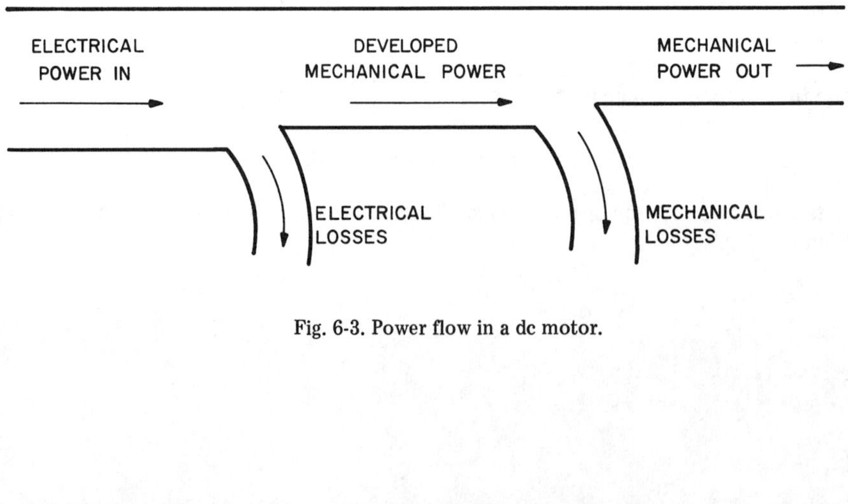

Fig. 6-3. Power flow in a dc motor.

$$+ \frac{K_e}{K_t} J\omega(t) \frac{d\omega(t)}{dt} + L_a i_a(t) \frac{di_a(t)}{dt} \qquad (6\text{-}26)$$

The first term on the right-hand side of Eq. (6-26) represents the i^2R loss in the armature winding as a result of armature current flow. The second term represents the power loss due to the viscous friction, and the third term represents the power loss due to all other friction torques, and together the second and third terms represent the mechanical losses of the motor. The fourth term represents the actual mechanical output of the motor. The last two terms represent instantaneous power flows into or out of the energy storage devices. Note that the last term is the derivative with respect to time of the quantity $1/2(L_a i_a^2)$ and, as such, it can be interpreted as the rate of change of the energy stored in the inductance L_a. In a similar way, the second to last term is interpreted as the rate of change of the kinetic energy. If, at the beginning and the end of any finite time period, the energy stored in the inertial and inductive elements is the same, there is no net change in energy stored, hence no net power dissipated. Thus, the average power dissipated by inertial and inductive elements is zero.

The efficiency of the motor as an energy converter can be calculated by taking the ratio of the output mechanical power to the total input average power.

$$\text{Efficiency (power)} = \frac{T_L}{\frac{R_a I_a^2 K_t}{K_e \omega} + B\omega + T_f + T_L} \times 100\% \qquad (6\text{-}27)$$

where all the parameters are considered to be constants.

The efficiency of mechanical power conversion can be obtained by taking the ratio of the mechanical power output to the developed mechanical power.

$$\text{Mechanical conversion efficiency} = \frac{T_L}{B\omega + T_f + T_L} \times 100\% \qquad (6\text{-}28)$$

As an example, a small dc permanent-magnet motor operates at an armature voltage of 16 volts and draws 2 amperes armature current while supplying 0.08 N-m torque to a load which turns at a speed of 200 radians/second. All of the friction is assumed to be viscous making $T_f = 0$ and $B = 100 \times 10^{-6}$ N-m. The armature resistance is 3 ohms. We calculate both the total power efficiency and the mechanical power conversion efficiency for such a motor.

$$\text{Efficiency (power)} = \frac{0.08}{\frac{3 \times 2^2}{200} + 100 \times 10^{-6} \times 200 + 0.08} \times 100\%$$

$$= \frac{0.08}{0.06 + 0.02 + 0.08} \times 100\%$$

$$= 50\% \qquad (6\text{-}29)$$

$$\text{Mechanical conversion efficiency} = \frac{0.08}{100 \times 10^{-6} \times 200 + 0.08} \times 100\%$$

$$= \frac{0.08}{0.1} \times 100\%$$

$$= 80\% \tag{6-30}$$

Notice that the major portion of the power lost is due to the resistance of the armature.

6.5 A THERMAL MODEL FOR A DC MOTOR

As indicated in the example above, half of the power consumed by a small dc motor in steady-state operation is dissipated as heat in the motor itself. In incremental motion systems, most of the generated torque is used for accelerating and decelerating the load. This mode of operation results in very little output power, and, therefore, most of the input power is converted to heat in the motor itself. Thus, it is important to understand and model this heat dissipation, at least in a gross manner so as to predict temperature rises within the motor. In order to arrive at a simple thermal model for the motor, we represent the motor by two elements only—the armature with its associated shaft and bearings as a rotating element and the stator (see Fig. 6-4).

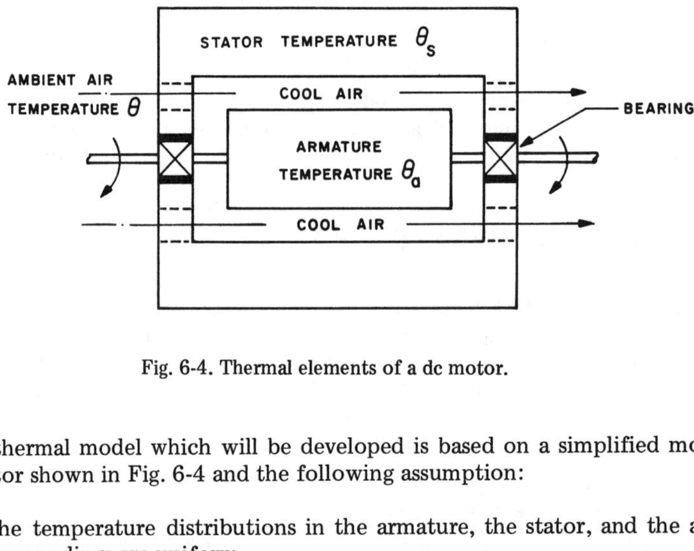

Fig. 6-4. Thermal elements of a dc motor.

The thermal model which will be developed is based on a simplified model for the dc motor shown in Fig. 6-4 and the following assumption:

1. The temperature distributions in the armature, the stator, and the ambient surroundings are uniform.
2. The heat transfer between surfaces at different temperatures is proportional to the temperature differential.
3. The total power dissipated and converted to heat (the first, second, and third terms on the right-hand side of Eq. (6-26) is all dissipated in the armature and the bearings, and is labeled p_L.

The amount of power (in the form of heat) that is transmitted from the armature to the stator is given by

$$p_{as}(t) = G_{as}[\theta_a(t) - \theta_s(t)] \tag{6-31}$$

where $\theta_a(t)$ and $\theta_s(t)$ are the relative-to-ambient temperatures of the armature and the stator, respectively, and G_{as} is the heat conductance between the armature and the stator. The heat power transferred between the stator and the ambient air surrounding the stator is

$$p_{so}(t) = G_{so}\theta_s(t) \tag{6-32}$$

and the heat power transferred from the armature to the ambient is

$$p_{ao}(t) = G_{ao}\theta_a(t) \tag{6-33}$$

In both the armature and the stator, the change in temperature with respect to time is proportional to the total heat power dissipated in that body, hence,

$$K_a \frac{d\theta_a(t)}{dt} = p_L(t) - p_{as}(t) - p_{ao}(t) \tag{6-34}$$

where the right-hand side represents the total net heat power dissipated in the armature. For the stator, the heat equation is

$$K_s \frac{d\theta_s(t)}{dt} = p_{as}(t) - p_{so}(t) \tag{6-35}$$

Substitutions of Eqs. (6-31), (6-32), and (6-33) into Eqs. (6-34) and (6-35) produce two differential equations in two unknowns with the input quantity being the total power dissipated in the armature p_L.

$$K_a \frac{d\theta_a(t)}{dt} = p_L(t) - G_{as}[\theta_a(t) - \theta_s(t)] - G_{ao}\theta_a(t) \tag{6-36}$$

$$K_s \frac{d\theta_s(t)}{dt} = G_{as}[\theta_a(t) - \theta_s(t)] - G_{so}\theta_s(t) \tag{6-37}$$

Transfer functions relating either the armature temperature or the stator temperature to the total input power dissipated p_L can be obtained by taking the Laplace transforms of Eqs. (6-36) and (6-37) and then solving for either θ_a or θ_s as functions of p_L. These transfer functions are given below.

$$\frac{\Theta_a(s)}{P_L(s)} = \frac{sK_s + G_{as} + G_{so}}{K_aK_ss^2 + [K_s(G_{as} + G_{ao}) + K_a(G_{as} + G_{so})]s + G_{as}G_{ao} + G_{as}G_{so} + G_{ao}G_{so}} \tag{6-38}$$

$$\frac{\Theta_s(s)}{P_L(s)} = \frac{G_{as}}{K_aK_ss^2 + [K_s(G_{as} + G_{ao}) + K_a(G_{as} + G_{so})]s + G_{as}G_{ao} + G_{as}G_{so} + G_{ao}G_{so}} \tag{6-39}$$

Equations (6-38) and (6-39) make it possible to theoretically calculate the armature and stator temperatures as functions of time if the time history of the power dissipated in the armature is known and if the thermal constants for the machine are known. Such calculations become rather unwieldy as the time history for the power dissipation becomes complicated. Moderate effort is needed to compute θ_a and θ_s for the case in which the power dissipated is constant. Several alternatives to the complete time history computation for the armature and stator temperatures may prove useful where approximate answers may be sufficient for the engineering analysis.

Steady-State Simplification

If the power dissipated in the armature reaches a constant, the armature and stator temperatures will eventually approach a steady-state value, and this steady-state value as a function of the magnitude of the power dissipated is easily computed by assuming the applied power to be the step function and invoking the final-value theorem in the Laplace domain. Thus, the steady-state armature temperature as a function of the constant power dissipated is

$$\theta_a(\text{steady-state}) = p_L\left[\frac{G_{as} + G_{so}}{G_{as}G_{ao} + G_{as}G_{so} + G_{ao}G_{so}}\right] \tag{6-40}$$

and the equation for steady-state stator temperature is

$$\theta_s(\text{steady-state}) = p_L\left[\frac{G_{as}}{G_{as}G_{ao} + G_{as}G_{so} + G_{ao}G_{so}}\right] \tag{6-41}$$

where p_L is the magnitude of the applied step power.

Forced Air Cooling

If, as a result of extra forced air cooling and reduced thermal coupling between the armature and the stator, some machines have $G_{as} \ll G_{ao}$ such that the heat power transferred to the stator can be ignored, then Eq. (6-36) reduces to

$$K_a \frac{d\theta_a(t)}{dt} = p_L(t) - G_{ao}\theta_a(t) \tag{6-42}$$

and the transfer function between the armature temperature and the power dissipated becomes

$$\frac{\Theta_a(s)}{P_L(s)} = \frac{1/G_{ao}}{\dfrac{K_a}{G_{ao}}s + 1} \tag{6-43}$$

with $1/G_{ao}$ being called the *thermal resistance* and K_a/G_{ao} being called the *thermal time constant*. Note that as the air flow is increased, the heat conductance, G_{ao}, will increase also. This results in a reduced thermal resistance and a longer thermal time constant. Both of these effects are favorable for the operation of the motor.

Parallel Method of Analysis

The transfer functions between the armature and stator temperatures and the dissipated power both contain the same polynomials in s in the denominator [see Eqs. (6-38) and (6-39)]. This quadratic may be factored into its two separate factors and the transfer function written as a partial-fraction expansion as in the following expression:

$$\frac{\Theta_a(s)}{P_L(s)} = \frac{sK_s + G_{as} + G_{so}}{C(s\tau_1 + 1)(s\tau_2 + 1)} = \frac{R_1}{s\tau_1 + 1} + \frac{R_2}{s\tau_2 + 1} \qquad (6\text{-}44)$$

where C is the constant $G_{as}G_{ao} + G_{as}G_{so} + G_{ao}G_{so}$. Thus, for a given power dissipation p_L, the armature temperature may be computed as the sum of two temperature components corresponding to the terms on the right-hand side of Eq. (6-44). This form of the transfer function may lead to simplified computations, assuming that R_1, R_2, τ_1, and τ_2 can be computed. In cases where one time constant differs from another by at least an order of magnitude, computations can be simplified by approximating one term as a constant in comparison with the other. A similar parallel model can be created for the stator temperature transfer function.

6.6 SUMMARY

Both the permanent-magnet and wound-field dc motors with constant current have similar transfer functions between velocity and applied armature voltage, and between velocity and load torque changes. Each transfer function is shown to be of the second order and to have the same denominator, although the numerators are not similar. If, however, armature inductance is negligible, both transfer functions reduce to first-order expressions differing from each other by a constant only.

Energy losses in smaller dc machines are high with power efficiencies of 50 percent not uncommon. All the energy lost is transformed to heat and may cause excessive heating particularly in the armature. Transfer functions between either stator or armature temperatures and the power lost in the armature have second-order denominators, although under some conditions, e.g., high forced air cooling, they simplify to first-order expressions.

The transfer characteristics derived in this chapter are all deterministic, linear, and time invariant, and require knowledge of machine parameters. Techniques for measuring such parameters are described in Chapter 9.

REFERENCES

1. *DC Motors, Speed Controls, Servo Systems*, 3rd ed., Engineering Handbook, Electro-Craft Corporation, 1975.

2. A. E. Fitzgerald, C. Kingsley and A. Kusko, *Electric Machinery*, 3rd ed., McGraw-Hill Book Company, New York, 1971.

3. L. W. Matsch, *Electromagnetic and Electromechanical Machines*, Intext Educational Publisher, College Division of Intext, San Francisco, California, 1972.

7 Torsional Resonance In High-Performance Incremental Motion Systems

J. TAL and H. WAAGEN

7.1 INTRODUCTION

High-performance incremental motion servos are used in systems where the key performance criteria are rapid motion times and high-speed slew capabilities. These servo systems must have a high gain-bandwidth product in order to maintain rise time values as small as 1 millisecond (msec). Since many system applications are satisfied with closed velocity loops, and since velocity loops are relatively easy to stabilize, it would seem that the problem of designing velocity control systems is easily solved by selecting a motor with a short mechanical time constant. This, in theory, would result in a wide bandwidth servo system. Unfortunately, this does not hold in actual practical applications.

In actuality, the design problem is complicated by the presence of torsional resonances that occur between the motor armature, the dc tachometer, and the driven load. These resonances arise from the fact that the motor shaft, having a finite stiffness, deflects torsionally when torque is applied to it. The shaft deflection in turn results in different angular positions at various parts of the mechanical system. The relative angular positions depend on the excitation frequency and on the system parameters. At some frequencies, it may occur that the angular deflections of two parts are in opposite phase; this condition allows the mechanical system to store kinetic energy, and, as a result, to increase the amplitude of the angular deflection. This phenomenon is called torsional resonance, and the frequencies at which it occurs are called resonance frequencies.

In order to simplify the model of the motor, this resonance phenomenon is usually ignored in most discussions of motor modeling. Instead, the common assumption is that all of the system mass is lumped at one point. According to this assumption, the relationship between the angular position, θ, and the generated torque, T_g, is given by

$$T_g = J\frac{d^2\theta}{dt^2} + B\frac{d\theta}{dt} \tag{7-1}$$

where J is the system inertia and B denotes the viscous friction coefficient. Equation (7-1) can be written as a differential equation in ω, the angular velocity:

$$T_g = J\frac{d\omega}{dt} + B\omega \tag{7-2}$$

or it can be expressed by the transfer function:

$$\frac{\Omega(s)}{T_g(s)} = \frac{1}{sJ + B} \tag{7-3}$$

where $\Omega(s)$ and $T_g(s)$ are the Laplace transforms of $\omega(t)$ and $T_g(t)$, respectively. The torsional resonance alters the transfer function of the motor and adds to the difficulty of stabilizing the loop. The resulting tendency toward oscillations limits the servo performance and may cause damage to the mechanical structure and links, thereby affecting the reliability of the system.

The objective of this chapter is to analyze the torsional resonance phenomenon and to forward some suggestions for minimizing its effects. We will start by discussing the operation of a velocity control system and the attendant effects of resonance on it. We will later discuss a simplified two-body structure before we analyze the more general case of three-body structures. Finally, techniques for minimizing the effects of resonance will be discussed with the objective in mind of enhancing the transient performance of the servo system.

7.2 THE EFFECTS OF TORSIONAL RESONANCE ON THE SYSTEM RESPONSE

In order to view the effects of torsional resonance, consider the control system shown in Fig. 7-1. This system is designed to generate a trapezoidal velocity profile for the motor. Figure 7-2 illustrates typical waveforms of the system.

When the start command is initiated in the system shown in Fig. 7-1, the motor command flip-flop is set and the error signal attains its maximum value. At this time, the motor voltage and current are at their maximum values because of the negligible electrical time constant. As a consequence, the motor will begin accelerating to some predetermined speed dependent on the displacement requirements. When the dwell velocity is reached, the motor will run at this speed as long as the motor command is high. At some point in time, a stop command will be generated and the motor command will drop to the low state. The input signal to the servo will then be determined solely by the magnitude of the tachometer output voltage. Since this voltage is negative, the drive voltage will be negative and large, causing the motor to decelerate to a stop. This cycle will be repeated upon initiation of the next start command.

It is important to increase the loop gain in incremental motion systems because increased gain improves the speed regulation. The increase in gain also widens the system bandwidth, allowing a rapid response to command. However, as the loop gain is increased, the resonance phenomenon is elicited. This results in sinusoidal vibration

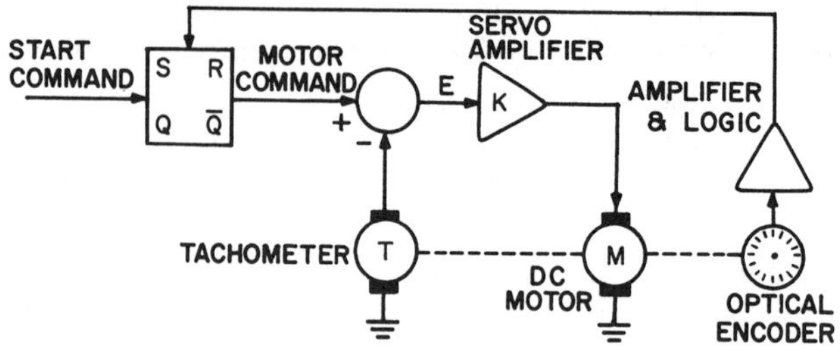

Fig. 7-1. DC servo incremental motion system.

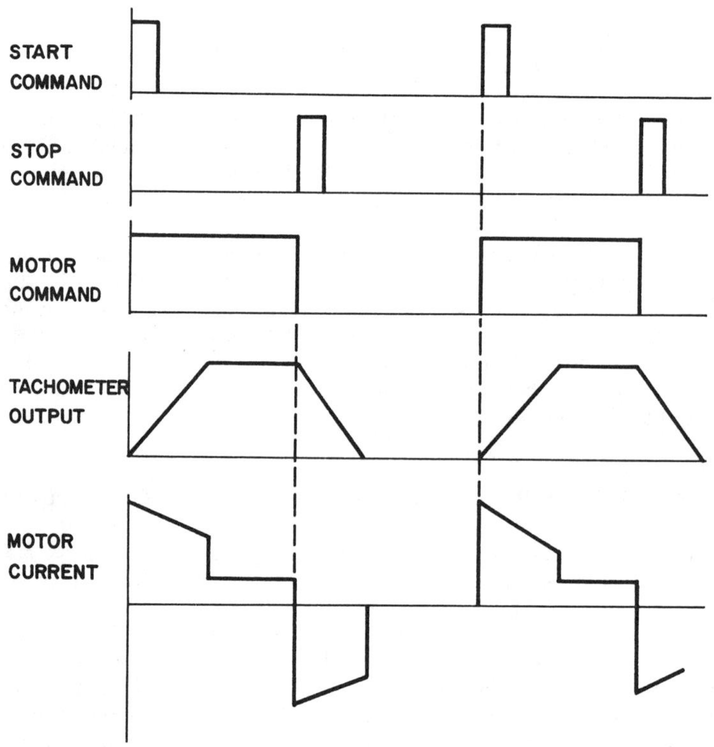

Fig. 7-2. Typical waveforms of the control system in Fig. 7-1.

at the resonance frequency, as illustrated in Fig. 7-3. Further increase in loop gain will lead to system instability.

Since the effect of torsional resonance is highly undesirable, it is necessary to understand and to analyze this phenomenon in order to minimize its effect. We start the analysis with a model of the mechanical system as a two-body structure.

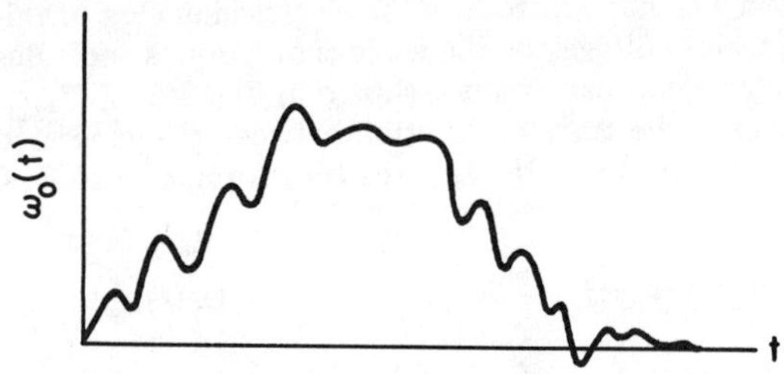

Fig. 7-3. Tachometer distortion caused by torsional resonance.

7.3 TORSIONAL RESONANCE IN TWO-BODY STRUCTURES

Consider the mechanical structure shown in Fig. 7-4. This is a simplified representation of a motor and tachometer. The angular positions of the motor and the tachometer are denoted by θ_m and θ_1, respectively. Similarly, the moments of inertia of the motor and the tachometer are denoted by J_m and J_1, respectively. The shaft linking the two bodies is assumed to have zero moment of inertia, a stiffness K_1, and damping B_1. Note that B_1 represents the frictional damping due to the angular velocity differences between the motor and the shaft. This is in addition to the frictional damping due to the motor velocity, which is represented by B_m. Clearly, the damping is generated in various parts of the system; but, for the sake of simplicity, we will assume that the damping is lumped.

In order to write the dynamic equations of the system, let the torque generated by the motor be T_g and let the torque applied to the shaft be T_1. The resulting dynamic equations are

$$T_g = J_m \ddot{\theta}_m + B_m \dot{\theta}_m + T_1 \tag{7-4}$$

$$T_1 = J_1 \ddot{\theta}_1 \tag{7-5}$$

The torque T_1 causes the shaft to deflect. The deflection is governed by the equation

$$T_1 = K_1(\theta_m - \theta_1) + B_1(\dot{\theta}_m - \dot{\theta}_1) \tag{7-6}$$

Equations (7-4), (7-5), and (7-6) can be combined together to yield

$$T_g = J_m \ddot{\theta}_m + B_m \dot{\theta}_m + J_1 \ddot{\theta}_1 \tag{7-7}$$

$$J_1 \ddot{\theta}_1 = K_1(\theta_m - \theta_1) + B_1(\dot{\theta}_m - \dot{\theta}_1) \tag{7-8}$$

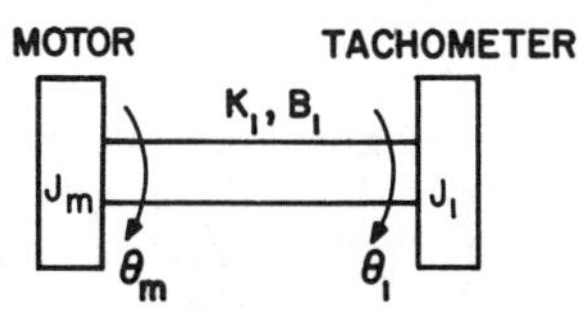

Fig. 7-4. Two-body structure.

At this point it is convenient to introduce the electrical analog of this mechanical system. Let the current and voltages be the analogs of torques and velocities, respectively. The resulting analog electrical circuit is shown in Fig. 7-5.

In order to develop the transfer function between the generated torque, T_g, and the angular position, θ_1, we apply the Laplace transformation to Eqs. (7-7) and (7-8). This yields

$$T_g(s) = (s^2J_m + sB_m)\Theta_m(s) + s^2J_1\Theta_1(s) \tag{7-9}$$

$$s^2J_1\Theta_1(s) = (sB_1 + K_1)\Theta_m(s) - (sB_1 + K_1)\Theta_1(s) \tag{7-10}$$

All initial conditions have been set to zero.

Equations (7-9) and (7-10) can be combined together to form

$$T_g(s) = \left[\frac{(s^2J_m + sB_m)(s^2J_1 + sB_1 + K_1)}{sB_1 + K_1} + s^2J_1\right]\Theta_1(s) \tag{7-11}$$

The mathematical model is simplified if we determine the transfer function between the torque T_g and the angular velocity ω_1. Since the Laplace transform of ω_1, $\Omega_1(s)$, is given by

$$\Omega_1(s) = s\Theta_1(s) \tag{7-12}$$

the resulting transfer function is

$$\frac{\Omega_1(s)}{T_g(s)} = \frac{sB_1 + K_1}{(sJ_m + B_m)(s^2J_1 + sB_1 + K_1) + sJ_1(sB_1 + K_1)} \tag{7-13}$$

Note that if the shaft stiffness, K_1, were infinite, Eq. (7-13) would become

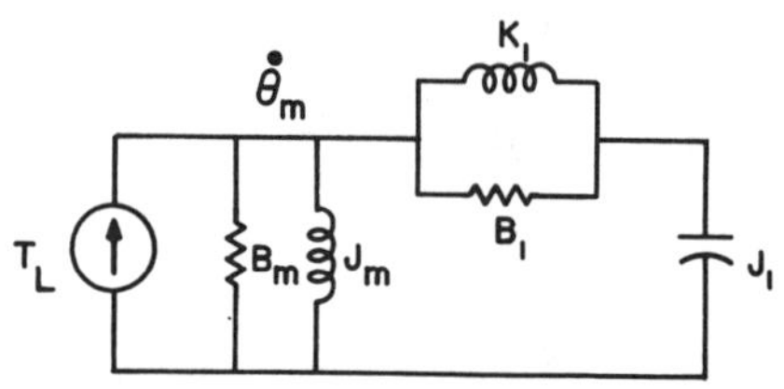

Fig. 7-5. Electrical analog of a two-body structure.

$$\lim_{K_1 \to \infty} \frac{\Omega_1(s)}{T_g(s)} = \lim_{K_1 \to \infty} \frac{K_1}{(sJ_m + B_m)K_1 + sJ_1K_1} = \frac{1}{s(J_m + J_1) + B_m} \tag{7-14}$$

which is exactly the transfer function of Eq. (7-3) derived under the assumption that the inertia is lumped at one point, with $J = J_m + J_1$.

Comparing Eq. (7-13) with Eq. (7-3) reveals that the torsional resonance adds one real zero and two poles to the transfer function. Since the zero is negative and real and generally of large magnitude, it has little effect on the system, and, therefore, it may be ignored. The poles, on the other hand, have a significant effect since they are always complex and usually have very little damping. The system poles are the roots of the characteristic equation which is obtained by equating the denominator of Eq. (7-13) to zero.

$$(sJ_m + B_m)(s^2J_1 + sB_1 + K_1) + sJ_1(sB_1 + K_1) = 0 \tag{7-15}$$

Equation (7-15) has three roots. One root is the result of the mechanical pole; it has a small magnitude. The other two roots, caused by the torsional resonance, have large magnitudes. In order to determine the first root, note that at low frequencies K_1 is much greater than the magnitude of sB_1 or s^2J_1. Therefore, we may assume that

$$K_1 \gg |sB_1|$$

and that

$$K_1 \gg |s^2J_1|$$

Consequently, Eq. (7-15) can be simplified to the form

$$(sJ_m + B_m)K_1 + sJ_1K_1 = 0 \tag{7-16}$$

The mechanical pole, which is the root of Eq. (7-16), is equal to

$$p_1 = -\frac{B_m}{J_m + J_1} \tag{7-17}$$

This pole is the same as the one in the transfer function given by Eq. (7-3).

In order to determine the resonance poles, we assume that B_m is very small in relation to the magnitude of sJ_m at frequencies near the resonance frequencies. Therefore, we may write

$$|sJ_m| \gg B_m \tag{7-18}$$

In view of Eq. (7-18), we may approximate Eq. (7-15) by

$$sJ_m(s^2J_1 + sB_1 + K_1) + sJ_1(sB_1 + K_1) = 0 \tag{7-19}$$

Next divide Eq. (7-19) by s, which represents the low-frequency pole, p_1,

$$s^2J_1J_m + sB_1(J_1 + J_m) + K_1(J_1 + J_m) = 0 \tag{7-20}$$

and define the equivalent inertia, J_e, as

$$J_e = \frac{J_1J_m}{J_1 + J_m} \tag{7-21}$$

Combining Eq. (7-20) with Eq. (7-21) results in the equation

$$s^2J_e + sB_1 + K_1 = 0 \tag{7-22}$$

The last equation has two roots:

$$p_{2,3} = \frac{-B_1 \pm \sqrt{B_1^2 - 4K_1J_e}}{2J_e} \tag{7-23}$$

However, since

$$4K_1J_e \gg B_1^2 \tag{7-24}$$

the damping associated with these poles is very small.

We may approximate Eq. (7-23) by

$$p_{2,3} = \frac{-B_1}{2J_e} \pm j\sqrt{\frac{K_1}{J_e}} \tag{7-25}$$

Thus, the resonance poles are complex and their imaginary parts are dominant. Furthermore, the imaginary parts, which give the resonance frequency, are determined by the stiffness K_1 and by the equivalent moment of inertia J_e.

Although the negative real part of the poles is very small, it plays an important role in stabilizing the system because it determines the amount of damping associated with the poles.

The motor transfer function in Eq. (7-13) may be represented by its poles and zeros, as shown in Fig. 7-6. This representation is readily applicable to the root locus analysis.

An alternative way of representing the torsional resonance effect is by the frequency response of the motor. This is a natural representation since it describes the

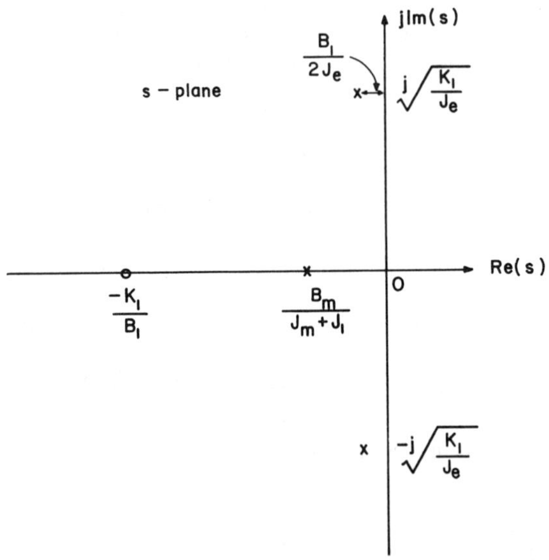

Fig. 7-6. Pole-zero representation of the transfer function of a two-body structure.

measurement results. The advantage of the frequency response representation is that it allows a quantitative description of resonance peaks. A typical frequency response curve is shown in Fig. 7-7.

The analysis given in this section is based on the assumption that the mechanical system may be adequately modeled as a two-body structure. However, in some cases, the two-body model is not sufficient and a more general form is needed. In such cases we may assume the adequacy of a three-body structure. An analysis of three-body structures is given in the following section.

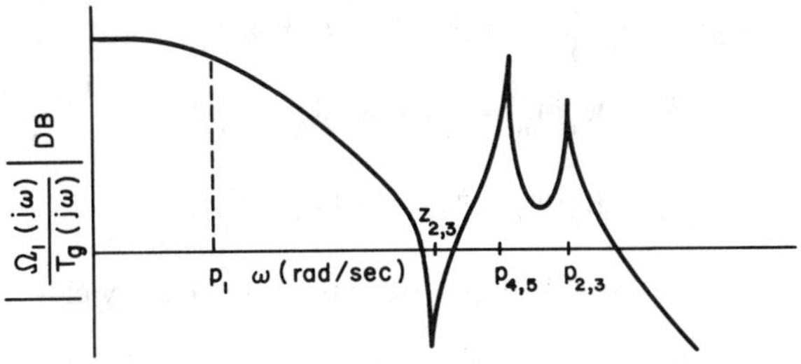

Fig. 7-7. Frequency response of a two-body structure.

7.4 TORSIONAL RESONANCE IN THREE-BODY STRUCTURES

The mechanical structure considered in this section consists of three bodies coupled by shafts. This is a more realistic representation of incremental motion systems, because they often include a motor, a load, and a tachometer. This mechanical system is shown in Fig. 7-8. Let the moments of inertia of the three bodies be denoted by J_m, J_1, and J_2. Subscript m refers to the motor, subscript 1 refers to the tachometer, and subscript 2 refers to the load. Similarly, let their angular positions be denoted by θ_m, θ_1, and θ_2. Denote the torque which is applied to the two bodies by T_1 and T_2. Let the stiffness and the damping of the shafts be denoted by K_1, K_2 and B_1, B_2.

Next, consider the dynamic equations of the system in response to the generated torque, T_g. These are given by

$$T_g = J_m \ddot{\theta}_m + B_m \dot{\theta}_m + T_1 + T_2 \tag{7-26}$$

$$T_1 = J_1 \ddot{\theta}_1 \tag{7-27}$$

$$T_2 = J_2 \ddot{\theta}_2 \tag{7-28}$$

The two shafts deflect according to the equations

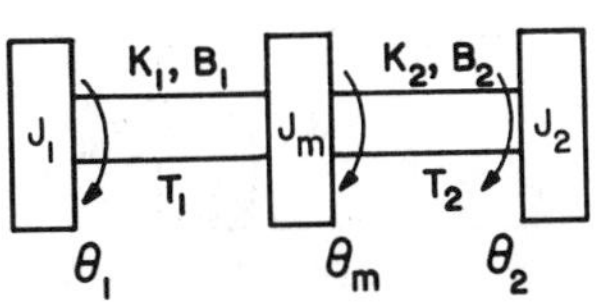

Fig. 7-8. Three-body mechanical system.

$$T_1 = K_1(\theta_m - \theta_1) + B_1(\dot{\theta}_m - \dot{\theta}_1) \tag{7-29}$$

$$T_2 = K_2(\theta_m - \theta_2) + B_2(\dot{\theta}_m - \dot{\theta}_2) \tag{7-30}$$

The electrical analog of this mechanical system is shown in Fig. 7-9. As in Fig. 7-5, the analogs of torque and velocity are current and voltage, respectively.

Equations (7-26) through (7-30) may be simplified by eliminating T_1 and T_2. This yields

$$T_g = J_m\ddot{\theta}_m + B_m\dot{\theta}_m + J_1\ddot{\theta}_1 + J_2\ddot{\theta}_2 \tag{7-31}$$

$$J_1\ddot{\theta}_1 = K_1(\theta_m - \theta_1) + B_1(\dot{\theta}_m - \dot{\theta}_1) \tag{7-32}$$

$$J_2\ddot{\theta}_2 = K_2(\theta_m - \theta_2) + B_2(\dot{\theta}_m - \dot{\theta}_2) \tag{7-33}$$

Applying the Laplace transform to Eqs. (7-31) through (7-33) yields

$$(s^2J_m + sB_m)\Theta_m(s) + s^2J_1\Theta_1(s) + s^2J_2\Theta_2(s) = T_g(s) \tag{7-34}$$

$$(sB_1 + K_1)\Theta_m(s) - (s^2J_1 + sB_1 + K_1)\Theta_1(s) = 0 \tag{7-35}$$

$$(sB_2 + K_2)\Theta_m(s) - (s^2J_2 + sB_2 + K_2)\Theta_2(s) = 0 \tag{7-36}$$

These can be written in a matrix form:

$$\begin{bmatrix} s^2J_m + sB_m & s^2J_1 & s^2J_2 \\ sB_1 + K_1 & -(s^2J_1 + sB_1 + K_1) & 0 \\ sB_2 + K_2 & 0 & -(s^2J_2 + sB_2 + K_2) \end{bmatrix} \begin{bmatrix} \Theta_m(s) \\ \Theta_1(s) \\ \Theta_2(s) \end{bmatrix} = \begin{bmatrix} T_g(s) \\ 0 \\ 0 \end{bmatrix} \tag{7-37}$$

In order to simplify the notations, define the following polynomials in s:

$$P_i(s) = s^2J_i + sB_i + K_i \qquad i = 1, 2 \tag{7-38}$$

$$Q_i(s) = sB_i + K_i \qquad i = 1, 2 \tag{7-39}$$

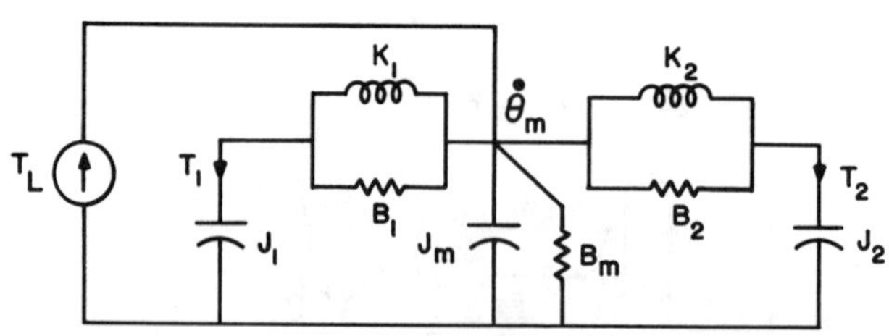

Fig. 7-9. Electrical analog of three-body structures.

Then we may write the dynamic equations of Eq. (7-37) as

$$\begin{bmatrix} s^2J_m + sB_m & s^2J_1 & s^2J_2 \\ Q_1 & -P_1 & 0 \\ Q_2 & 0 & -P_2 \end{bmatrix} \begin{bmatrix} \Theta_m(s) \\ \Theta_1(s) \\ \Theta_2(s) \end{bmatrix} = \begin{bmatrix} T_g(s) \\ 0 \\ 0 \end{bmatrix} \tag{7-40}$$

The matrix equations of Eq. (7-40) allow us to derive the transfer function between the input T_g and any angular position. Since the angular velocity of the tachometer, ω_1, is used for feedback, it is important to develop its transfer function. The same method employed below may be used to determine the armature velocity or the load velocity.

We can solve Eq. (7-40) for $\Theta_1(s)$. This yields

$$\Theta_1(s) = \frac{P_2Q_1}{(s^2J_m + sB_m)P_1P_2 + s^2J_1Q_1P_2 + s^2J_2P_1Q_2} T_g(s) \tag{7-41}$$

In order to determine the transfer function between $T_g(s)$ and $\Omega_1(s)$, note that

$$\Omega_1(s) = s\Theta_1(s) \tag{7-42}$$

Now combine Eq. (7-41) with Eq. (7-42) to yield

$$\frac{\Omega_1(s)}{T_g(s)} = \frac{P_2Q_1}{(sJ_m + B_m)P_1P_2 + sJ_1P_2Q_1 + sJ_2P_1Q_2} \tag{7-43}$$

The transfer function of Eq. (7-43) is a ratio of polynomials in s. As the orders of the numerator and the denominator are three and five, respectively, the transfer function has three zeros and five poles. We will start by analyzing the poles; later, we will turn our attention to the zeros. Since one of the poles is real and has a small magnitude, and since the other four are generally complex with large magnitudes, we may use the following approximation: at low frequencies we assume that

$$P_i = Q_i = K_i \qquad i = 1, 2 \tag{7-44}$$

Similarly, we assume that, at higher frequencies,

$$|sJ_m| \gg B_m \tag{7-45}$$

In order to determine the system poles, we start with the characteristic equation

$$(sJ_m + B_m)P_1P_2 + sJ_1P_2Q_1 + sJ_2P_1Q_2 = 0 \tag{7-46}$$

In order to find the low frequency pole, substitute Eq. (7-44) into Eq. (7-46), we get

$$[s(J_m + J_1 + J_2) + B_m]K_1K_2 = 0 \tag{7-47}$$

The first pole is found to be

$$p_1 = -\frac{B_m}{J_m + J_1 + J_2} \tag{7-48}$$

This is of the same form as the pole of the transfer function of Eq. (7-3).

For the resonance poles, apply Eq. (7-45) to Eq. (7-46). This simplifies the equation. Next, divide by s to eliminate the effect of p_1. The resulting equation is

$$J_m P_1 P_2 + J_1 P_2 Q_1 + J_2 P_1 Q_2 = 0 \tag{7-49}$$

As can be seen, Eq. (7-49) is a fourth-order equation which does not have an analytical closed-form solution. In order to obtain an approximate solution, assume that the tachometer inertia, J_1, is much smaller than J_m. This is a valid assumption in most cases. Therefore, we can ignore the term $J_1 P_2 Q_1$ in Eq. (7-49). This results in

$$J_m P_1 P_2 + J_2 P_1 Q_2 = 0 \tag{7-50}$$

or

$$(J_m P_2 + J_2 Q_2) P_1 = 0 \tag{7-51}$$

Since Eq. (7-51) is a product of two polynomials, we may determine the roots of each term separately. Starting with $P_1(s)$, we write the equation

$$P_1(s) = s^2 J_1 + sB_1 + K_1 = 0 \tag{7-52}$$

and the corresponding roots are

$$p_{2,3} = \frac{-B_1 \pm \sqrt{B_1^2 - 4J_1 K_1}}{2J_1} \tag{7-53}$$

Since B_1^2 is generally much smaller than $4J_1 K_1$, we may write Eq. (7-53) as

$$p_{2,3} = -\frac{B_1}{2J_1} \pm j\sqrt{\frac{K_1}{J_1}} \tag{7-54}$$

Note that this pair of complex poles is completely determined by the parameters of the tachometer and its coupling, and that it is almost independent of the load J_2. Therefore, these two poles are referred to as the tachometer resonance poles. The imaginary part of these poles is typically in the range of 3 to 5 kilohertz.

The other pair of complex poles is determined from the equation

$$J_m P_2 + J_2 Q_2 = 0 \tag{7-55}$$

or explicitly

$$J_m (s^2 J_2 + sB_2 + K_2) + J_2 (sB_2 + K_2) = 0 \tag{7-56}$$

This can be written as

$$s^2 J_m J_2 + s(J_m + J_2)B_2 + (J_m + J_2)K_2 = 0 \tag{7-57}$$

Define the equivalent moment of inertia, J_e, as

$$J_e = \frac{J_m J_2}{J_m + J_2} \tag{7-58}$$

and rewrite Eq. (7-57) as

$$s^2 J_e + sB_2 + K_2 = 0 \tag{7-59}$$

The resulting poles are

$$p_{4,5} = \frac{-B_2 \pm \sqrt{B_2^2 - 4J_e K_2}}{2J_e} \tag{7-60}$$

These can be approximated by

$$p_{4,5} = \frac{-B_2}{2J_e} \pm j\sqrt{\frac{K_2}{J_e}} \tag{7-61}$$

Equation (7-61) indicates that this pair of poles is determined by the parameters of the motor and the load. Therefore, they are referred to as the motor-load resonance. The resonant frequency of this pair depends on the system parameters and is usually smaller than the tachometer resonant frequency.

Equations (7-54) and (7-61) indicate the dependence of the resonance poles on the system parameters. An increase in shaft stiffness results in an increase in resonance frequencies, whereas an increase in the moments of inertia decreases the resonance frequencies. The damping does not affect the resonance frequency, but it does affect the real part of the poles, and as a result influences the dynamic behavior of the system.

The zeros of the transfer function can be found by setting the numerator of Eq. (7-43) to zero.

$$P_2 Q_1 = (s^2 J_2 + sB_2 + K_2)(sB_1 + K_1) = 0 \tag{7-62}$$

The roots of Eq. (7-62), which are the zeros of the system, are

$$z_1 = -\frac{K_1}{B_1} \tag{7-63}$$

and

$$z_{2,3} = \frac{-B_2 \pm \sqrt{B_2^2 - 4K_2 J_2}}{2J_2} \cong -\frac{B_2}{2J_2} \pm j\sqrt{\frac{K_2}{J_2}} \tag{7-64}$$

Thus, one of the zeros is real and the other two are complex. The complex zeros are determined by the load parameters and the coupling. Since it follows from Eq.

(7-58) that $J_e < J_2$, the frequency of the complex zeros is always smaller than that of the complex poles. However, in cases where the load inertia J_2 is small, J_e approaches J_2, bringing the frequency of the zeros closer to that of the motor-load poles.

In concluding the analysis of three-body structures, we summarize the discussion by indicating the values of the poles and the zeros of the transfer function. These are given by the following equations:

$$p_1 = -\frac{B_m}{J_m + J_1 + J_2} \tag{7-65}$$

$$p_{2,3} = -\frac{B_1}{2J_1} \pm j\sqrt{\frac{K_1}{J_1}} \tag{7-66}$$

$$p_{4,5} = -\frac{B_2}{2J_2} \pm j\sqrt{\frac{K_2}{J_e}} \tag{7-67}$$

$$z_1 = -\frac{K_1}{B_1} \tag{7-68}$$

$$z_{2,3} = -\frac{B_2}{2J_2} \pm j\sqrt{\frac{K_2}{J_2}} \tag{7-69}$$

The poles and zeros of the system may be described in the complex plane, as shown in Fig. 7-10.

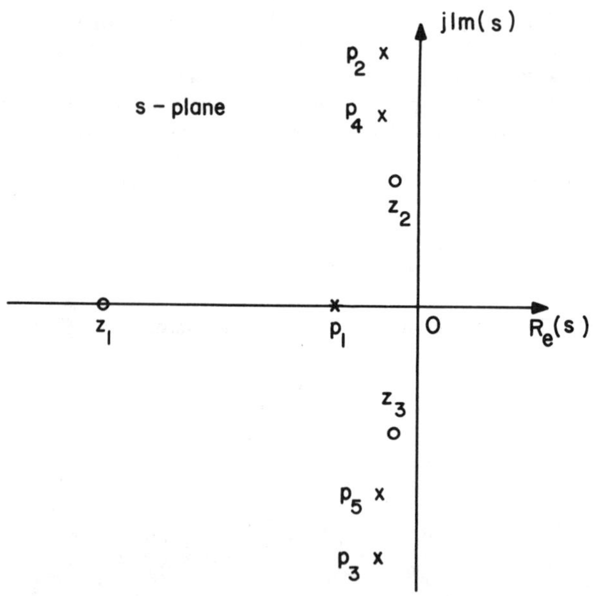

Fig. 7-10. Pole-zero representation of the transfer function of a three-body structure.

The frequencies of the poles and the zeros can be found from the frequency response measurement of the system. Since z_1 is negative with a large magnitude, it has little effect on the frequency response. However, the effects of all the other poles and zeros can be clearly seen on the Bode plot of the frequency response. A typical curve is shown in Fig. 7-11.

The three-body model is a generalization of the two-body model described in the previous section. Consequently, all results of Sec. 7.3 can be obtained from the results of this section by setting $J_2 = 0$. The removal of the load inertia, J_2, reduces Eq. (7-43) to the form of Eq. (7-13). Similarly, the poles $p_{4,5}$ and the zeros $z_{2,3}$ will disappear, and the remaining poles and zero will be identical to those found in Sec. 7.3.

The measured frequency response curve may have more poles than what is described here. If that is the case, it indicates that the three-body structure is not an adequate approximation for the electromechanical system, and a higher-order model is needed to describe the system more accurately. Having analyzed the torsional resonance phenomenon, we can now examine its effect on the system stability. This examination follows.

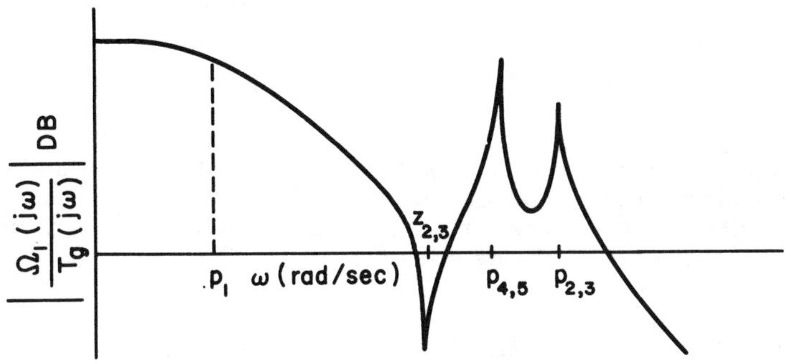

Fig. 7-11. Frequency response curve of a system with three-body structure.

7.5 EFFECT OF TORSIONAL RESONANCE ON SYSTEM STABILITY

The effect of the torsional resonance is represented by the addition of at least one pair of complex poles to the motor transfer function. Since these poles have very little damping, they have an adverse effect on the system stability. This can be seen by considering the root-locus diagram of the system. Let us suppose that the motor can be represented as a two-body structure, and that the amplifier has an additional pole, p_0. The open-loop transfer function, L(s) is

$$L(s) = \frac{K(s + z_1)}{(s + p_0)(s + p_1)(s + p_2)(s + p_3)} \tag{7-70}$$

The root-locus diagram for this system is shown in Fig. 7-12. Since the poles $p_{2,3}$ are very close to the imaginary axis, their root-loci branches may cross the imaginary axis even when the gain is moderate. This is certainly undesirable; it places a

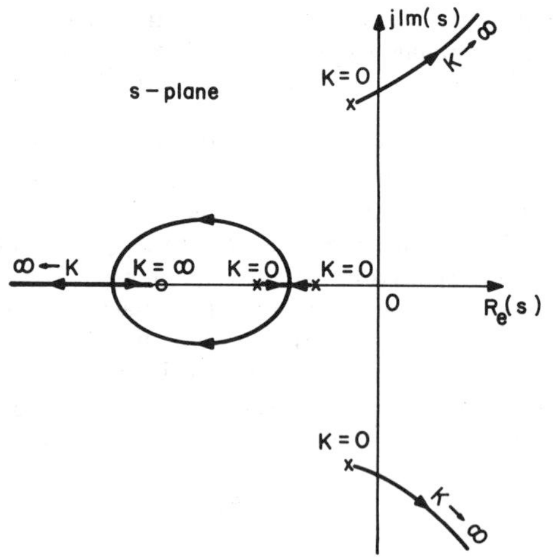

Fig. 7-12. Root-locus diagram of Eq. (7-70).

severe restriction on the loop gain and, therefore, on the system bandwidth.

The root-locus diagram for the case of three-body structures is more complex, but the net effect is similar. There, too, the torsional resonance limits the loop gain, placing a severe limitation on the system performance.

7.6 TECHNIQUES FOR MINIMIZING RESONANCE EFFECTS

Several approaches may be used to minimize the effects of system resonances on servo performance. They are as follows.

1. Increasing The Shaft Diameters.

An examination of the expressions for the system poles indicates that the resonance frequencies have the general form of $\sqrt{K/J_e}$ where K is the shaft stiffness and J_e is the equivalent moment of inertia. It can also be shown that the stiffness of a straight cylindrical shaft is $K = GJ_p/L$ where G is the shear modulus of elasticity, J_p is the polar moment of inertia, and L is the distance between the inertial elements. Substituting $\pi d^4/32$ for J_p, with d being the shaft diameter, we obtain the expression for the resonance frequency:

$$\sqrt{\frac{K}{J_e}} = \sqrt{\frac{\frac{\pi d^4 G}{32}}{LJ_e}} = d^2\sqrt{\frac{\frac{\pi G}{32}}{LJ_e}} \tag{7-71}$$

It can be seen that the significant parameter in determining the resonant frequency is the shaft diameter, and that increasing its size will result in an attendant improvement in system performance. (This assumes that the armature-shaft coupling introduces negligible compliance into the system.) If the shaft has n steps, each having a length L_i and a diameter d_i, the resulting polar moment of inertia is given by

$$J_p = \frac{\sum_{i=1}^{n} L_i}{\sum_{i=1}^{n} \frac{32L_i}{\pi d_i^4}} \tag{7-72}$$

2. Reducing The Physical Separation Between Inertial Elements.

As can be seen from Eq. (7-71), a decrease in the length L will increase the resonance frequency. This, of course, has a less significant effect on resonance than does increasing the shaft diameter.

3. Minimizing System Inertias.

A reduction in the moment of inertia of the mechanical components will result in a smaller value of J_e, as shown by Eq. (7-21). This decrease in J_e will increase the resonance frequency, as can be seen from Eq. (7-71).

4. Use Of A Ramp Velocity Command.

An examination of the Fourier transform of a step function with a sharp leading edge reveals that the frequency content of such a step is very wide. When such a step is applied to a motor, it excites the torsional resonance modes. The effect of the torsional resonance is reduced if we replace the step function by a ramp, because the frequency content of the ramp is narrower. However, the use of a ramp as a command signal may cause a delay in the velocity response.

5. Use Of A Notch Filter.

Another solution to the resonance problem is to alter the movement of the root-loci branches by adding zeros to attract those branches. However, since it is impossible to add zeros without adding poles, we select the zeros to be imaginary and the poles to be real and negative. A circuit that can generate these poles and zeros is called a "notch filter" circuit; it can be designed in various ways. One possible design for such a filter is shown in Fig. 7-13. Since the output terminal of the filter should not be loaded, it is connected to the positive input terminal of an operational amplifier. The transfer function of this filter is

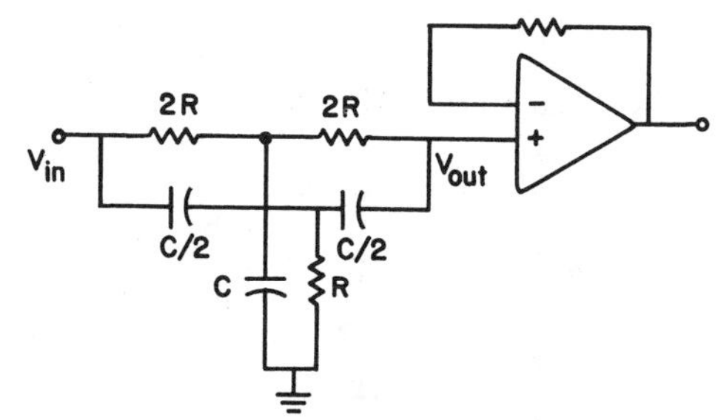

Fig. 7-13. Notch filter.

$$\frac{V_{out}(s)}{V_{in}(s)} = \frac{R^2C^2s^2 + 1}{R^2C^2s^2 + 4RCs + 1} \qquad (7\text{-}73)$$

This filter has zeros at

$$z_{1,2} = \pm j \frac{1}{RC} \qquad (7\text{-}74)$$

and poles at

$$p_1 = -\frac{0.27}{RC} \qquad (7\text{-}75)$$

$$p_2 = -\frac{3.73}{RC} \qquad (7\text{-}76)$$

Thus, one can select the value of RC such that the frequency of the zeros coincides with that of the resonance poles, thereby eliminating their effect. The cost of this approach is the addition of some real poles, the most significant being p_1 which has a relatively small magnitude.

6. Use Of A Derived Velocity Signal.

In low friction systems, filtering of the resonance can be achieved by summing the tachometer voltage with a voltage proportional to the motor current. This voltage is then filtered by a capacitor and the resultant output is used as the velocity feedback. The circuit is shown in Fig. 7-14. The tachometer voltage, V_T, is proportional to the velocity

$$V_T = K_g \omega \qquad (7\text{-}77)$$

where K_g is the tachometer constant.

When the friction is negligible, the motor current is proportional to the acceleration. Accordingly, we may write V_s as

$$V_s = A \frac{d\omega}{dt} \qquad (7\text{-}78)$$

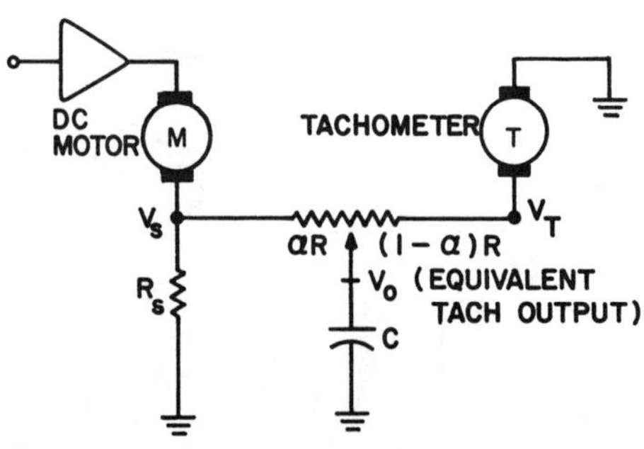

Fig. 7-14. A circuit for generating a derived velocity signal.

where A is given by

$$A = \frac{R_s J}{K_T} \tag{7-79}$$

Now the node equation in the Laplace transform domain for the given circuit at the node V_0 is

$$\frac{V_s(s) - V_0(s)}{aR} + \frac{V_T(s) - V_0(s)}{(1-a)R} - CsV_0(s) = 0 \tag{7-80}$$

If we combine Eq. (7-80) with Eqs. (7-77) and (7-78) and rearrange the terms, we obtain

$$V_0(s) = \frac{sA(1-a) + aK_g}{a(1-a)sRC + 1}\,\Omega(s) \tag{7-81}$$

In order for V_0 to be proportional to ω, both the pole and the zero in Eq. (7-81) must have the same value. This requires

$$a^2 = \frac{A}{K_g RC} = \frac{R_s J}{K_g K_T RC} \tag{7-82}$$

Equation (7-82) can be satisfied by an appropriate adjustment of the potentiometer. For the value of a given by Eq. (7-82), Eq. (7-81) becomes

$$V_0(s) = aK_g\Omega(s) \tag{7-83}$$

As can be seen from Eq. (7-83), the voltage V_0 is proportional to the angular velocity ω. While using a derived velocity signal is advantageous in that summing the tachometer signal with the motor current tends to minimize the resonance effect, it also has a disadvantage in that the feedback signal is not proportional to the velocity, and, therefore, may have an adverse effect on the speed regulation. This is especially true when the friction load is significant.

7.7 OTHER CONSIDERATIONS

As the motor temperature rises, the motor's mechanical characteristics change. A temperature rise results in a slight decrease in the effective stiffness and in a significant increase in the damping. This shifts the resonance poles to the left and slightly decreases their imaginary parts. As expected, the increased damping has a favorable effect on the loop stability; therefore, it is sufficient to stabilize the system with the motor at room temperature. Furthermore, the notch filter compensation will be effective even if the resonance poles "move" with the temperature change, since the zeros attract the root-loci branches in their vicinity.

Also of importance is the fact that bellows couplings (which are used to compensate for shaft misalignments) are unacceptable for use in a high-performance intermittent motion system; the resultant compliance which they introduce is very low, and this results in low resonant frequencies.

7.8 SUMMARY AND CONCLUSIONS

In this chapter we have examined the phenomenon of torsional resonance in incremental motion servo systems. We have seen how torsional resonance can affect the system, and how it poses problems of system instability for the designer. We characterized torsional resonance mathematically and employed practical models for the mechanical system based on both two- and three-body structures. We looked at the problem in terms of pole and zero placement, and derived six methods for reducing torsional resonance in practical applications. Each of these methods was examined in terms of its specific advantages and shortcomings.

The designer's choice of which method(s) to employ will ultimately depend on the actual system application, its various physical limitations, and the system's motion time requirements.

REFERENCES

1. H. Waagen, "Reduce Torsional Resonance in Incremental-Motion Servos," *Control Engineering*, pp. 85-88, Apr. 1969.
2. *DC Motors, Speed Controls, Servo Systems*, 3rd ed., Engineering Handbook, Electro-Craft Corporation, 1975.
3. B. C. Kuo, *Automatic Control Systems*, 3rd ed., Prentice-Hall Inc., Englewood Cliffs, N.J., 1975.
4. V. W. Eveleigh, *Control Systems Design*, McGraw-Hill Book Company, New York, 1972.

8 DC Tachometers As Velocity Sensors

J. TAL and E. K. PERSSON

8.1 INTRODUCTION

Dc tachometers, because of their ability to provide shaft velocity information, play an important role in incremental-motion control systems. Dc tachometers can function in various ways; they can be used as velocity indicators to provide visual shaft speed readout, or they can provide feedback information needed for stabilizing position control systems. But perhaps their most significant contribution to servo systems is in providing shaft velocity feedback for speed control systems. In this chapter we will examine the different functions of the dc tachometer as well as different tach designs, mathematical modeling in both dc tachometers and dc motor-tachometer units, and general tachometer characteristics.

The dc tachometer is an electromechanical device capable of generating an electrical signal which is proportional to the angular velocity of the motor shaft. The function of the dc tachometer is directly opposite to that of the dc motor: a dc motor converts electrical energy into mechanical energy, while a dc tachometer converts mechanical energy into electrical. Although the structure and principles of operation of dc motors and tachometers are very similar, there are important differences between the two: because dc motors operate at high power levels, attention must be given to their power generation and dissipation characteristics. Dc tachometers, on the other hand, always operate at low power levels; as a result, their critical characteristics have to do with the linearity of the relationship between shaft velocity and dc voltage, and the minimization of ripples in the output voltage.

The terms "dc tachometer" and "dc generator" have, in the past, been used interchangeably to describe the same object. To the authors' knowledge, this ambiguity of terms may have arisen from early applications of tachometers, when their use was limited to giving visual readouts of shaft velocity (usually in conjunction with a voltmeter calibrated in RPM). Because tachometers were used in industrial environments of high electrical noise, and because they were often required to carry velocity information over considerable distances, it was necessary for their output impedance to be low. As a result of this requirement, early tachometers were capable of generating a considerable amount of power; thus, they became known as "generators." The term seems to have carried over into servo systems through the popularity of the ac servo systems of the early '50s; product literature at that time usually referred to ac tachometers as "tachometer-generators." Thus, we find today that "dc generator" is sometimes used instead of the perhaps more accurate term, dc tachometer.

As mentioned before, dc tachometers today are most widely used in velocity control systems where the shaft velocity information they provide is used for velocity feedback purposes. The tachometer signal in such systems is compared to the motor's command voltage; the difference between the two is then amplified and used to drive the servo system motor (see Fig. 8-1). Because of this direct correlation between the tachometer and the motor, the accuracy of the actual shaft velocity is largely dependent on the accuracy of the tachometer. (For further discussion of velocity control systems, see Chapter 12.)

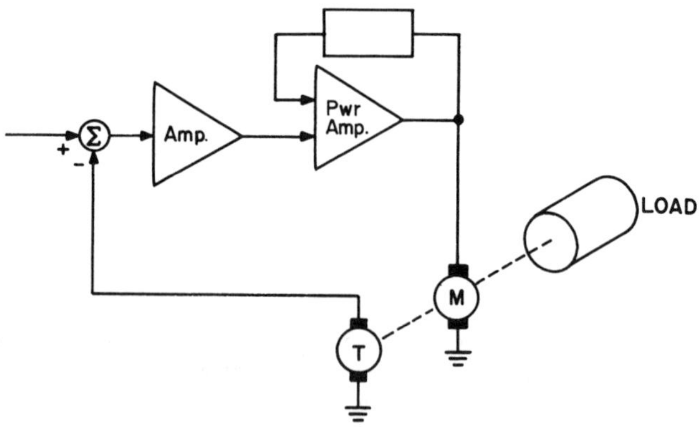

Fig. 8-1. Block diagram of velocity servo with tachometer feedback.

Another use for velocity sensors is in position control systems, as described in Chapter 13. In these systems, the tachometer signal is again used for velocity feedback. The feedback provides damping to the position servo loop, thereby enhancing the overall stability of the system (see Fig. 8-2). The tachometer accuracy in position control systems is not as critical as in velocity servo systems.

The third and most traditional use of dc tachometers is in providing visual shaft speed readout. Tachometers functioning in this capacity are generally used in conjunction with a voltmeter calibrated directly in RPM. Since the voltage output of the tachometer is directly related to the shaft velocity, a visible indication of that velocity is achieved by connecting the tachometer output voltage to the voltmeter (see Fig. 8-3).

It is possible to obtain a velocity-proportional signal without using a dc tachometer. One way is to use an ac tachometer in conjunction with a rectifier and filter system; this gives a voltage proportional to shaft velocity over a limited speed range.

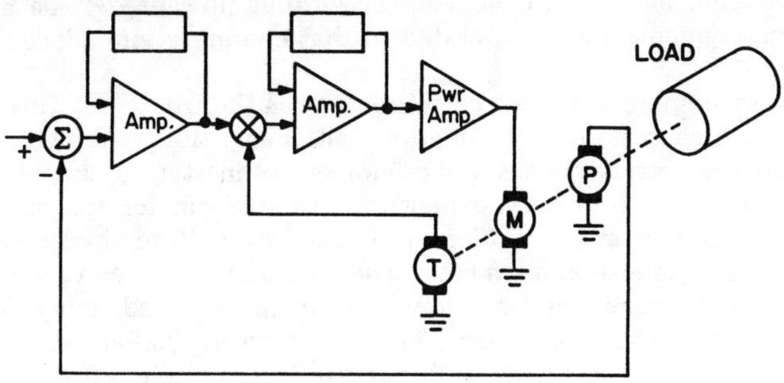

Fig. 8-2. Block diagram of position sensor with tachometer damping.

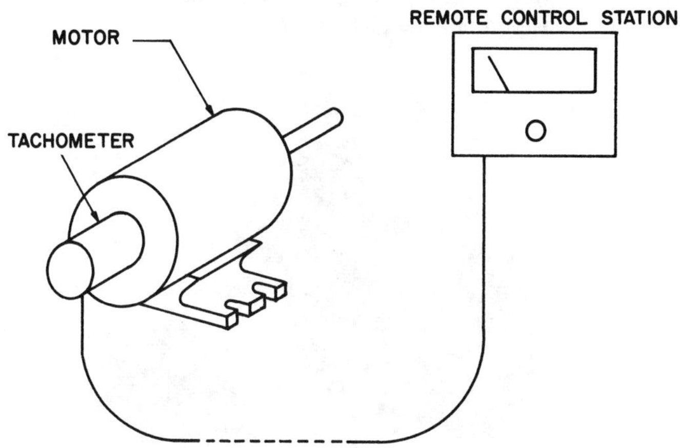

Fig. 8-3. Tachometer providing shaft velocity readout.

The ac tachometer, however, does not indicate the direction of shaft rotation. (The polarity of the dc tachometer is dependent on the direction of the shaft rotation.) Another alternative is to use a frequency demodulator to generate a velocity-proportional voltage from the frequency of an optical encoder signal. Methods such as the above, while possessing some limited advantages in certain applications, cannot begin to approach the dc tachometer in its ability to provide velocity information over a very broad range of conditions. Therefore, in this chapter, we will limit the discussion of velocity sensors to an examination of true dc tachometers.

8.2 TYPES OF DC TACHOMETERS

As mentioned above, the dc tachometer and the dc motor share similar structures and principles of operation. The brush-type dc tachometer, like its dc motor counterpart, has an armature, a commutator, a brush rigging, and a permanent-magnet field. Unlike the dc motor, however, its output impedance can be relatively high due to its low output level. In addition, because the design of the dc tachometer is geared specifically toward producing an accurate dc voltage without unwanted "side effects," many commutation segments are incorporated so that commutation-induced voltage ripple is minimized.

The most common type of dc tachometer is the iron core. This brush-type tachometer uses high grade rotor laminations with many slots to establish a path of uniform magnetic reluctance. It has a high moment of inertia and is subject to relatively high voltage ripples. It is the least expensive of the tachometer designs.

Where both low moment of inertia and low voltage ripples are desired, the moving-coil tachometer may be a better choice than the iron core. It is very light, has a low moment of inertia, and has low voltage ripples. Predictably, it is also more expensive than the iron core. Figure 8-4 shows a moving-coil armature and a housing-and-magnet assembly. The armature diameter in this example is 0.7 inches; the length, including the 19 bar silver commutator, is 1.4 inches. It has a moment of inertia of 9×10^{-5} oz-in-sec^2. Both the iron-core and the moving-coil tachometers use brushes and commutators designed and selected to provide high conductivity with a minimum of electrical noise and mechanical wear. Typically a 50-50 composition of silver and graphite is used for the brushes; the commutator material can be either a silver-bearing copper composition or a hard silver alloy.

Fig. 8-4. A low-inertia moving-coil tachometer armature, housing-and-magnet assembly.

The third type of dc tachometer is the brushless tachometer. Like its brushless dc motor counterpart, it uses no mechanical contacts to commutate the active windings; for this reason, the commutation system must be synchronized with the shaft angular position. Optical or magnetic encoders are generally used to provide this information to the commutation circuit. The switching of the various voltage-producing windings can be accomplished in a variety of ways, depending on the system requirements for output dc voltage linearity and voltage ripple content. The main advantage of brushless dc tachometers is that the lack of brushes (and, therefore, of mechanical contact) means longer component life. An example of a brushless tachometer output is shown in Fig. 8-5, where an oscilloscope picture of the three trapezoidal phase waveforms is displayed. A block diagram of the system follows in Fig. 8-6.

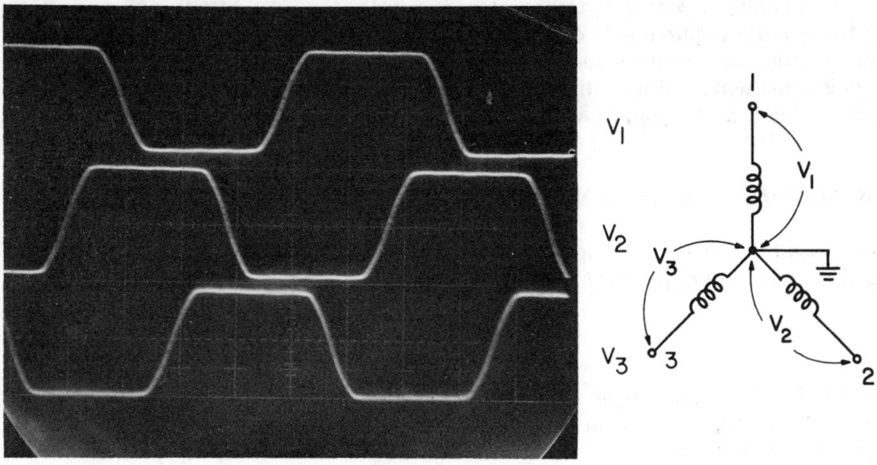

Fig. 8-5. Example of a three-phase brushless tachometer output voltage waveform.

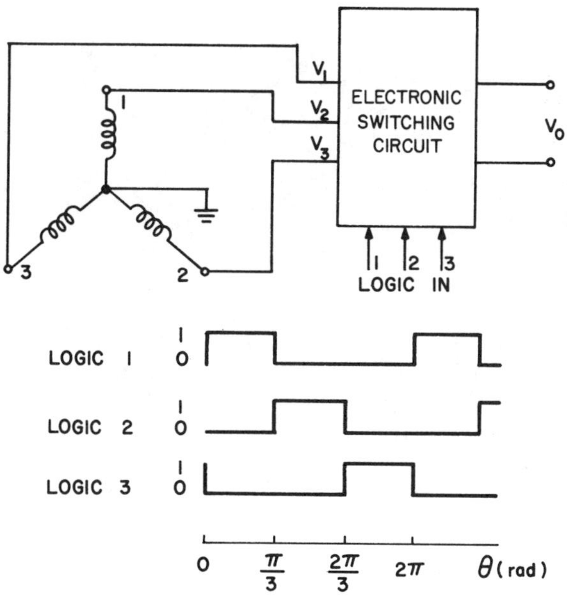

Fig. 8-6. Block diagram of commutation system for a three-phase brushless tachometer.

This diagram demonstrates the switching logic function controlling the electronic switching in the commutating circuit. The design of the electronic switching circuit requires special care to prevent switching transient voltages from causing significant output voltage ripples.

The choice of which dc tachometer to use in any given application is determined by the specific requirements of the system, especially in terms of electrical specifications relating to temperature-dependent or time-dependent properties. We will now examine mathematical modeling of the dc tachometer in order to more fully understand its limits and capabilities.

8.3 MATHEMATICAL MODELING OF DC TACHOMETERS

The output signal of a dc tachometer, V_g, consists of two components: the desired signal, E_g, and the ripple, n(t). This can be expressed by the equation

$$V_g = E_g + n(t) \qquad (8\text{-}1)$$

A typical tachometer voltage is illustrated by Figs. 8-7 and 8-8. Since the voltage E_g is directly proportional to the angular velocity, ω, the relationship between the two can be given by the equation

$$E_g = K_g \omega \qquad (8\text{-}2)$$

where K_g is the tachometer voltage constant. (A common value for this constant is K_g = 3 volts/KRPM.)

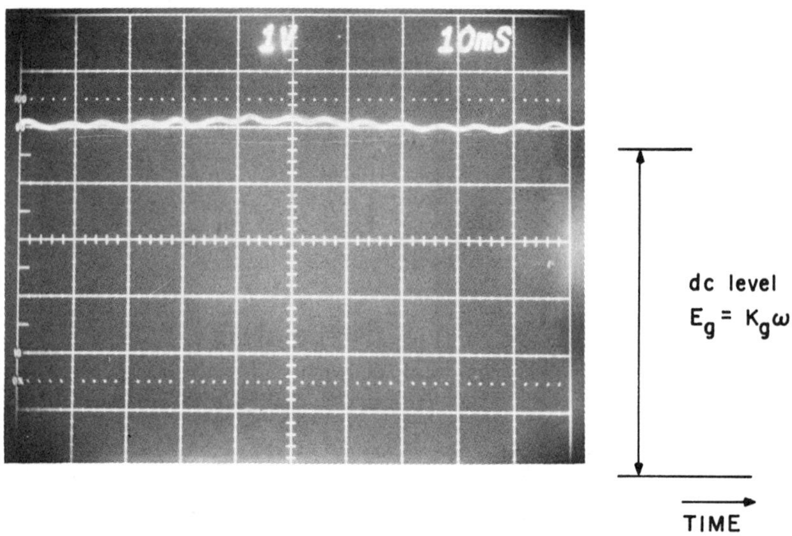

Fig. 8-7. Typical dc tachometer signal under constant-speed conditions.

In some applications, the shaft angular position, θ, is considered to be the input, in which case the transfer function is

$$\frac{E_g(s)}{\Theta(s)} = sK_g \qquad (8\text{-}3)$$

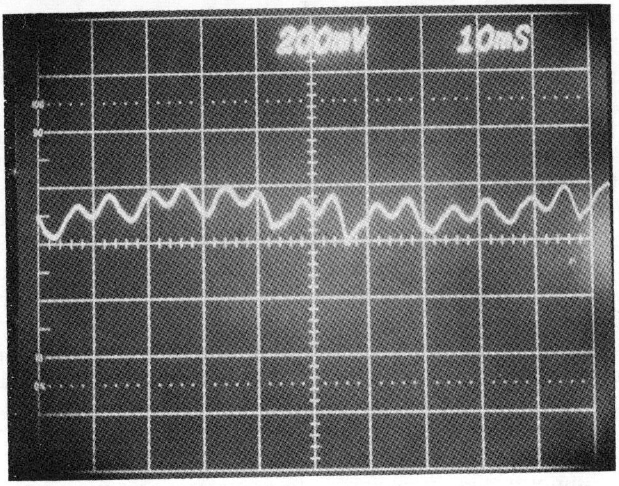

Fig. 8-8. Ripple voltage in Fig. 8-7 expanded in time and amplitude.

This is the form most commonly used in textbooks and in applications involving position control systems.

The ripple, or undesired signal, n(t), is sometimes referred to as tachometer "noise." It can be caused by a variety of factors, including the following:

Commutation Ripples. Commutation ripples, which make up the most significant part of the noise, are caused by the transition of the tachometer brushes between commutator segments. The signal generated is periodic with a fundamental frequency f_c:

$$f_c = \frac{100}{6} kun \tag{8-4}$$

where u is the number of commutator segments; k is equal to 1 when u is even, and to 2 when u is odd; and n is the shaft velocity in KRPM. The resulting frequency is expressed in Hertz. The amplitude of the commutation noise increases with the shaft velocity; this increase, however, is not proportional. Because the control system responds most effectively to signals at low frequencies, the effect of commutation ripples will be greatest at low speeds (when the noise frequency is also low).

Armature Eccentricity. The second-most important element contributing to noise is eccentricity in the tachometer armature. The effect of this condition is the generation of an undesired signal which is periodic and whose fundamental frequency is equal to the angular frequency of the tachometer shaft. Because the frequency of this signal is relatively low, it can have an adverse effect on the system. However, the magnitude of this signal is often small in comparison to commutation ripples.

High Frequency Noise. The third element contributing to n(t) is high frequency noise, or "white noise." White noise is caused by various factors, the most significant being electromagnetic induction. White noise has the least critical effect on the system; because the frequency of the signal is high, it can be filtered out without affecting the overall system behavior.

8.4 MATHEMATICAL MODELING OF DC MOTOR-TACHOMETER UNITS

In order to avoid backlash problems which are created when the motor and the tachometer are not tightly coupled, most incremental motion systems employ tachometers mounted directly on the motor shaft. In fact, quite often the motor is built with an integral tachometer; this allows the motor and tachometer to share the same shaft, resulting in reduced torsional resonance between the two. While both of the above configurations virtually eliminate backlash, they also introduce an electromagnetic coupling between the motor and tachometer caused by the proximity of their magnetic fields. This condition necessitates dealing with the motor and the tachometer as a single unit. The motor-tachometer unit is examined in the following discussion.

The dc motor-tachometer unit may be represented by the block diagram of Fig. 8-9. As we can see from Fig. 8-9, the tachometer voltage V_g is affected by the motor voltage V_m in two ways: directly, by means of the velocity ω and the signal V_{g1}; and indirectly, as a result of electromagnetic coupling.

In order to simplify the discussion of tachometer-motor units, we will consider only the desired component of the tachometer voltage, E_g; the undesirable component, n(t), is not affected by electromagnetic coupling and can be ignored for the time being. In order to illustrate the coupling effect in simple terms, we will assume in the following discussion that the motor may be modeled as a transfer function with a single pole:

$$M(s) = \frac{\Omega(s)}{V_m(s)} = \frac{K_m}{s\tau_m + 1} \tag{8-5}$$

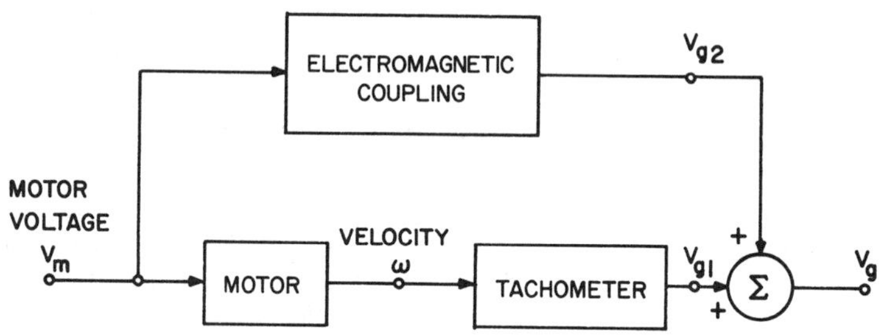

Fig. 8-9. Block diagram representation of the coupling between the motor voltage and the tachometer voltage.

Since the output voltage V_g is proportional to the velocity, ω, we may write

$$V_{g1} = K_g\omega \tag{8-6}$$

where K_g is the tachometer voltage constant.

Now we can combine Eq. (8-5) with Eq. (8-6) to form the transfer function for the direct branch of the model:

$$\frac{V_{g1}(s)}{V_m(s)} = \frac{K_g K_m}{s\tau_m + 1} \tag{8-7}$$

The electromagnetic coupling depends on the geometric structure of the motor and the tachometer. (In most cases, the coupling effect may be represented as a high-pass filter.) Furthermore, the relative position of the motor brushes and the tachometer brushes will introduce a phase shift a between the motor voltage and the tachometer voltage. As a consequence, the transfer function representing the electromagnetic coupling is given by

$$\frac{V_{g2}(s)}{V_m(s)} = \frac{K_2 s}{s\tau_2 + 1} e^{ja} \tag{8-8}$$

The total tachometer voltage, V_g, is the sum of V_{g1} and V_{g2}.

$$V_g = V_{g1} + V_{g2} \tag{8-9}$$

Therefore, the total transfer function of the motor-tachometer unit becomes

$$\frac{V_g(s)}{V_m(s)} = \frac{K_m K_g}{s\tau_m + 1} + \frac{K_2 s}{s\tau_2 + 1} e^{ja} \tag{8-10}$$

It can be seen that the combined transfer function depends on a, the relative position of the motor housing and the tachometer housing. This principle can be illustrated by the following example.

Example 8-1.

Consider a motor-tachometer unit in which the direct branch transfer function is

$$\frac{V_{g1}(s)}{V_m(s)} = \frac{10}{s + 10} \tag{8-11}$$

and whose electromagnetic coupling is given by

$$\frac{V_{g2}(s)}{V_m(s)} = \frac{0.2s}{s + 5} e^{ja} \tag{8-12}$$

We will now examine separately the combined transfer function for two conditions: when $a = 0$, and when $a = 180°$.

Case a: $a = 0$

The combined transfer function in this case is equal to:

$$T_1(s) = \frac{10}{s + 10} + \frac{0.2s}{s + 5} = \frac{0.2(s^2 + 60s + 250)}{(s + 5)(s + 10)} \tag{8-13}$$

This transfer function has two real poles (at $p_1 = -5$ and $p_2 = -10$) and two real zeros (at $z_1 = -4.5$ and $z_2 = -55.5$). The poles and zeros of $T_1(s)$ are shown in Fig.

8-10. The frequency response of this system is shown in Fig. 8-11 (magnitude) and Fig. 8-12 (phase). The magnitude of $T_1(s)$ is approximated by its asymptotic curve.

Case b: a = 180°

The combined transfer function in this case is given by

$$T_2(s) = \frac{10}{s+10} - \frac{0.2s}{s+5} = \frac{0.2(-s^2+8s+250)}{(s+5)(s+10)} \tag{8-14}$$

This transfer function has the same poles as $T_1(s)$, but the zeros have changed. The zeros in this case are $z_1 = -5.5$ and $z_2 = 45.5$. The poles and zeros of $T_2(s)$ are shown in Fig. 8-13. Note that one of the zeros is in the right-half plane; this zero will tend to attract the branches of the root locus, resulting in an unstable system. Thus, the electromagnetic coupling effect in this case is detrimental to the system stability. The frequency response of $T_2(s)$ is shown in Fig. 8-11 (magnitude) and Fig. 8-12 (phase). The magnitude is approximated by the asymptotes as shown in Fig. 8-11.

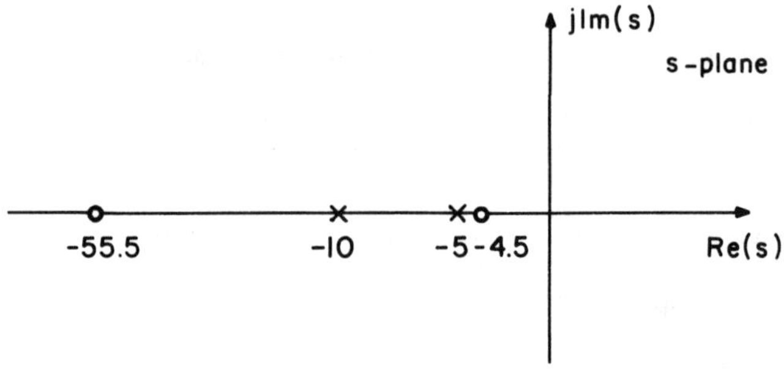

Fig. 8-10. Pole-zero configuration of $T_1(s)$.

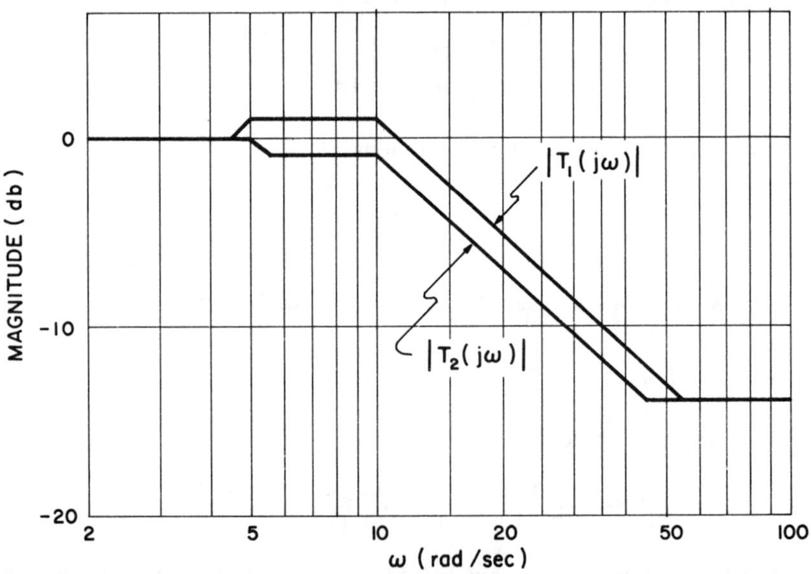

Fig. 8-11. Asymptotic approximations for the magnitude of $T_1(j\omega)$ and $T_2(j\omega)$.

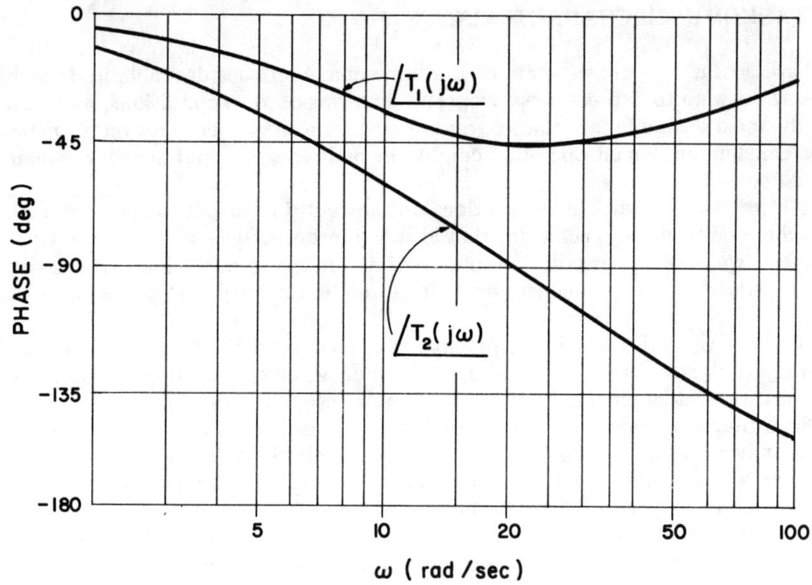

Fig. 8-12. Phase curves of $T_1(j\omega)$ and $T_2(j\omega)$.

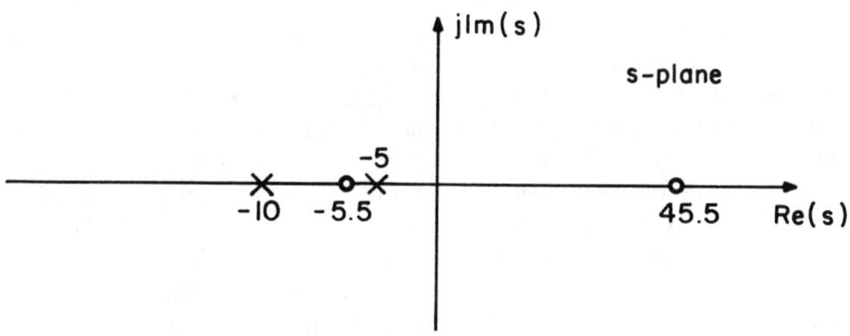

Fig. 8-13. Pole-zero configuration of $T_2(s)$.

The above example illustrates the simple fact that the combined transfer function of the motor and the tachometer may change drastically as a function of the relative angular position of the motor and the tachometer. This change may have a significant overall effect on the system as the motor-tachometer zeros move into the right-half of the s-plane.

Another significant fact is illustrated by Figs. 8-11 and 8-12; that is, while a change in the zeros of the transfer function has little effect on the magnitude of the transfer function, a change in zeros has a significant effect on the phase characteristics of the transfer function. Therefore, when the electromagnetic coupling is analyzed, it is more useful to measure the phase than it is to measure the gain.

8.5 TACHOMETER CHARACTERISTICS

We have so far mentioned some of the basic characteristics desirable in dc tachometers. If we were to express these properties in terms of ideal conditions, we could say that the most desirable tachometer would be one whose voltage constant K_g remained truly constant under all operating conditions and whose output signal V_g emanated ripple-free.

Unfortunately, such a device does not currently exist (although some tachometers come quite close to achieving that ideal). In order to deal with real world devices, therefore, we must understand the sources of deviations from the ideal. The deviations can be divided into three categories: voltage ripple, linearity, and temperature stability.

Voltage ripple is caused by inherent design characteristics of the tachometer; it is generally (but not always) a function of the shaft velocity. As we saw in Sec. 8-3, the basic causes of tachometer "noise" are commutation ripples, armature eccentricity, and high frequency noise. Commutation-induced noise can be caused by voltage variations resulting from a change in flux as the tachometer brushes traverse from commutator bar to commutator bar, by commutation of the tachometer current during the commutation event, and by surface noise of the commutator-brush interface. Perhaps the most dominant ripple component in quality tachometers is commutation-induced ripple, which can have cyelic effects at commutation frequency (see Sec. 8-3). The inductively caused noise due to the commutation event is usually of high frequency and can be easily filtered by a simple RC low-pass network. The surface noise is often small enough that it can simply be ignored. Other causes of voltage ripple include magnetic anomalies in the field, rotor eccentricities, and externally-induced stray fields in the armature. The stray-field pick-up can be minimized by magnetic shields, but the rotor eccentricity, which can cause a twice-around effect, is not so easily detected or dealt with. Because rotor eccentricity is usually a second-order effect, it is often obscured by commutation ripple. It can, however, be observed in low-ripple tachometers (those with peak-to-peak/average ripple levels below 1 percent).

Because the sources of tachometer ripple are not always proportional to the shaft velocity, it may be necessary to examine the full velocity range of interest for voltage ripple if the system is to be used in critical applications. In low-inertia incremental motion control systems, well-designed moving-coil tachometers are often found to have more favorable voltage-ripple characteristics than corresponding iron-core units. This is because the moving-coil tachometer has a lower inductance and operates in a magnetic field free of rotor-induced reluctance changes. Chapter 9 includes a discussion of measurements of voltage gradients and tachometer ripples. (Typical values of voltage ripples range from 6 percent to 1 percent peak-to-peak average.)

The linearity characteristics of dc tachometers are generally very good (less than 1/2 percent) when operating at "normal" speeds (3000 RPM or less). Some nonlinearities may be noticed at higher velocities, however. The particular characteristics of the nonlinearities will depend on the individual tachometer design. Causes of nonlinearities include commutation noise due to brush bounce at high speeds and slightly eccentric commutators, eddy current and hysteresis loss effects in the armature assembly, and brush position change caused by a change in the direction of shaft rotation.

Temperature stability problems are mainly caused by changes in the magnetic field resulting from a change in magnet temperature. Some magnet materials are more sensitive to temperature change than others. Barium ferrite, for instance, has a temperature coefficient of −0.2%/°C, while ALNICO 5 has a coefficient of 0.015%/°C; ALNICO 8 has a coefficient of less than −0.01%/°C. Some new rare-earth compounds have been produced recently which are claimed to have essentially a zero temperature coefficient. Magnetic temperature coefficients may be compensated for by means of either magnetic shunts (using a material which has a coefficient opposite to that of the magnet) or electrical compensation (using resistive elements with a known temperature coefficient circuit). In the latter case, it may be necessary to use a specific termination resistance recommended by the manufacturer in order to achieve the specified temperature stability.

All that remains to be considered is component wear. In a well-designed brush-type dc tachometer, a minimal amount of brush and commutator wear occurs. While life expectancies depend on the type of environmental and operational conditions present, a quality brush-type dc tachometer's components should give from 10,000 to 50,000 hours of service. The brushless dc tachometer will have even longer life.

8.6 CONCLUSIONS

Dc tachometers as velocity sensors find wide applications in incremental motion control systems. Having expanded from their limited initial use of supplying visual shaft speed readout, tachometers now play an integral role in providing velocity feedback for both speed and position control systems. In this chapter, we discussed the nature of the dc tachometer and the different types of tachometers currently available. We investigated mathematical modeling of dc tachometers both individually and as part of the dc motor-tachometer unit. We looked at the effects of electromagnetic coupling and discussed tachometer "noise." Finally, we studied ideal tachometer properties and how to compensate for deviations from the ideal by tracing problem areas to their possible sources. The information presented herein should aid the system designer in selecting the right tachometer for his particular application.

REFERENCES

1. *DC Motors, Speed Controls, Servo Systems*, 3rd ed., Engineering Handbook, Electro-Craft Corporation, 1975.

2. B. C. Kuo, *Automatic Control Systems*, 3rd ed., Prentice-Hall, Inc., Englewood Cliffs, N.J., 1975.

3. S. A. Davis and B. K. Ledgerwood, *Electromechanical Components for Servomechanisms*, McGraw-Hill Book Company, New York, 1961.

9 Testing DC Motors And Tachometers

E. K. PERSSON and J. TAL

9.1 INTRODUCTION

In this chapter we will discuss the different methods used to determine the performance parameters of dc motors and generators. The objective of such testing is to isolate and measure the various factors which affect the performance of a dc motor or generator, thereby providing the information needed to determine that motor or generator's suitability for operating in a particular application. The following factors must be taken into account when evaluating dc motors: armature resistance and inductance, torque and voltage constants, motor friction and damping, and armature moment of inertia. Methods for measuring these effects will be discussed in Sec. 9.2. In determining the parameters of dc tachometer performance, output resistance, voltage gradient, and tachometer ripples are the effects which must be considered. These will be discussed in Sec. 9.3.

In describing the test methods used, we have adopted the viewpoint of accommodating the user who has limited access to sophisticated testing equipment. Therefore, in cases where several testing methods could be used, we have narrowed the discussion to include only those methods which are the simplest to use or which do not require high specialized equipment.

Since the specific motor application can indicate the probable output performance limitation (be it thermal, magnetic, electrical, or mechanical), we must first decide what parameters are of concern to us. In many cases it is entirely adequate to have a motor characterized simply by its resistance and its torque constant. From these two values alone, one can determine whether the motor can handle the application requirements, and what the internal power dissipation in the motor will be.

However, a more thorough investigation is often required to obtain a more complete picture of the suitability of the motor or generator in question. While we will be concerned with both mechanical and electrical measurements in this chapter, we will focus our attention on measuring only the major (i.e., the most important) parameters of dc motors and generators as listed above. For a more complete treatment of the subject, including a discussion of additional test procedures, we refer the reader to the ECC Handbook [1].

9.2 TESTING DC MOTORS

Armature Resistance

Any testing of armature resistance in a brush-type dc motor must take into account the fact that brush resistance is not always equivalent to "pure" resistance. The brush-commutator interface usually develops a nonlinear resistance, the parameters of which are determined by the metal-graphite composition of both the brush and the commutator material. Because of this influence, ohmmeters may give an erroneous indication of armature resistance. In addition, armature resistance tends to vary with shaft position, especially in motors having few commutation points.

The most convenient alternative to using a ohmmeter in measuring armature resistance is to use the volt-ammeter method, as shown in Fig. 9-1. The motor armature is mechanically locked at some angular position, and a current I is then applied to the motor. The voltage V is measured across the motor, and the armature resistance is determined from the ratio of the voltage to current according to Ohm's law. By making several measurements at typical operating current levels and at several shaft angle positions, an average value can be obtained. Since armature resistance is specified at 25°C, care should be taken that the selected current value does not cause critical heating of the armature.

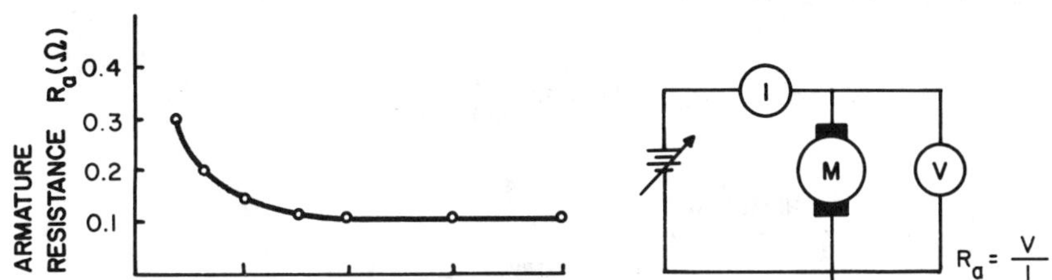

Fig. 9-1. Armature resistance measurement as a function of test current.

Armature Inductance

Armature inductance in dc motors may be most easily measured by using an impedance bridge. Since the measured value of armature inductance may be a function of the shaft position, a series of measurements should be taken. If the measurements for shaft position are taken at four 90-degree intervals, a reliable average value for armature inductance can be determined.

Torque And Voltage Constants

Since, as we learned in Chapter 5, the torque constant K_t and the voltage constant K_e are directly proportional in dc motors, it is sufficient to measure only one of them in order to derive both values. Since the choice of which constant to measure depends on the equipment available, methods for testing both constants will be presented. The one exception to this interchangeability rule applies to measuring the torque constant when a dynamometer is available. Determining the torque constant by using a dynamometer is more informative than deriving it from the motor's voltage constant because the motor's behavior under load conditions (revealing brush sparking, motor heating, and possible demagnetization) can be observed.

Measuring The Back EMF Constant

The back emf constant K_e can be measured by running the motor as a generator (that is, driven by another motor) and measuring the generated voltage E_g. The setup for this test is shown in Fig. 9-2. The shaft velocity, ω, is then measured and compared to the generated voltage. The voltage constant is then obtained as the ratio of the two measurements:

$$K_e = \frac{E_g}{\omega} \tag{9-1}$$

For example, if the motor voltage is 7 volts at a velocity of 1000 RPM, the voltage constant is

$$K_e = 7 \text{ volts/1000 RPM} \tag{9-2}$$

Using the SI units, K_e is equal to

$$K_e = 0.06685 \frac{\text{volts}}{\text{rad/sec}} \tag{9-3}$$

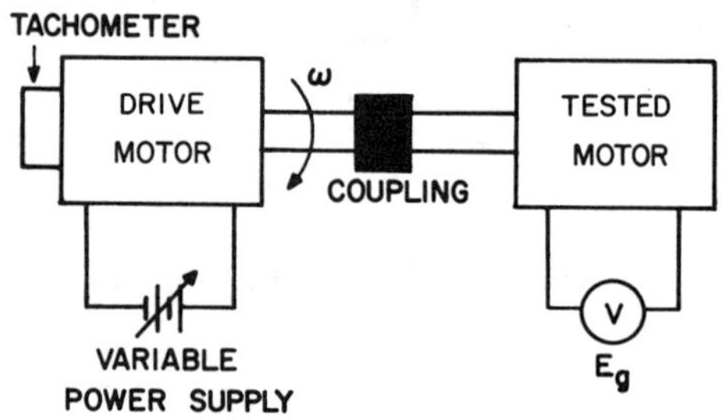

Fig. 9-2. Method of measuring the back emf constant of a dc motor.

Knowledge of K_e enables us to determine K_t directly; K_e and K_t are the same when expressed by the SI system. For the example given above, we obtain

$$K_t = 0.06685 \text{ Newton-meter/ampere} \tag{9-4}$$

which is equal to

$$K_t = 9.47 \frac{\text{oz-in}}{\text{ampere}} \tag{9-5}$$

Note that when K_e is expressed in volts/1000 RPM, and when K_t is expressed in oz-in/ampere, the relationship between the two is given by

$$K_t = 1.35K_e \tag{9-6}$$

It is important to note that, in a well-designed dc motor, the relationship between the motor velocity and voltage is linear; when the motor velocity is constant in such a motor, the voltage is constant as well. However, all practical motors will generate a voltage which varies somewhat with the angular position of the shaft. Therefore, it may prove worthwhile to connect an oscilloscope to the output voltage of the motor in order to observe the voltage function with respect to shaft rotation. A voltage which varies with shaft position indicates the possibility of torque ripple in the motor.

Measuring The Torque Constant

Obtaining the torque constant requires the use of a dynamometer or an equivalent device capable of indicating torque output and shaft velocity. A typical test setup is shown in Fig. 9-3. In this example, a dc power supply energizes the motor being tested, which is in turn coupled to the dynamometer. The motor is run at a constant speed, and the load of the dynamometer is then raised to the rated load of the motor. A plot of the measurement results should fall in a straight line; the slope of that plotted line equals K_t; that is,

$$K_t = \frac{\Delta T}{\Delta I} \tag{9-7}$$

where ΔI denotes the incremental change in the motor current, and ΔT denotes the corresponding incremental change in motor torque. If the plotted line is not straight, it indicates a nonlinear characteristic of the motor which may be due to lamination or magnet saturation. (It is important to note here that the rated torque output of a motor is usually defined as a continuous torque output. It is common in incremental motion systems for the rated torque conditions to be exceeded during acceleration and deceleration; therefore, under high torque conditions, the torque measurement should be made as quickly as possible so that excessive motor heating does not occur.)

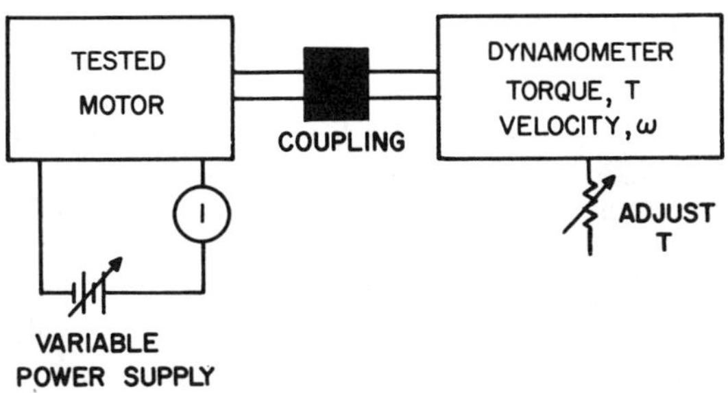

Fig. 9-3. Torque constant measurement of a dc motor.

Motor Friction And Damping

Friction in the motor can be divided into two categories: constant friction and viscous friction. Viscous friction occurs as a result of such factors as bearing lubrication and electromagnetic losses (eddy current and hysteresis losses). The magnitude of viscous friction is proportional to the motor velocity: the higher the speed, the greater the friction. Viscous frictional torque is represented by the term $B\omega$, where ω is the angular velocity of the motor shaft. The constant B is called the viscous damping coefficient. Constant friction is dependent on the motion state of the motor. When the motor is at rest, the constant friction is called static friction, and is denoted by T_s. Static friction is basically that frictional component which prevents initial rotation. As the motor begins to rotate, the constant friction drops to a new value called the dynamic friction, T_d. The value of T_d is generally lower than the value of T_s. In order to determine the three quantities T_s, T_d, and B, the following methods can be used.

The static friction, T_s, can be measured simply be observing the armature current needed to make the armature rotate. This can be done by connecting a variable-voltage power supply to the motor and increasing the current until the shaft begins to rotate. The highest current reading can then be multiplied by the known torque constant in order to derive T_s.

The dynamic friction, T_d, can be measured by noting the current needed to run the motor at a constant speed. A plot of the no-load current versus speed shaft velocity over the range being tested will give not only the value of T_d, but also the value of

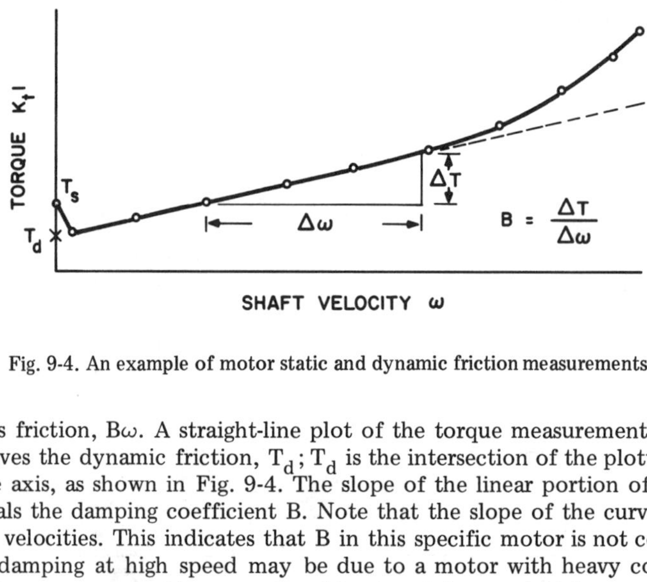

Fig. 9-4. An example of motor static and dynamic friction measurements.

the viscous friction, $B\omega$. A straight-line plot of the torque measurement versus shaft velocity gives the dynamic friction, T_d; T_d is the intersection of the plotted line with the torque axis, as shown in Fig. 9-4. The slope of the linear portion of the plotted curve equals the damping coefficient B. Note that the slope of the curve changes at high shaft velocities. This indicates that B in this specific motor is not constant. The increased damping at high speed may be due to a motor with heavy conductors or high iron losses, causing eddy current and hysteresis losses at high commutation frequencies. (Such a motor would be less suitable for a high-speed operation than a motor with straight-line characteristics over the usable speed range.)

Moment Of Inertia

In this section we will discuss two ways of measuring motor armature moment of inertia: the direct method and the indirect method. The direct measurement technique requires that the motor be disassembled (not recommended for the amateur); the moment of inertia is then measured by using a torsional pendulum. The indirect method determines the moment of inertia from the frequency response of the motor. This is a much simpler technique, but it requires previous knowledge of the motor torque constant, K_t.

The Direct Method Using The Torsional Pendulum

When an object with a moment of inertia J is mounted on a torsional pendulum, as shown in Fig. 9-5, the resulting period of oscillation of the pendulum is proportional to the square root of the total moment of inertia.

To find the armature's moment of inertia using this method, the armature must be removed from the motor structure. (For very precise measurements, the inner races of the ball bearings should be retained in order to include the bearing contribution to the inertia.) The armature is then suspended along its axis of rotation by a torsional spring such as a piano wire. The spring constant is chosen such that the period is easily measurable.

We start the test by measuring the period of oscillation when a homogeneous cylinder of simple proportions is suspended from the piano wire. The moment of inertia of the known cylinder is denoted by J_1, and the period of oscillation corresponding to it is denoted by T_1. Next the test object (the armature of unknown inertia) is

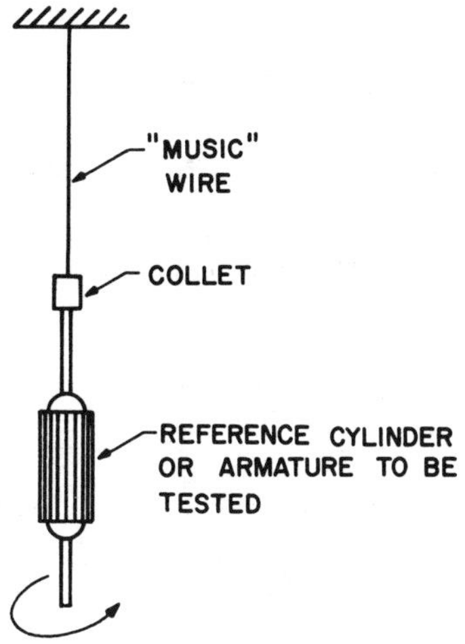

Fig. 9-5. Torsional pendulum test of armature.

suspended. Let the oscillation period in this case be denoted by T_2, and let the unknown armature inertia be denoted by J_2. The collet, which is used to connect the torsional pendulum to the subject of the test, is assumed to be the same for both experiments; denote its inertia by J_3. The unknown moment of inertia, J_2, can be found from the formula

$$J_2 = \frac{(J_1 + J_3)T_2^2}{T_1^2} - J_3 \tag{9-8}$$

where

J_1 is the moment of inertia of the reference cylinder

and

J_3 is the moment of inertia of the collet.

The Indirect Method Applying Sinusoidal Current

The indirect method of measuring motor armature moment of inertia is based on applying a current which includes a dc bias and a sinusoidal component to the motor, and measuring the resulting acceleration. This method is limited to the measurement of the motor armature and its inertial load, and requires knowledge of the torque constant K_t. This test also requires the utilization of a velocity sensor such as a tachometer. The test setup is shown in Fig. 9-6.

The motor is driven by a current which has a dc component and a sinusoidal component. The dc component of the current is adjusted so that the motor runs in the same direction, although its velocity is modulated sinusoidally. Let the sinusoidal component of the current be

$$i(t) = I_m \sin 2\pi ft \tag{9-9}$$

The resulting torque is

$$T_g(0) = K_t I_m \sin 2\pi ft \tag{9-10}$$

and the armature acceleration which results from the sinusoidal torque is

$$a(t) = \frac{K_t I_m}{J} \sin 2\pi ft \tag{9-11}$$

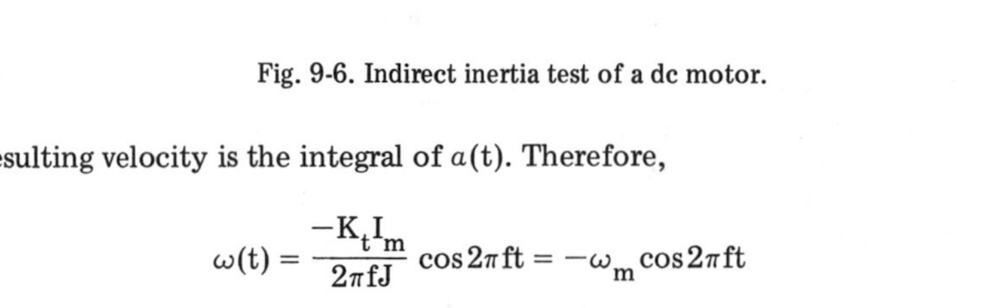

Fig. 9-6. Indirect inertia test of a dc motor.

The resulting velocity is the integral of $a(t)$. Therefore,

$$\omega(t) = \frac{-K_t I_m}{2\pi fJ} \cos 2\pi ft = -\omega_m \cos 2\pi ft \tag{9-12}$$

Now, if we compare the amplitudes of the sinusoidal components of the velocity $\omega(t)$ and the current $i(t)$, we find that

$$\omega_m = \frac{K_t I_m}{2\pi f J} \tag{9-13}$$

Thus, the moment of inertia is given by

$$J = \frac{K_t I_m}{2\pi f \omega_m} \tag{9-14}$$

Note that friction was ignored in this analysis. This is permissible since the motor current has a dc component large enough to keep the motor running in the same direction. The effect of the viscous damping, B, may be minimized by reducing the amplitude of the velocity changes.

In order to increase the accuracy of this test, we may run several measurements at different modulating frequencies f. Note that the ratio I_m/ω_m is proportional to f; therefore, if we plot the test results of I_m/ω_m versus the frequency on log-log paper, the resulting points should fall on a straight line with a slope of 45 degrees. This method allows us to identify the faulty measurements and to average the test results.

An important point to consider is that, if the velocity sensor/tachometer moment of inertia is of a significant value compared to that of the motor, the tachometer moment of inertia must be subtracted from the total inertia—unless the tachometer is actually going to be used in the intended application.

Additional Measurements

Mechanical measurements such as shaft run-out and shaft end play are straightforward measurements which can be easily made if the necessary indicating instruments are available. With knowledge of the armature inductance and resistance, the electrical time constant can be calculated. This value can also be measured by applying a step voltage across the motor with the shaft locked. If the current value is then displayed on an oscilloscope, the exponential rise in relation to time can be observed, thus providing us with the electrical time constant.

Likewise, the mechanical time constant can be measured by the application of a voltage step. This and other measurements are described in detail in [1].

9.3 TESTING TACHOMETERS

In this section we will describe the test methods used for establishing the more common tachometer parameters of output resistance, voltage gradient, and voltage ripple.

Output Resistance

The output resistance can be easily measured by using the half-voltage method. The tachometer is run at a constant speed and is connected, through a switch S_1, to a variable load resistance R_L as shown in Fig. 9-7. The load resistance R_L is adjusted so that when the switch S_1 is closed, the output voltage V_g drops to half of its no-load value. At that point, R_L equals the output resistance.

Voltage Gradient

The generator voltage gradient (or voltage constant) is the ratio between the output voltage and the generator angular velocity. Measuring the voltage gradient can be

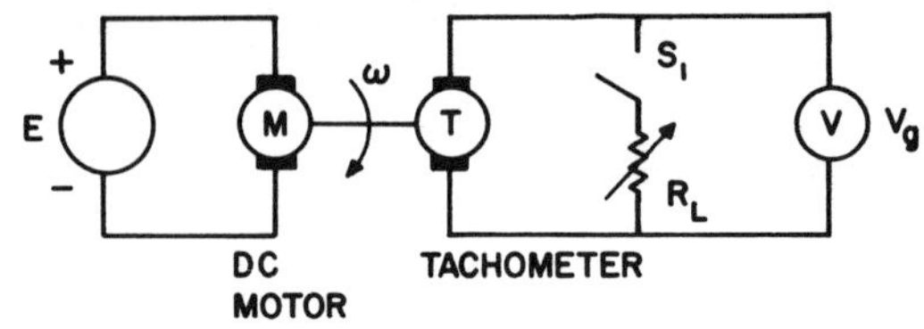

Fig. 9-7. Half-voltage method for measuring dc tachometer output resistance.

accomplished by running the generator at a known velocity, measuring the voltage, and then determining the ratio between the two.

For example, if the generator voltage is 3 volts at 1,000 RPM, the voltage gradient is

$$K_g = 3 \text{ volts}/1000 \text{ RPM} \tag{9-15}$$

or

$$K_g = 0.0286 \frac{\text{volt}}{\text{rad/sec}} \tag{9-16}$$

Tachometer Ripple

Dc tachometer-generator ripple is generally specified in one of two ways. In non-critical applications, tachometer ripple is given as the ratio between the RMS value of the output voltage ac components and its average dc value. In more critical applications (where a high degree of ripple control is essential), it is given as the ratio between the peak-to-peak values of ac components and its average dc value. The latter measurement requires an oscilloscope; the former requires true RMS and dc voltmeters.

The fundamental frequency of the generator ripples is determined by the number of commutator segments, and can be calculated by using the following expression:

$$f_r = \frac{100}{6} \text{ kun} \tag{9-17}$$

where u is the number of commutator segments; k is equal to 1 when u is even and to 2 when u is odd, and n is the shaft speed in kiloRPM. The resulting frequency, f_r, is expressed in Hertz.

Another aspect of tachometer-voltage ripple is a twice-around component (or twice-per-pole-pair in a multipole) tachometer. This is due to an eccentricity in the mounting of the tachometer armature, causing the voltage gradient to be higher during the instant when one part of the rotor is closer to a pole than the other. This effect is observed mostly in low-ripple tachometers, where this second-order component becomes visible due to the generally lower-ripple picture. This effect, which can cause problems in a servo loop, is more difficult to minimize by RC filtering since twice-around ripple always has a lower frequency than commutation ripple.

The RMS ripple measurement is made by simply connecting a dc voltmeter and a true RMS meter with an ac input network to the tachometer terminals while the tachometer is driven at a constant speed in the region of interest.

Peak-to-peak ripple measurements are made in a similar manner; by connecting the tachometer output voltage to an oscilloscope, a direct measurement of the desired

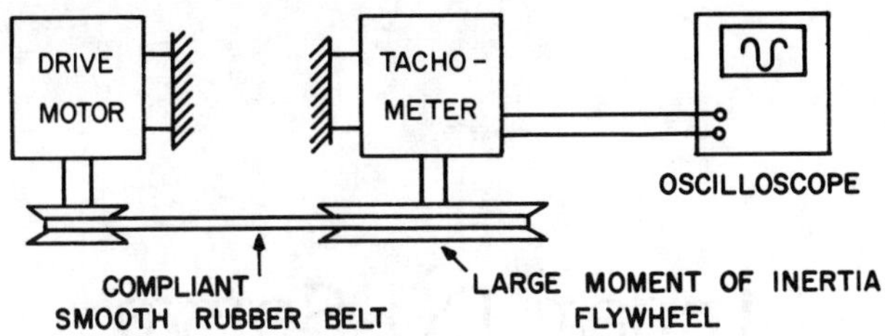

Fig. 9-8. A mechanical low-pass filter drive system for testing tachometer ripple.

voltage components can be taken.

Testing for tachometer ripple becomes more difficult with high quality tachometers in which the ripple value amounts to less than 2 percent of the dc voltage. Most test methods contribute their own inherent shaft velocity ripples through the drive mechanisms employed in the test itself, and these ripples are generally of the same magnitude. Therefore, it is important to insure that the tachometer velocity is indeed constant. This can be done by attaching a flywheel to the tachometer, as shown in Fig. 9-8. The large moment of inertia of the flywheel provides the mechanical filtering needed to generate a constant velocity.

9.4 CONCLUSION

In this chapter we have discussed different methods used in testing dc motors and generators to determine their performance parameters. With a view toward the least complicated methodology, we have taken a separate look at dc motors and generators. For dc motors, we have described tests which can be used to measure armature resistance and inductance, torque and voltage constants, motor friction and damping, and armature moment of inertia. For generators, we looked at ways to measure voltage gradient and tachometer ripples. Finally, we discussed the special considerations involved in dealing with high quality tachometers. These testing methods can prove highly valuable in determining the correct choice of a dc motor or generator for use in any incremental motion servo system.

REFERENCES

1. *DC Motors, Speed Controls, Servo Systems*, 3rd ed., Engineering Handbook, Electro-Craft Corporation, 1975.

2. B. C. Kuo, *Automatic Control Systems*, 3rd ed., Prentice-Hall, Inc., Englewood Cliffs, N.J., 1975.

3. B. D. Wedlock and J. K. Roberge, *Electronic Components and Measurements*, Prentice-Hall, Inc., Englewood Cliffs, N.J., 1969.

10 Linear DC Servo Amplifiers

R. SCHMIDT

10.1 INTRODUCTION

Linear servo amplifiers are usually constructed using a linear gain element such as an operational or differential amplifier to drive a power stage which, in turn, drives the motor. The power stage is included in the local feedback loop to linearize a desired parameter such as motor voltage, motor current, or a combination of both. In this fashion nonlinearities and temperature drift in the power stage are reduced by the loop gain of the system. An additional op-amp is often used as a compensation or summing amplifier and serves as a summing point for the command and major system feedback, such as an analog tachometer (in velocity control systems), and generates an error signal which causes the power amplifier to correct the controlled parameter.

Linear amplifiers control motor voltage or current by controlling the voltage applied to the motor and thus must drop a voltage equal to the difference between the supply voltage and the required motor voltage while delivering the appropriate current. A significant amount of power is dissipated in the output transistors, especially while running the motor under low-speed high-torque conditions where the motor back emf is low and the current is high. This is in marked contrast to switching amplifiers. Switching amplifiers control motor voltage by varying the duty cycle of voltage applied to the motor and operate in either a saturated or off mode and dissipate little power in either state, giving rise to efficient operation. Switching amplifiers, however, introduce new problems of their own. In order to avoid unduly high switching frequencies (with attendant switching losses), a minimum amount of motor (or external) inductance is required. This increases the electrical time constant of the system and may lead to reduced system bandwidth. In addition, electromagnetic interference (EMI) problems are often severe, and the systems tend to be more complex and failure prone. Thus, there is no clear-cut superiority of one type of amplifier over the other.

In general, linear amplifiers are preferred in wide-bandwidth low-power systems (less than several hundred watts). They are ideal in driving low-inertia moving-coil motors where high acceleration currents are required for short time intervals (several milliseconds) where output transistor peak current ratings may be used to advantage.

Conversely, switching amplifiers are usually used in larger systems, especially those which require extended operation at low speed and high torque where the power dissipation of the linear amplifier is high.

Linear amplifiers may be classified according to function: unidirectional or bidirectional (Sec. 10.2 and Sec. 10.3), according to configuration: bipolar or bridge (Sec. 10.3), or according to transfer function: voltage source or current source (Sec. 10.3 and Sec. 10.7).

Linear power stage design considerations, including predominant failure modes, are discussed in Sec. 10.4, Sec. 10.5, and Sec. 10.6, and a linear bipolar amplifier design example is given in Sec. 10.8.

10.2 UNIDIRECTIONAL SERVO AMPLIFIERS

Unidirectional amplifiers are the simplest form of linear amplifiers and are used in systems where a motor is driven in only one direction; that is, they provide a linear transfer function in only the first quadrant of operation (Fig. 10-1), such as positive motor voltage and positive motor current, and they are not capable of reversing the motor voltage or current. Such systems are characterized by rapid acceleration capability but slow deceleration capability since the motor is decelerated only by system friction and viscous damping.

The addition of dynamic braking capability to a unidirectional amplifier results in two-quadrant operation (Fig. 10-2). It still allows rotation in only one direction but has the capability of reversing motor current by removing drive current and providing a current path for current delivered by the motor back emf generator in the opposite direction. Deceleration current is a function of the back emf and, thus, is a function of speed, providing greater deceleration capability at high speed and reducing to zero near zero speed.

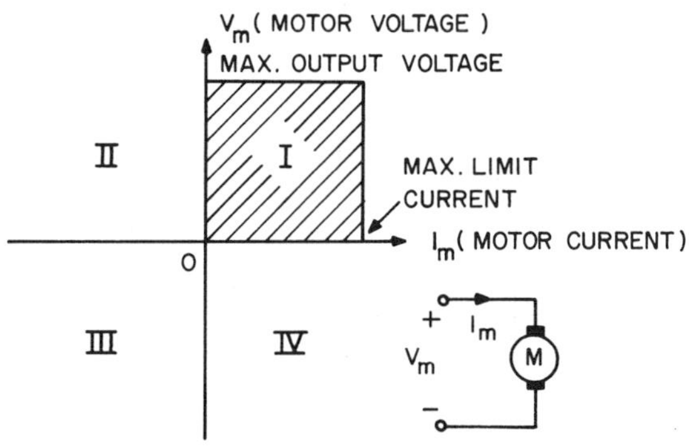

Fig. 10-1. Unidirectional amplifier—single-quadrant control.

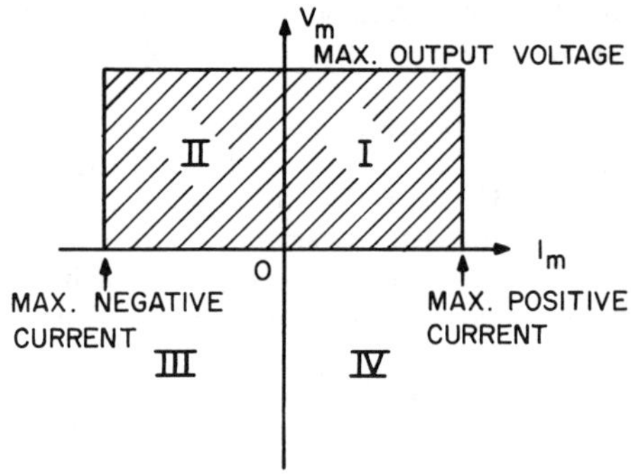

Fig. 10-2. Unidirectional amplifier with dynamic braking—two-quadrant control.

It should be remembered that operation in quadrant II is transient in nature and is only applicable while decelerating a system from sufficient speed to produce the desired deceleration current.

Figure 10-3 depicts a typical system deceleration profile for a unidirectional system with and without dynamic braking. Observe that the effectiveness of dynamic braking is reduced at low speed.

Figure 10-4 shows a simplified schematic of a unidirectional amplifier with dynamic braking.

When the motor is being driven, Q_2 conducts through diode D into the motor. Q_3 does not conduct since its base is one diode voltage drop more positive then the emitter. When Q_2 ceases to conduct, resistor R_d establishes base current in Q_3 turning it on. Current then flows in a reverse direction from the motor through Q_3. In this case, braking current is limited only by the back emf and motor armature resistance R_a. To limit braking current to a lower value, an additional resistance may be connected from the collector of Q_3 to ground. Q_1 serves as a current limiter for acceleration only, shunting base current from Q_2 when the drop across R_3 reaches approximately 0.7 volt, thus, limiting the acceleration current to: $I_{lim} = 0.7/R_3$ amp.

Note that the system configuration is that of a noninverting op-amp with a power booster with a voltage gain of $(R_2/R_1) + 1$. The noninverting configuration is compatible with single-supply operations.

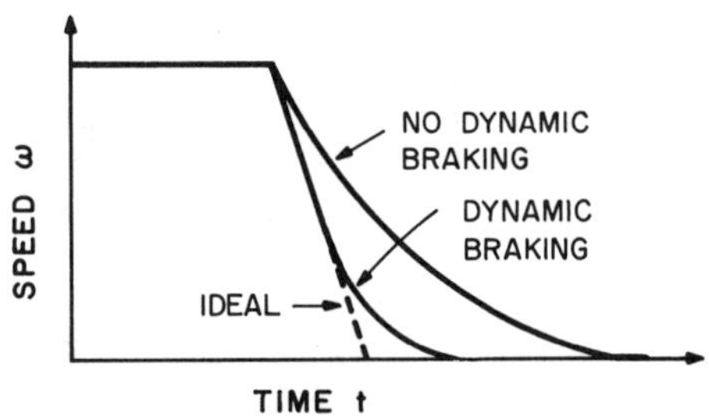

Fig. 10-3. A typical deceleration profile of a unidirectional system with and without dynamic braking.

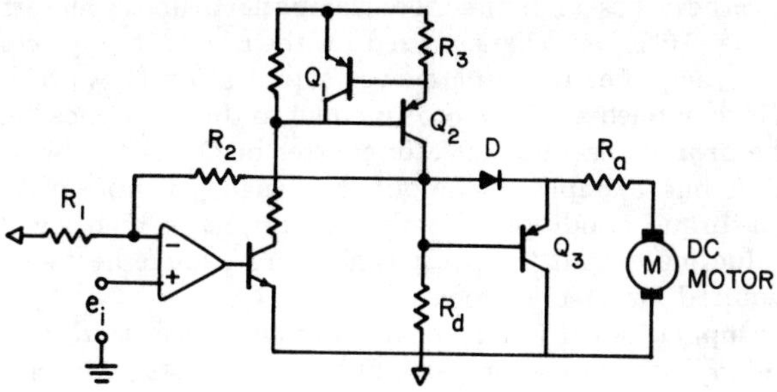

Fig. 10-4. Unidirectional amplifier with dynamic braking.

10.3 BIDIRECTIONAL SERVO AMPLIFIERS

Bidirectional servo amplifiers are capable of operating a motor in either direction and have a linear transfer function through zero speed. They are capable of full four-quadrant control (Fig. 10-5) and can apply voltage and current of either polarity to the motor.

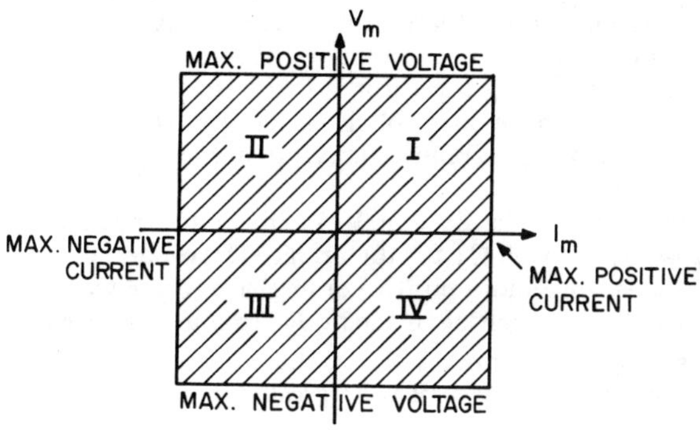

Fig. 10-5. Bidirectional servo amplifier showing full four-quadrant control.

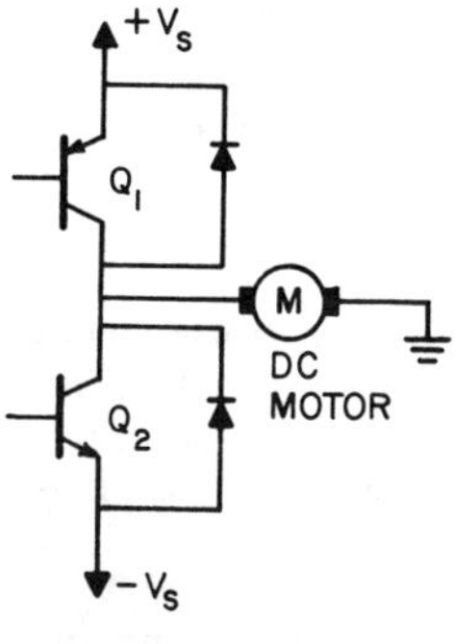

Fig. 10-6. Linear bipolar output stage.

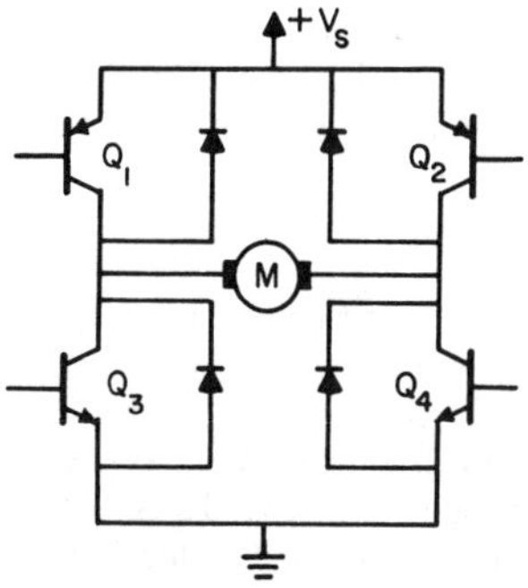

Fig. 10-7. Linear bridge output stage.

The two basic types of bidirectional amplifiers, linear bipolar (Fig. 10-6) and linear bridge (Fig. 10-7), are characterized by their output stage configurations [1].

The bipolar amplifier uses dual-power supplies for forward and reverse operations. The voltage applied to the motor is equal to the difference between the supply voltage and the drop across the collector-emitter of Q_1 or Q_2, whichever is conducting. As with all linear amplifiers, amplifier efficiency is not great, especially under low-speed high-torque conditions where the current is high and the voltage drop across the conducting output transistor is also large. Large heat sinks, often with fan cooling, are required for larger systems.

While the bipolar linear is simpler in form and easier to design than the bridge, the bipolar design does have several significant drawbacks which should be evaluated at the system level.

1. Bipolar power supplies are required.
2. The $V_{ceo(sus)}$ rating of the power transistors must be greater than the sum of the two supply voltages, usually $2V_s$.
3. Linear bipolar amplifiers are particularly susceptible to forward-bias second-breakdown* failure of the conducting power transistor(s) when decelerating a large inertial load from high speed.

The linear bridge requires only a single supply in order to apply a positive or negative voltage to the motor that is nearly equal in magnitude to the supply voltage.

For the bridge, Q_1 and Q_4 conduct when motor voltage is applied in one direction, while Q_2 and Q_3 conduct for the opposite direction of motor voltage. The bridge is capable of connecting the motor to the supply bus in one direction or the other depending on which pair of output transistors is conducting. In this case, the voltage across each conducting transistor is equal to one half the difference between the supply voltage and the motor voltage assuming equal transistor voltage drops.

The bridge configuration only requires that the $V_{ceo(sus)}$ rating of the power transistors be greater than V_s.

Free-wheeling diodes (Fig. 10-6 or 10-7) are connected across Q_1 and Q_2 for the bipolar and across Q_1 through Q_4 for the bridge. The diodes clamp the output voltage to one diode voltage drop more positive or negative than $\pm V_s$ for the bipolar. For the bridge the diodes clamp either leg of the bridge to one diode drop more positive than V_s to one diode drop below ground. Thus, the output transistors are never subjected to voltage stresses more than several volts greater than V_s for the bridge and $2V_s$ for the bipolar.

A more subtle advantage of the bridge is reduced voltage stress on a given conducting transistor. If the bridge is properly designed, the difference between the supply voltage and the required motor voltage is equally split between two conducting transistors. This greatly increases the current capability of the output transistors before forward-bias second-breakdown failure is encountered. The increased safe output current capability for equivalent motor performance using identical power transistors in the bridge configuration is demonstrated by the following example.

Example

A linear amplifier must apply plus or minus 25 volts to a motor having an armature resistance of 1 ohm. A current limit of ±5 amp and a supply voltage of 30 volts is assumed. It is assumed that the motor back emf is at 25 volts whereupon motor current is abruptly reversed to decelerate the motor. First, consider a linear bipolar amplifier, as shown in Fig. 10-8.

First the motor is accelerated to a speed such that the back emf, $V_e = 25$ volts; Q_1 is conducting with Q_2 off. The motor current is then reversed to the limit value of

---

*Secondary breakdown is discussed later in this section.

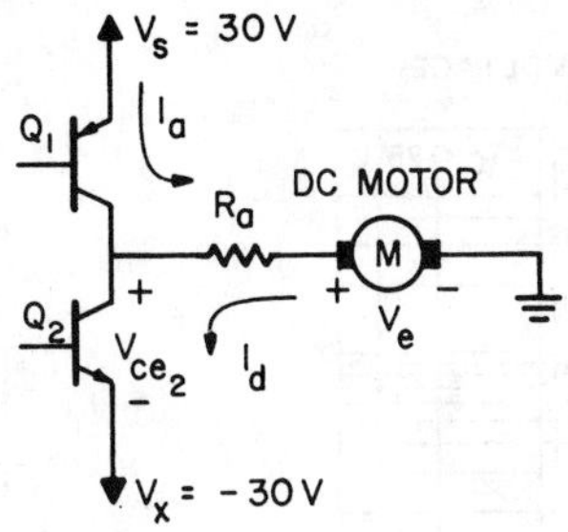

Fig. 10-8. Forward-bias second-breakdown stress on linear bipolar amplifier.

I_d = 5 amps; Q_2 is now conducting. The current limiter allows Q_2 to conduct only to the extent that current limit is not exceeded, i.e., V_{ce_2} must be such that the voltage across R_a (1 ohm) does not exceed 5 volts.

Then,

$$-30 + V_{ce_2} + I_d R_a - V_e = 0 \tag{10-1}$$

or

$$V_{ce_2} = V_e + 30 - I_d R_a \tag{10-2}$$

Inserting the values selected, the last equation becomes

$$V_{ce_2} = 25 + 30 - 5 = 50 \text{ volts} \tag{10-3}$$

The conducting transistor, then, must be able to sustain 50 volts at 5 amps, or, a peak power of 250 watts. The peak power dissipation decreases as the back emf of the motor decreases during deceleration. The time duration for which the transistor must dissipate high power is dependent on the system moment of inertia and other factors.

Figure 10-9 shows the safe operating area of a typical 100-volt 30-amp power transistor. At 50-volts and 5-amps the transistor is clearly out of its continuous or dc safe operating region and can sustain only 1.5 amps before entering forward-biased second breakdown. The system is unsafe. Four output transistors in parallel on each side would be required to bring the system within the dc safe operating area (SOA). The dc SOA is usually used for deceleration times greater than the thermal time constant of the chip, usually about 10 milliseconds.

A linear bridge, on the other hand, presents a more favorable solution (Fig. 10-10).

We first assume that Q_1 and Q_4 are conducting and that the back emf, V_e, is 25 volts. The motor current is then reversed by turning Q_1 and Q_4 off and Q_2 and Q_3 on. The system is designed so that $V_{ce_2} = V_{ce_3}$.

Thus,

$$V_s = V_{ce_2} - V_e + IR_a + V_{ce_3} \tag{10-4}$$

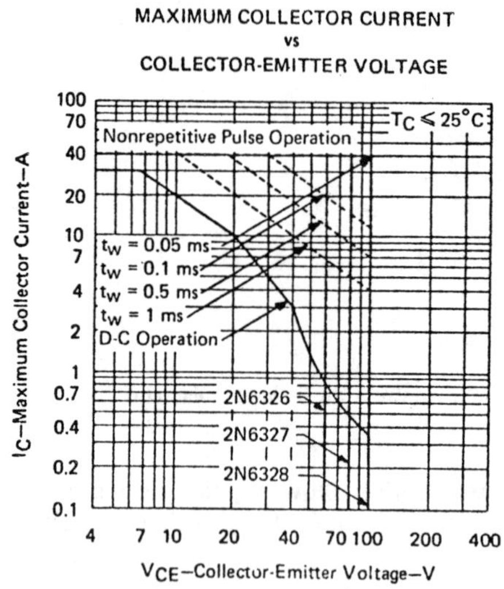

Fig. 10-9. Power transistor safe operating area.

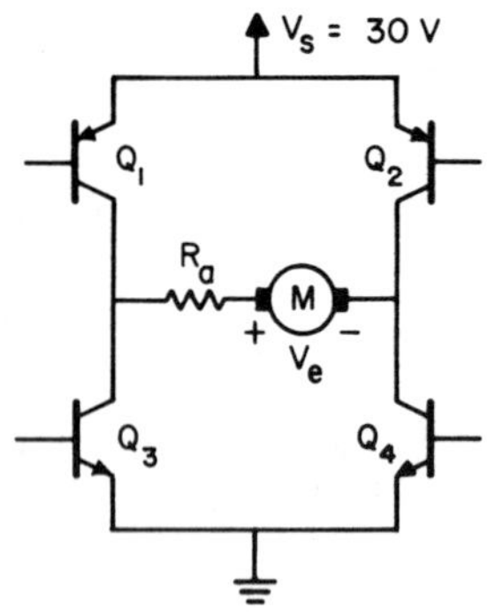

Fig. 10-10. Linear bridge amplifier.

Since

$$V_{ce_2} = V_{ce_3} \tag{10-5}$$

$$V_s = 2V_{ce} - V_e + IR_a \tag{10-6}$$

or

$$V_{ce} = \frac{V_s + V_e - IR_a}{2} \tag{10-7}$$

Thus,

$$V_{ce} = \frac{30 + 25 - 5}{2} = 25 \text{ volts} \tag{10-8}$$

For the linear bridge each power transistor sees a maximum of 25 volts at 5 amps at the instant of deceleration. This is within the safe operating region of the transistor, which is rated at 7 amps at $V_{ce} = 25$ volts. Thus, a single output transistor (rather than four in parallel) suffices for a total of four in the bridge rather than the eight required in the bipolar.

Linear bipolar amplifiers, however, continue to be the most popular owing mainly to their ease of design. They function well if operated within their safe limits. A significant advantage of the bipolar amplifier is that since one end of the motor is connected to ground, a low-valued resistor (0.1 ohm) connected between the motor and ground may be used to obtain a voltage proportional to motor current for current limiting or current feedback.

Since the motor has no ground connection in the bridge, motor current information is more difficult to extract. One method, if bipolar low voltage (±15 volts) supplies are available for an op-amp, is shown in Fig. 10-11.

In this case,

$$I_0 = -\frac{e_o R_1}{R_2 R_s} \tag{10-9}$$

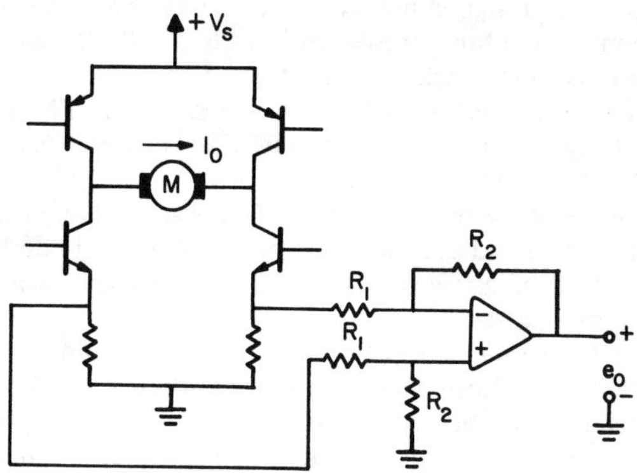

Fig. 10-11. Obtaining a voltage proportional to motor current in a linear bridge.

10.4 POWER AMPLIFIER DESIGN CONSIDERATIONS

A considerable amount of care must be taken in selecting output transistors and the configuration in which they are used. The following points should be given careful consideration.

1. Output Device Selection

The following points must be considered before output transistor selection may be made in a given application.

a. What is the maximum current and voltage required at the motor? The power supply voltage must be a minimum of several volts higher than the maximum motor voltage. For unregulated power supplies the difference between the full-load low-line and no-load high-line power supply voltages can be significantly greater than the normal ±10 percent line variations, so abnormally high $V_{ceo(sus)}$ ratings for the power transistors are required.
b. For what time duration is peak acceleration or deceleration current required? May the pulsed safe operating area (SOA) of the devices be utilized or must the system remain within dc SOA? Second-breakdown stress during deceleration often dictates larger output transistors (or more in parallel) than does simple peak current requirements [2].
c. What power amplifier bandwidth is required? This may be dictated by current limiter speed requirements. In general, the slowest transistors that meet the bandwidth requirements are preferred since they usually are more rugged in terms of second-breakdown characteristics.

d. Will the design be bipolar or a bridge? Are dual supplies available? Do forward-bias second-breakdown considerations during deceleration make the bridge more attractive? The bridge requires about half the $V_{ceo(sus)}$ rating for output devices than the bipolar does.

2. Output Stage Design

In general, the designer has two basic choices in power amplifier output stage design. The complementary configuration, Fig. 10-12, uses NPN and PNP output transistors in a fully symmetrical arrangement.

The quasi-complementary output stage is shown in Fig. 10-13. It usually features an inverted Darlington arrangement of an NPN output transistor and a PNP driver to simulate a PNP output transistor.

The complementary output stage offers ease of design and symmetry of performance, especially gain and bandwidth, whereas the quasi-complementary output stage allows the use of all NPN transistors in the output stage which are usually less expensive and often more rugged.

It is easier to parallel output transistors in the complementary design requiring only emitter resistors whose value is such that they drop about 0.7 volts at the maximum collector current. The increased availability of economical complementary pair output transistors has enhanced the use of the complementary configuration.

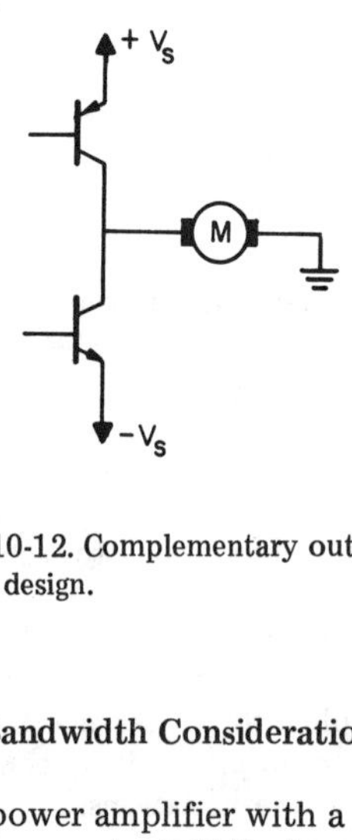

Fig. 10-12. Complementary output stage design.

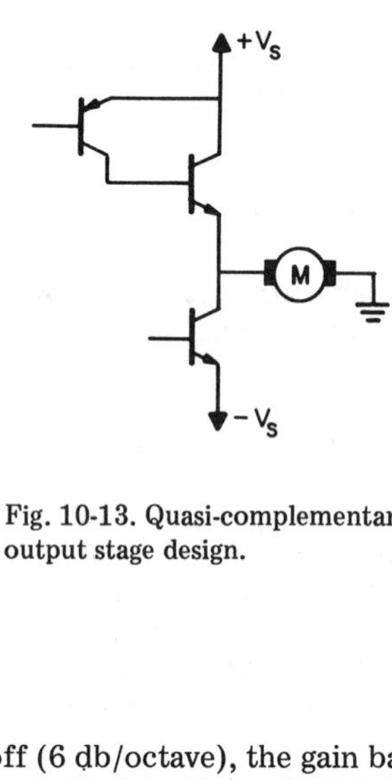

Fig. 10-13. Quasi-complementary output stage design.

3. Gain Bandwidth Considerations

In a power amplifier with a single pole roll-off (6 db/octave), the gain bandwidth product is a constant. If the bandwidth is to be controlled (as it should be), it follows that the gain must also be controlled. Power amplifiers are, therefore, usually designed with one or more internal feedback loops to stabilize amplifier voltage or current gain.

The following technique (Fig. 10-14) offers controlled closed-loop voltage gain as well as controlled current gain. The current mirror configuration of the drivers and output stage also improves the power supply ripple rejection.

The transistors Q_1 and Q_2 comprise a current mirror (as do Q_3 and Q_4). If adequate drive current, I_1, is available from Q_1, current I_0 is directly proportional to I_1.

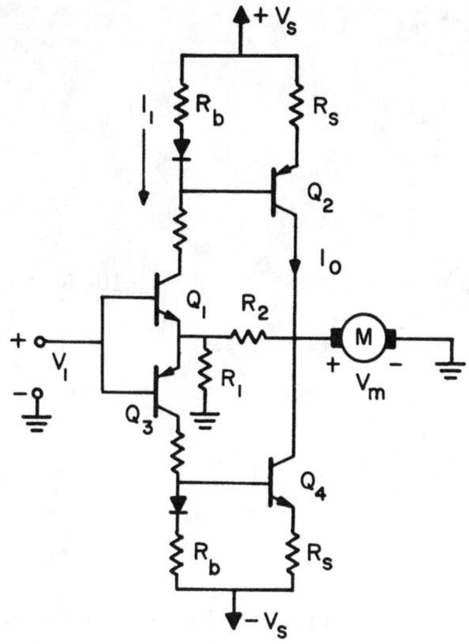

Fig. 10-14. Controlled-gain output stage configuration.

$$I_0 = I_1 \frac{R_b}{R_s} \tag{10-10}$$

The system voltage gain V_m/V_1 is controlled by R_1 and R_2 (ignoring the V_{be} drops of Q_1 and Q_3). The resistors R_1 and R_2 are equivalent to the feedback resistors in a noninverting op-amp configuration. Therefore,

$$\frac{V_m}{V_1} = 1 + \frac{R_2}{R_1} \tag{10-11}$$

Thus, the closed-loop voltage and current gains of the output stage are well controlled in themselves making the system less susceptible to transistor parameter and supply voltage variations.

Since the current mirror insures that output current is a function of drive current only, the system is relatively immune to supply variations and β variations. Hence, it performs more predictably with the wide parameter variations exhibited by power and driver transistors.

10.5 CROSSOVER DISTORTION IN POWER AMPLIFIERS

Crossover distortion or "deadband," as it is commonly called, is defined as a low-gain (or zero-gain) region about the zero output voltage point caused by low conduction (or lack of conduction) of any of the output transistors. The problem usually arises due to emitter-base drops of the driver or output transistors which must be exceeded by the driving signal before conduction can begin (Fig. 10-15).

As voltage V_1 is varied, neither Q_1 nor Q_3 conducts so long as $-V_{be} \leqslant V_1 \leqslant V_{be}$. Therefore, output transistors Q_2 and Q_4 do not conduct while V_1 is within this

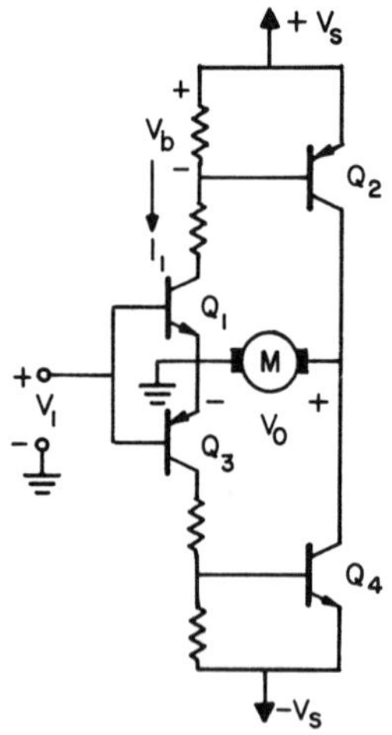

Fig. 10-15. Crossover distortion.

region. Furthermore, even when V_1 exceeds the V_{be} drop of Q_1 or Q_2, the respective output transistors do not conduct until the driver current is sufficient to cause $I_1 R_b$ to exceed the V_{be} (0.7 volts) of the output transistors.

Although the above discussion has centered around the bipolar design, it applies equally to the linear bridge. Crossover distortion in the bridge, however, occurs at half the power supply voltage where the bridge output voltages are equal and the motor voltage is zero. The driver circuitry usually translates the drive so that this corresponds to a zero voltage input signal.

Figures 10-16(a) and 10-16(b) show the effect of crossover distortion on the amplifier transfer function and the resulting output distortion for a sine wave input. Note that the system gain, $\Delta V_0/\Delta V_1$, decreases markedly near the zero point.

Severe problems may be encountered if crossover distortion is not minimized. Systems properly compensated (critically damped) for small velocity variations near zero have a tendency to overshoot (underdamped) or oscillate during larger velocity excursions due to the increased system gain away from zero.

Crossover distortion is a critical problem in positioning systems. Here the system must provide holding torque to the motor at the zero point (zero velocity). Since the amplifier gain is low at this point due to crossover distortion, gains elsewhere in the system must be increased to provide adequate position loop gain. This results in potential stability problems since once a position error signal is sufficiently high to overcome crossover distortion, the system gain increases dramatically causing an underdamped or oscillatory condition.

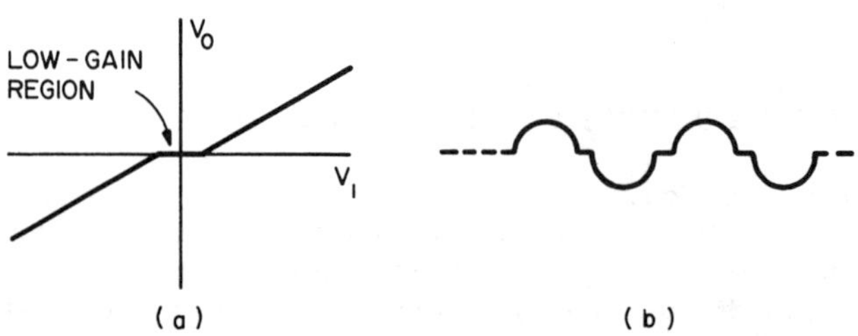

Fig. 10-16. (a) Transfer function showing crossover distortion.
(b) Output distortion for a sinusoidal input.

Numerous methods have been devised to overcome the problem. Some work well, while others cause new problems.

A common technique used to reduce crossover distortion in the drivers is shown in Fig. 10-17. Diodes D_1 and D_2 offset the emitter-base drops of Q_1 and Q_2, while resistors R_1 and R_2 provide base drive current. Resistors R_3 and R_4 establish a maximum error current caused by any mismatch in V_{D1} versus V_{beQ1} or V_{D2} versus V_{beQ2}. The error current is $(V_D - V_{be})/R_{3(4)}$.

With $V_1 = 0$, a steady-state bias current I_1 could flow (with a corresponding bias current I_2 on the opposite side) depending on the mismatch of V_D to V_{be} of the drivers.

Diodes D_3 and D_4 perform a similar function at the bases of Q_3 and Q_4 deflecting most of the driver current I_1 and I_2 to the bases of Q_3 and Q_4 until sufficient current is present to break down D_3 and D_4. Thus, even small drive currents are directed to the bases of the power transistors rather than the base resistors (R_5 and R_6) which results in output current for very small drive currents and reduced crossover distortion.

With large output currents, the variation in possible bias current in the output transistors due to errors in mismatch of emitter-base drops with diode drops becomes severe resulting in large steady-state power dissipation in the output transistors.

Schottky diodes used for D_1 and D_2 and silicon diodes for D_3 and D_4 can minimize this problem while still enhancing crossover distortion.

Another approach is to elminate D_1 and D_2 and rely on closed-loop feedback (voltage or current) to reduce crossover distortion by the loop gain of the system. It should be remembered that the loop gain decreases with frequency and that the crossover distortion increases with increasing frequency.

It is generally desirable to minimize open-loop crossover distortion by techniques similar to those described above before attempting to reduce it further by

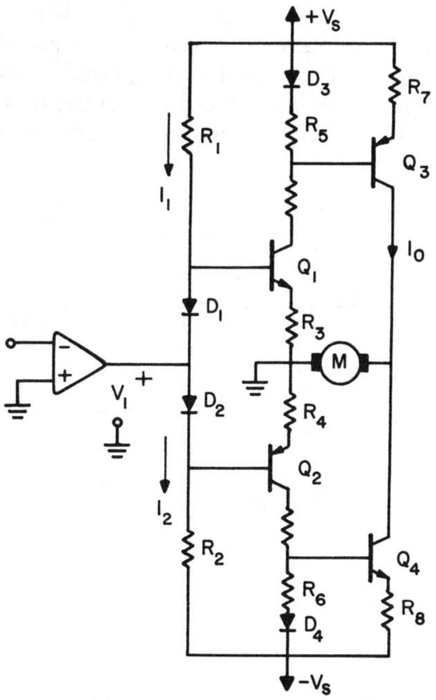

Fig. 10-17. Reduction of driver crossover distortion.

feedback. Failure to do this may result in stability problems due to the excessive loop gain required for crossover distortion reduction. For large power amplifiers, at least, it is usually advisable to reduce open-loop crossover distortion only to the point where it can be guaranteed that no output transistor bias current flows. Output transistor bias current may result in excessive heating and is difficult to control.

10.6 CURRENT LIMITING TECHNIQUES

A reliable (and fast) current limiter is mandatory when driving permanent-magnet dc motors with transistor power amplifiers. Both the motor and the amplifier can be damaged by even short duration overcurrent conditions. In general, current limiters should have the following characteristics:

1. The current limiter should maintain relatively flat limiting characteristics, even with large amounts of overdrive.
2. It should be fast enough to prevent current overshoot when a motor is decelerating from high speed.
3. It should be closed loop in form to prevent large summing amplifier error voltages when in current limit.
4. Ideally, the current limiter should be located as close as possible to the output transistors.

Other characteristics such as adjustability, two-stage current limit (higher values of pulsed current are allowed for short times to accelerate or decelerate), or foldback current limit may be required in certain applications.

Some compromise is often required when desired current limiter characteristics are selected.

A transconductance power amplifier* is desirable for two reasons: (1) Because the input drive voltage is clamped, the amplifier is a closed-loop form of current limiter; (2) Only the voltage necessary to produce the required current is present at the input to the amplifier. Since the summing amplifier drives the power amplifier, a clamp across the feedback points of the summing amplifier limits the summing amplifier output voltage and, hence, motor current to a predetermined value (Fig. 10-18).

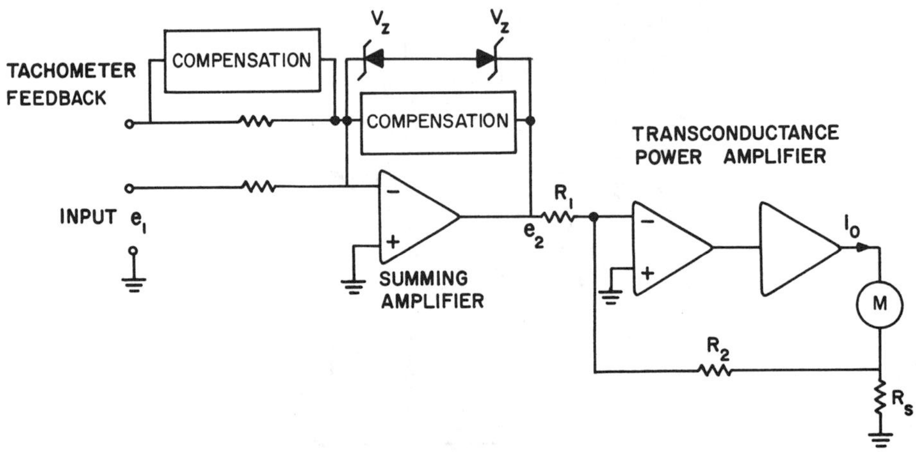

Fig. 10-18. Transconductance power amplifier current limiter.

*A transconductance amplifier is an amplifier whose output current is proportional to input voltage.

The power amplifier transconductance is given by

$$\frac{I_0}{e_2} = -\frac{R_2}{R_1 R_s} \tag{10-12}$$

Since voltage e_2 is limited to $\pm(V_z + V_d)$ where V_z is the zener voltage and $V_d = 0.7$ volts, the current limit is given by:

$$I_{lim} = -e_{2max}\frac{R_2}{R_1 R_s} \tag{10-13}$$

$$I_{lim} = \pm(V_z + V_d)\frac{R_2}{R_1 R_s} \tag{10-14}$$

The speed required for the current limiter is dependent on the maximum possible rate of rise of motor current, which occurs at the instant of deceleration from maximum speed. The maximum applied voltage to the motor could approach $2V_s$ under these conditions in the absence of current limiting.

To calculate the required speed, let I_{lim} equal the selected value of current limit. Assuming a deceleration condition, the motor current may be written as:

$$I_m(t) = \frac{2V_s}{R}(1 - e^{-t/\tau}) \tag{10-15}$$

where

$$\tau = L_a/R_a$$

and

L_a = armature inductance

R_a = armature resistance

Now

$$I_m(T) = I_{lim} = \frac{2V_s}{R_a}(1 - e^{-T/\tau}) \tag{10-16}$$

where T is the time required for current to reach the limit value.

Solving for T from Eq. (10-16), we obtain

$$T = \tau ln\left[\frac{2V_s}{2V_s - R_a I_{lim}}\right] \tag{10-17}$$

The current limiter response time, or amplifier response time in the case of transconductance amplifiers, should be substantially less than T.

For the transconductance amplifier we may relate the required response time T to the power amplifier bandwidth.

The 10-90 percent rise time of the amplifier is given by [3]

$$T_{10\text{-}90} = \frac{0.35}{f_o}$$

where f_o is the amplifier 3 db bandwidth. We require that $T_{10\text{-}90} \ll T$. Therefore,

$$f_o \gg \frac{0.35}{T}$$

For example, consider a transconductance amplifier driving a motor having the following armature resistance and inductance:

$R_a = 2$ ohms

$L_a = 6$ millihenries

Assume $V_s = 30$ volts. Let I_{lim}, the current limit, equal 5 amperes. Then,

$$T = \frac{6 \times 10^{-3}}{2} \, ln\left[\frac{60}{60 - 10}\right]$$

$$= 547 \, \mu\text{sec}$$

We then require that the amplifier bandwidth be

$$f_o \gg \frac{0.35}{547 \times 10^{-6}} = 640 \text{ Hz} \tag{10-18}$$

A simpler, alternative method is to assume that the maximum rate of rise of the motor current is

$$\frac{\Delta I}{\Delta T} = \frac{2V}{L} \tag{10-19}$$

Then,

$$\Delta T = \frac{L\Delta I}{2V}$$

For the above case, we obtain

$$\Delta T = \frac{5 \times 6 \times 10^{-3}}{60} = 500 \, \mu\text{sec} \tag{10-20}$$

We again demand that the amplifier 10-90 percent rise time be less than ΔT and obtain

$$\frac{0.35}{f_o} < \Delta T$$

Therefore,

$$f_o > \frac{0.35}{\Delta T} = 700\ \text{Hz} \tag{10-21}$$

The transconductance amplifier current limiter, however, requires the satisfactory operation of all of the components in the power amplifier. An op-amp failure, for instance, could cause the destruction of the output transistors or the motor.

It is often desirable to back up the transconductance current limiter with a simple fixed current limiter located at the output transistor drive circuitry. The limiter is set slightly above the maximum limit current range of the transconductance amplifier to provide maximum protection to the system. Several such direct current limit techniques are shown below.

If the driver and output transistors are configured as a current mirror (Fig. 10-19), then the relationship of output current to drive current is given by:

$$I_0 = I_1 R_1 / R_2 \tag{10-22}$$

It is also required that current I_1 be much greater than the base current of Q_2 and that diode D_1 offsets the emitter-base drop of Q_2.

It can be seen that the voltage across R_1 equals the voltage across R_2. If the voltage across R_1 is limited (as in the example, to 0.7 volts by D_2) to a certain value, then the voltage across R_2 is limited to the same value. Hence, the output current is limited in this case to:

$$I_{lim} = 0.7/R_2 \tag{10-23}$$

Another technique involves direct sampling of output transistor emitter current and the shunting of the drive current when the output current achieves a certain value (Fig. 10-20).

In this case, when the voltage $I_0 R_2$ exceeds the emitter-base drop of Q_2, Q_2 starts to conduct, shunting the base drive of Q_1 to that value necessary to maintain $I_0 R_2 \approx 0.7$ volts. Thus,

$$I_{0\,lim} \approx 0.7/R_2 \tag{10-24}$$

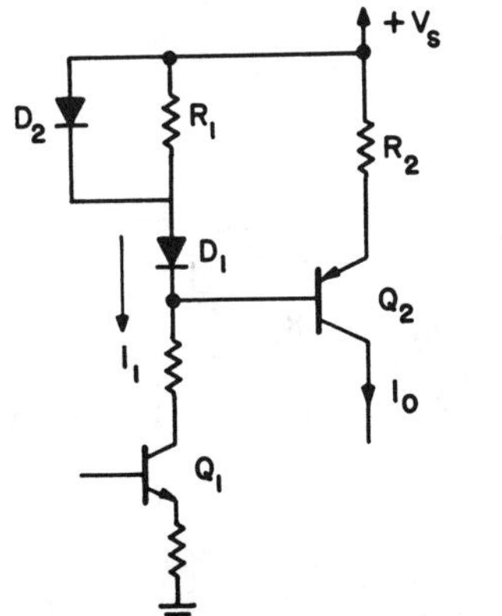

Fig. 10-19. Current limiter used with current mirror driver and output transistor.

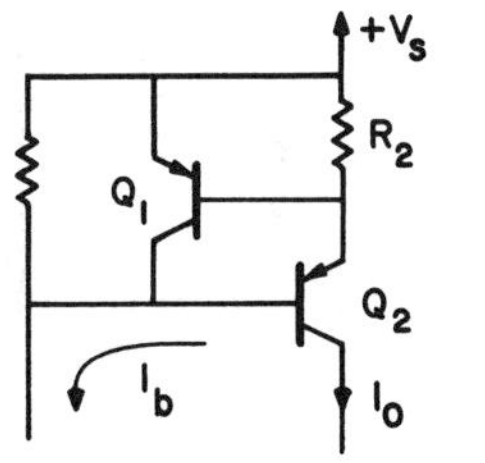

Fig. 10-20. Direct-sampling current limiter.

These techniques provide an inexpensive, yet effective, secondary current limit function and may be used in either the bipolar or bridge configurations.

10.7 INPUT-OUTPUT RELATIONSHIPS IN LINEAR AMPLIFIERS

Amplifiers can be configured as voltage amplifiers (volts out/volts in), current amplifiers (amps out/amps in), or a combination of the two depending on the type of feedback used in the minor feedback loop. System performance can be enhanced by careful consideration of the characteristics of the two types.

The current source configuration allows current limiting to be accomplished simply by limiting the input voltage to the power amplifier. Since the output current of the current source or transconductance amplifier is proportional to input voltage, a simple voltage clamp around the summing amplifier causes the motor current to be limited to the desired value.

The voltage source presents a low output impedance to the motor, which, without changing voltage, can damp motor speed variations since the amplifier can sink current supplied by the back emf generator. Hence, a certain amount of damping is present in a voltage amplifier which could enhance stability.

The current source, on the other hand, presents a high output impedance to the motor and, consequently, has almost no damping. This feature is useful in driving multiphase, brushless dc motors where mutual coupling between windings could cause excessive damping if a voltage source were used.

The transconductance (current source) amplifier eliminates the dependence of the system on motor inductance [4], effectively cancelling the electrical pole of the motor for small signal variations where the compliance of the current source is not exceeded (voltage saturation). On the other hand, the transfer function includes the integration effect between current and acceleration which has an adverse effect on system stability. This is illustrated in the following derivation. Consider the system shown in Fig. 10-21.

When the amplifier gain is high, the negative op-amp input may be considered virtual at ground potential, resulting in the following expression:

$$\frac{V_1}{R_1} + \frac{V_0}{R_2} + \frac{iR_s}{R_3} = 0 \tag{10-25}$$

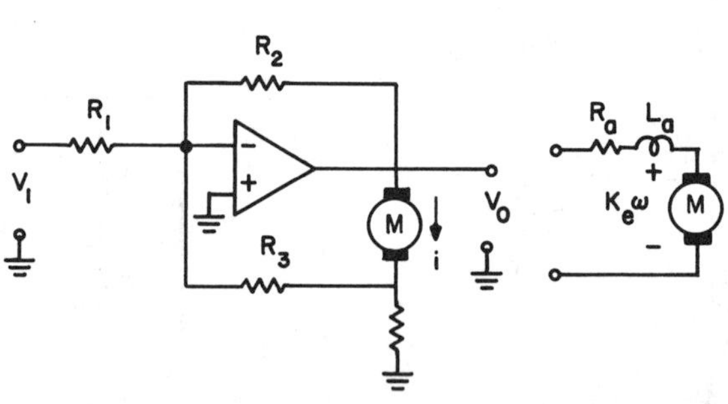

Fig. 10-21. Composite voltage source-current source amplifier.

As R_3 approaches an open circuit, we have

$$\frac{V_1}{R_1} + \frac{V_0}{R_2} = 0 \qquad (10\text{-}26)$$

or

$$\frac{V_0}{V_1} = -\frac{R_2}{R_1} \qquad (10\text{-}27)$$

Similarly, when the voltage feedback resistor, R_2, is made very large, Eq. (10-25) may be written as

$$\frac{V_1}{R_1} + \frac{iR_s}{R_3} = 0 \qquad (10\text{-}28)$$

or

$$\frac{i}{V_1} = -\frac{R_3}{R_1 R_s} \qquad (10\text{-}29)$$

which represents a current amplifier.

It is interesting to examine the effect of voltage and current feedback on the overall transfer function of the motor and amplifier.

Assuming that the voltage drop across R_s is low, we may assume that V_0 represents the motor voltage.

Using standard motor equations, we may write [5]

$$V_0 = K_e\omega + R_a i + L_a \frac{di}{dt} \qquad (10\text{-}30)$$

Also

$$K_t i = J\frac{d\omega}{dt} \qquad (10\text{-}31)$$

where K_t is the motor torque constant. Applying the Laplace transform to the last equation, we have

$$K_t I(s) = Js\Omega(s) \qquad (10\text{-}32)$$

or

$$I(s) = \frac{sJ}{K_t}\Omega(s) \qquad (10\text{-}33)$$

Applying the Laplace transform to Eq. (10-30), we have

$$V_0(s) = K_e\Omega(s) + (R_a + sL_a)I(s) \qquad (10\text{-}34)$$

Substituting these expressions into the Laplace transformed version of Eq. (10-25) yields

$$\frac{V_1(s)}{R_1} + \frac{K_eK_t\Omega(s) + sJ(R_a + sL_a)\Omega(s)}{R_2K_t} + \frac{sJR_s\Omega(s)}{K_tR_3} = 0 \qquad (10\text{-}35)$$

The transfer function between the input voltage V_1 and the velocity ω may now be written as

$$\frac{\Omega(s)}{V_1(s)} = -\frac{1}{R_1}\,\frac{R_2R_3K_t}{R_3[K_eK_t + sJ(R_a + sL_a)] + sJR_sR_2} \qquad (10\text{-}36)$$

In the special case of dominant current feedback where R_2 is large, we may approximate Eq. (10-36) by

$$\frac{\Omega(s)}{V_1(s)} = -\frac{R_3K_t}{sJR_1R_s} \qquad (10\text{-}37)$$

which represents a first-order transfer function with pure integration.

The opposite case of dominant voltage feedback (where R_3 is large) may be approximated by

$$\frac{\Omega(s)}{V_1(s)} = -\frac{R_2K_t}{R_1[K_eK_t + sJ(R_a + sL_a)]} \qquad (10\text{-}38)$$

or

$$\frac{\Omega(s)}{V_1(s)} = -\frac{\dfrac{R_2K_t}{R_1JL_a}}{s^2 + \dfrac{sR_a}{L_a} + \dfrac{K_eK_t}{JL_a}} \qquad (10\text{-}39)$$

which represents a second-order system.

The complete transfer function including both current and voltage feedback is given by rearranging Eq. (10-36),

$$\frac{\Omega(s)}{V_1(s)} = -\frac{\dfrac{R_2K_t}{R_1JL_a}}{s^2 + \left(\dfrac{R_a}{L_a} + \dfrac{R_sR_2}{L_aR_3}\right)s + \dfrac{K_eK_t}{JL_a}} \qquad (10\text{-}40)$$

A comparison of Eq. (10-40) with Eq. (10-39) which has voltage feedback only shows that the addition of current feedback increases the coefficient of s in the denominator of Eq. (10-40), thus, increasing the loop damping.

10.8 DESIGN EXAMPLE—LINEAR BIPOLAR AMPLIFIER

Consider the following specification for a linear power amplifier.

1. Bipolar type, ±28 volts supplies available
2. Output voltage ±24 volts
3. Output current ±10 amps peak
   2 amps RMS
4. Transconductance type $I_0/V_{in} = 2$ amps/volt
5. Motor armature resistance $R_a = 2$ ohms
6. Maximum back emf, $V_e = 25$ volts.

First, consider output device selection. Due to the possibility of high speed reversals, the output transistors may see collector-emitter voltages that approach $2V_s$. Thus, we see that $V_{ceo(sus)} > 2V_s = 56$ volts.

Assume that the high-line supply voltage may be ±30 volts, and select output transistors with $V_{ceo(sus)} = 80$ volts.

The output transistor current rating (or the number of transistors required in parallel) is usually determined by forward bias, I_{sb}, considerations. Maximum $V_{ce}I_c$ stress occurs at the instant of deceleration from high speed. Even if system specifications do not require operation near maximum speed, it is wise to assume it may happen due to the loss of tachometer feedback or other mishap.

The voltage across the conducting transistor during deceleration is given by

$$V_{ce} = V_s + V_e - IR_a \tag{10-41}$$

where V_s is the maximum power supply voltage; V_e, the maximum back emf; R_a, the motor armature resistance; and I, the motor current. An estimate of the maximum power stress point for I_{sb} considerations may be made by considering motor current to be at its limit value in Eq. (10-41). This approximation is used for an initial estimate of the second-breakdown safety margin. In general, the output transistors must remain within the SOA for all values of current from zero to I_{lim}. Therefore, Eq. (10-41) should be used to plot V_{ce} versus I directly on the SOA curves to verify that operation is always within the SOA for all possible values of current.

$$V_{ce\,max} = 30 + 25 - 20 = 35 \text{ volts} \tag{10-42}$$

Select for initial investigation the 2N6327, 2N6330 complementary pair of power transistors rated at 80 volts, 30 amps continuous. Regardless of the transistor type used and the number of transistors in parallel, if any, each must remain within the safe operating area (SOA) at the maximum junction temperatures to be encountered.

Since the I_{sb} rating of a transistor is relatively independent of the average junction temperature, and, in any case is usually specified at $T_j = 200°$ C, thermal ratings and I_{sb} ratings usually may be considered separately [6]. Transistor operation must remain within the thermal and I_{sb} safe areas.

Considering the I_{sb} problem first, recall that the peak current is 10 amps and $V_{ce\,max}$ is 35 volts during deceleration. Assume that deceleration times may exceed 10 milliseconds, which is longer than the thermal time constant of the collector junction (usually less than 10 milliseconds), so that pulsed SOA's may not be used. Referring to Fig. 10-22, follow the $V_{ce} = 35$ volts line upward until it crosses the dc SOA curve. The collector current corresponding to this point is approximately 4 amps. So it takes at least three transistors in parallel to safely obtain the 10 amps required. Next plot V_{ce} versus I_c on the SOA curve (Fig. 10-22) to verify that operation is within the SOA, for all values of motor current from I = 0 to $I = I_{lim}$ (Table I). In this case, the collector current in each transistor is given by $I_c = I/3$.

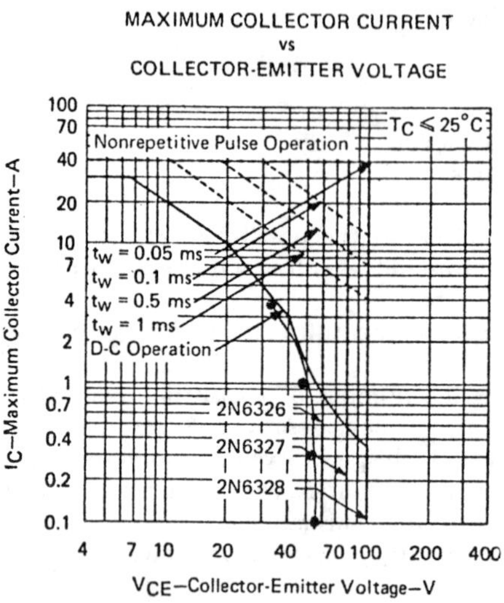

Fig. 10-22. Power transistor safe operating area.

Table I. V_{ce} versus I_c obtained from Eq. 10-41.

I	I_c	V_{ce}
0.3A	0.1A	54.4V
0.9	0.3	53.2
3.0	1.0	49.0
10.0	3.3	35.0

Next, a thermal analysis is performed to determine if the number of transistors in parallel must be increased due to thermal considerations. Maximum junction power dissipation must first be determined. The only accurate way to determine transistor power dissipation is to integrate the real-time product of $V_{ce}(t)i_c(t)$ over a worst-case motion profile. The average collector power dissipation is given by

$$P_{avg} = \frac{1}{T}\int_0^T V_{ce}(t)i_c(t)dt \tag{10-43}$$

Assume that $P_{avg} = 50$ watts in this case. Then, generate a thermal model for the system with the requirement that the maximum junction temperature for the system, T_{jmax}, be less than 200° C, as specified in the data sheet.

Include in the model the three collector junctions (assume in the model that either set of transistors may have to dissipate 50 watts although this may not always be the case); a thermal resistance, θ_{jc}, between junction and case; a thermal resistance, θ_{cs}, between case and heat sink; a heat sink temperature, T_s; a thermal resistance, θ_{sa}, between the sink and the outside world; and a maximum ambient temperature, T_a (Fig. 10-23). Model the power dissipation P as three separate current sources, P/3, assuming that each parallel transistor dissipates 1/3 the total power.

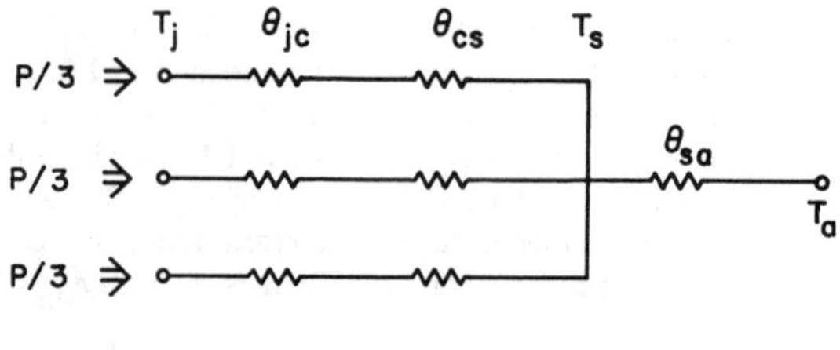

Fig. 10-23. Bipolar linear amplifier thermal model.

We may calculate T_j from

$$T_j = T_a + P\theta_{sa} + \frac{P}{3}(\theta_{jc} + \theta_{cs}) \tag{10-44}$$

If we assume

T_{jmax} = 150°C (derated for reliability considerations)

T_{amax} = 55°C

θ_{jc} = 0.875°C/watt (data sheet)

θ_{cs} = 2°C/watt (a thermal washer and a good thermal compound is used)

We may then determine the maximum allowable thermal resistance of the heat sink, θ_{sa}, from Eq. (10-44),

$$\theta_{sa} = \frac{1}{P}[T_{jmax} - T_a - \frac{P}{3}(\theta_{jc} + \theta_{cs})]$$

$$= \frac{1}{50}[150 - 55 - \frac{50}{3}(1.08)]$$

$$= 1.54^\circ\text{C/watt maximum}$$

The heat sink may be chosen from manufacturer's data for convection cooling. If fan cooling is used, a considerably smaller sink may be chosen. In any case, its thermal resistance must not exceed 1.54°C/watt if the desired 150°C maximum junction temperature is not to be exceeded.

In this case, we have deliberately chosen a heat sink large enough to maintain a safe junction temperature with three power transistors in parallel on each side. If the heat sink size were fixed by other considerations, more transistors could be required depending on the size of the sink.

Since three transistors in parallel are adequate, we may proceed to the design of the drivers. Each output transistor must deliver 3.3 amps maximum. The data sheet shows that these devices have a minimum static forward current transfer ratio h_{FE} of 25 at I_c = 5 amps, V_{ce} = 4 volts. We assume from this that 3.3 amps with a base current of 3.3/25 = 0.132 amps, V_{ce} will be somewhat less than 4 volts. We shall assume it is 3 volts and verify experimentally.

Figure 10-24 shows the drive requirements for the output transistors including emitter resistors, R_s, for paralleling. These resistors are chosen to drop approximately 0.7 volts at maximum collector current to reduce current sharing errors due to V_{be} mismatches and other parameter variations. The current sharing errors are approximately

$$\Delta I_c = \Delta V_{be}/R_s \tag{10-45}$$

We select $R_s \approx$ 0.7/3.3 amps = 0.21 ohms $\approx$ 0.22 ohms (a standard value). Thus, the mismatch in collector current is ΔI_c = 0.1/0.22 = 0.45 amps if a 0.1-volt V_{be} mismatch is the maximum expected.

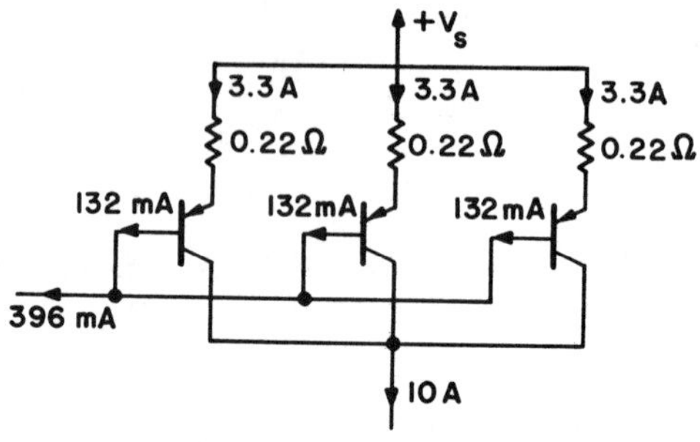

Fig. 10-24. Output transistor drive requirements.

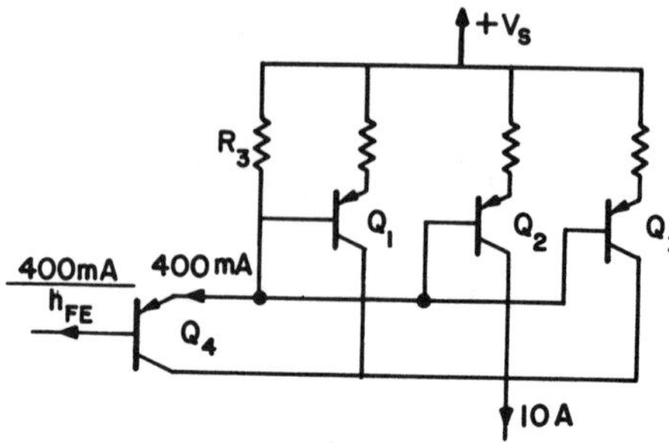

Fig. 10-25. Darlington output stage.

Base current for each transistor is 3.3 amp/25 = 132 milliamps for a total drive requirement of about 400 milliamps.

Although 400 milliamps may be obtained from the driver transistors, the power dissipation in the driver collector resistors is high since when the driver is fully on (as happens when the motor is disconnected), the power dissipation is at least 28 volts × 0.4 amps = 11.2 watts.

Furthermore, we would like to operate the system as a current mirror, which requires that the driver current be substantially greater than the required base current.

We, therefore, choose to operate the output stage in a Darlington configuration as in Fig. 10-25.

Transistor Q_4 may be substantially smaller than the output transistors but not so small that the $V_{ce(sat)}$ drop causes substantially increased collector-emitter drop in the output transistors. (The driver's saturation drop adds directly to the overall V_{ce} of the output stage.)

If we choose Q_4 to be an 80-volt transistor with an h_{FE} of 25 at 0.5 amps with V_{ce} = 1 volt, then the base current of Q_4 is approximately 0.4/25 = 16 milliamps. The small current through R_1 is neglected. The resistance R_1 is chosen to prevent leakage current in the output transistors from turning themselves on.

The design is continued in Fig. 10-26. The transistor Q_5 must sink 16 milliamps of base drive from Q_4 plus the current through R_1, D_1, and D_2. The current of Q_5 must be substantially greater than 16 milliamps for proper current mirror operation. Let the maximum collector current of Q_5 be 100 milliamps as limited by the collector resistor, R_3, and the power supply voltage, V_s.

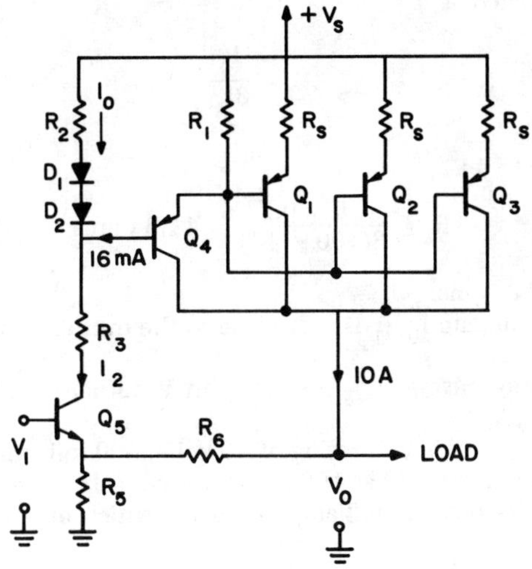

Fig. 10-26. Output stage design.

$$R_3 \approx \frac{28 \text{ volts}}{0.1 \text{ amps}} = 280 \text{ ohms}$$

Choose

$$R_3 = 270 \text{ ohms}$$

The power dissipated in R_3 is

$$P_{R3} = 28 \text{ volts} \times 0.1 \text{ amp} = 2.8 \text{ watts}$$

We choose

$$P_{R3} = 5 \text{ watts}$$

Diodes D_1 and D_2 offset the emitter-base drops of the output transistors and Q_4, and cause the voltage drop across R_2 to equal the drop across each R_s when Q_5 draws collector current.

The system current gain may be calculated by the following observation: Since

$$V_{R2} = V_{Rs}$$

then

$$I_1 R_2 = I_0 R_s/3$$

and

$$\frac{I_0}{I_1} = \frac{3R_2}{R_s}$$

Thus, R_2 may be selected as

$$R_2 = \frac{I_0}{3I_1}$$

and

$$R_s = \frac{10 \times 0.3}{3(100 - 16)} = 11.6 \text{ ohms}$$

We choose R_s to be 12 ohms.

The mirror current gain I_0/I_1 is very close to the overall current gain I_0/I_2 which is equal to 120.

Transistor Q_5 may have a V_{ceo} greater than V_s rather than $2V_s$ since it does not see voltages exceeding V_s.

The transistor Q_5 may have a rating of 40 volts, 500 milliamps such as 2N2222A which has an h_{FE} greater than 50 at 100 milliamps collector current.

Thus, Q_5 requires only 2 milliamps of drive which may be easily supplied by most op-amps.

Resistors R_5 and R_6 establish the closed-loop voltage gain. We design for a maximum output voltage of 30 volts and a reasonable input voltage that may be supplied by an op-amp such as 5 volts. The voltage gain, neglecting the offset voltage of Q_5, is given by

$$\frac{V_0}{V_1} = 1 + \frac{R_6}{R_5} = \frac{30}{5} = 6$$

Thus,

$$R_6 = 5R_5$$

The resistance of R_5 should be such that it drops only a small percentage of the drive voltage at 100 milliamps. We select R_5 to be 39 ohms. We can reduce R_3 by a similar amount to ensure that 100 milliamps of drive is maintained. The resistance of R_6 is then approximately 200 ohms. A wattage rating exceeding 4.5 watts is required for R_6.

This concludes the output stage design for the upper half. The lower half is entirely symmetrical. Figure 10-27 shows the complete design.

Op-amp U1 drives the input to the power amplifier. In this case, crossover distortion is adequately reduced by the system loop gain. Consequently, no special action is taken at the bases of Q_1 and Q_2 to reduce the distortion. Diodes D_1 through D_4, however, do help to reduce crossover distortion, deflecting all driver current to the base of Q_3 or Q_4 until the diodes conduct ($2V_d$) at which time Q_3 and Q_4 are capable of conduction.

Op-amp U1 is configured as a current source. Current feedback is taken from 0.1 ohm power resistor (R_s) between motor and ground.

The closed-loop current gain of the system is given by

$$\frac{I_0}{E_1} = -\frac{R_3}{R_1 R_s} = -2 \text{ amps/volt} \tag{10-46}$$

Since

$$R_s = 0.1 \text{ ohm}$$

$$\frac{R_3}{R_1} = 0.2$$

We let

$$R_1 = 10 \text{ kilohms}$$

Then

$$R_3 = 2 \text{ kilohms}$$

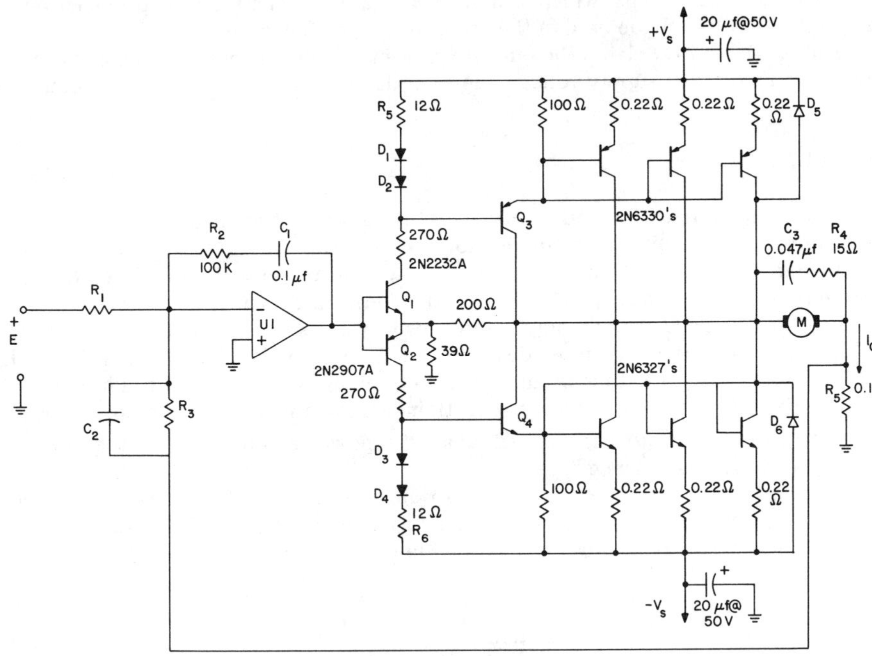

Fig. 10-27. Linear bipolar transconductance amplifier.

Several compensation networks are required to stabilize the system. While typical values are shown, it is necessary to test system compensation with the motor and op-amp to be used. This can be done by substituting a dummy load for the motor (R and L) and checking the response of output current I_0 to a step function of input voltage at E_1. This should be checked up to and including current limit. R_s is a convenient point for viewing a voltage proportional to I_0 on an oscilloscope. The system should be compensated for critical damping in the output current.

Lag/lead network, R_1, R_2, C_1, greatly influences the system and open-loop bandwidth. The system bandwidth is maximized by minimizing C_1, consistent with amplifier stability, so that the stability of the servo system in which the amplifier is used is not adversely affected. Other methods for compensating linear amplifiers, such as a series R-C from the (−) to (+) op-amp inputs, are given in references [7] and [8].

A high-frequency zero (several hundred kilohertz) is provided by R_4 and C_3 to compensate for a high-frequency pole in the power amplifier.

The capacitor C_2 is used to set the closed-loop bandwidth of the amplifier. This is done only to reduce the possibility of high-frequency motor excitation due to tachometer noise and other sources and, in any case, should be at least an order of magnitude higher in frequency than the closed-loop servo bandwidth.

As discussed in Sec. 10.6, input voltage E_1 may be clamped (bounded) to provide current limiting. In this case, if E_1 is limited to ±5 volts, then I_0 is limited to ±10 amps. Additional protection may be obtained by placing diodes across R_5 and R_6 to limit the output current to approximately $3 \times 0.7/0.22 = 9.55$ amps. A slight lowering of the emitter resistors to about 0.18 ohms sets this backup current limit to $3 \times 0.7/0.18 = 11.7$ amps, which would set it slightly higher than the prime current limiting and would only be used in the event of op-amp failure, etc.

Diodes D_5 and D_6 clamp the output voltage to a value no greater than one diode drop more than either supply voltage due to di/dt variations in the motor inductance.

10.9 CONCLUSION

Numerous choices must be made when considering which type and subtype of servo amplifier would best suit a given application.

Linear amplifiers are simpler and less noisy than pulsewidth modulated (PWM) amplifiers and are usually capable of wider bandwidths. Linear amplifiers are less efficient, however, and are suitable for low power systems.

In linear amplifiers one must determine if an unidirectional amplifier will fulfill his needs and, if so, whether or not dynamic braking is required. If a bidirectional amplifier is required—do the design constraints make the bridge or bipolar design more attractive? Once this is decided—shall the power amplifier be of the voltage amplifier or transconductance type?

Predominant failure modes such as forward-biased second breakdown have been discussed, and the need for a fast, and possibly redundant, current limiter stressed.

Finally, a design procedure for a bipolar linear amplifier has been given.

REFERENCES

1. *DC Motors, Speed Controls, Servo Systems*, 3rd ed., Electro-Craft Corporation, pp. 4-31, Oct. 1975.
2. W. D. Roehr, "Avoiding Second Breakdown," Motorola Semiconductor Products, Inc., AN415.
3. J. Millman and H. Taub, *Pulse, Digital and Switching Waveforms*, McGraw-Hill Book Company, New York, p. 44, 1965.
4. *DC Motors, Speed Controls, Servo Systems*, 3rd ed., Electro-Craft Corporation, pp. 4-36, Oct. 1975.
5. *DC Motors, Speed Controls, Servo Systems*, 3rd ed., Electro-Craft Corporation, pp. 2-17, Oct. 1975.
6. W. D. Roehr, "Avoiding Second Breakdown," Motorola Semiconductor Products, Inc., AN415.

7. "Integrated Circuit Operational Amplifiers," RCA, ICAN-5290.

8. J. G. Graeme, *Designing With Operational Amplifiers*, McGraw-Hill Book Company, New York, p. 16, 1977.

11 Pulsewidth Modulated Amplifier For DC Servo Systems

J. TAL and E. K. PERSSON

11.1 INTRODUCTION

The amplifier performs an important function in the dc servo system by providing the correct amount of current and voltage needed to drive the motor. In Chapter 10, we looked at amplifiers in which the voltages vary continuously; these are called linear amplifiers. These amplifiers are relatively simple and are capable of wide bandwidths, but they possess a characteristic which makes them unsuitable for use in many applications, especially those in which the power delivered by the amplifier is high. Their drawback is operational inefficiency caused by the dissipation of large amounts of power in the output transistors. Consider the following example: a dc motor and amplifier are connected in series, sharing the voltage from the power supply. If the supply voltage is 50 volts, the motor voltage 10 volts, and the current 20 amperes, the power delivered to the motor would be 200 watts. However, the power dissipated in the amplifier would amount to 800 watts — four times the power delivered. Clearly, dissipation of this magnitude is highly undesirable.

One way to solve the power dissipation problem is by using an amplifier which controls motor voltage by varying the duty cycle of the voltage applied to the motor. Amplifiers of this type are called switching amplifiers. They are particularly well-suited for use in large systems, especially those requiring extended operation at low speed and high torque (where the power dissipation in a linear amplifier would be very high).

Switching amplifiers derive their name from their operating characteristic: the transistors in amplifiers of this type are turned on and off like a switch. When the transistor is turned on, the voltage across it is negligible; when the transistor is turned off, the voltage across it is large, but the current is zero. In either case, the power dissipated in the output transistor is small.

The switching can be performed in various ways. One simple method is to switch the amplifier at a constant frequency, and to vary the "on" and "off" phases of the period according to the need. Such amplifiers are called pulsewidth modulated (PWM) amplifiers. A wider class of switching amplifiers utilizes a combination of pulsewidth and pulse-frequency modulation modes. Systems such as these possess an advantage in that their pulse frequencies can be adjusted, thereby enabling them to maintain a controlled current variation level. However, amplifiers of this type also have the considerable disadvantge of sweeping through a wide frequency spectrum, often causing oscillations and audible "chirps" when a mechanical resonance is met at a particular frequency. This factor makes the combined-mode switching amplifier unsuitable for many applications. In this chapter, the discussion of switching amplifiers will be focused on a description of the operation of the above-mentioned PWM amplifiers utilizing fixed switching frequencies.

PWM amplifiers can be operated in three modes: bipolar, unipolar, and limited unipolar. In the following chapter, we will discuss each mode in detail. We will also consider the problem of power dissipation in motors which are driven by PWM amplifiers, and we will discuss the selection of switching frequencies.

The issue of driver efficiency is not limited to dc motor systems; step motor drives and systems employing other electromechanical devices for the driving component are also susceptible to power dissipation problems. Thus, while this chapter will deal specifically with dc motor systems, the results can apply equally well to step motor drives and linear motor amplifiers.

11.2 MODES OF OPERATION

PWM amplifiers can be divided into three groups according to their configuration: bipolar, unipolar, and limited unipolar. The first and most basic mode of operation, the bipolar PWM amplifier, can be briefly described as follows. Consider the circuit shown in Fig. 11-1. Let the switching frequency be f_s and the switching period be t_f. Let the "on" interval occur during the first part of the period (between $t = 0$ and $t = t_1$), and let it be followed by the "off" interval (which occurs during the remainder of the switching period).

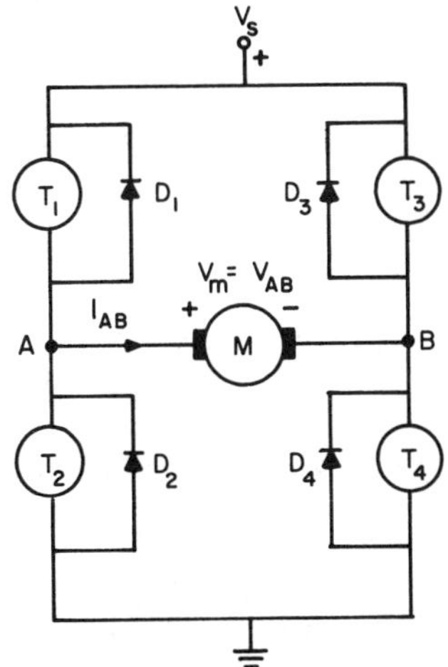

Fig. 11-1. Pulsewidth-modulated amplifier and a dc motor.

The bipolar PWM mode is such that transistors T_1 and T_4 are turned on during the "on" phase, and T_2 and T_3 are turned on during the "off" phase. Consequently, the motor voltage, V_m, is given by

$$V_m = V_{AB} = \begin{cases} V_s & 0 \leqslant t < t_1 \\ -V_s & t_1 \leqslant t < t_f \end{cases} \tag{11-1}$$

A second mode of operation, the unipolar mode, also reduces the number of transistors to be switched. In the unipolar PWM amplifier, the switching sequences depend on the sign of the amplifier input voltage, V_{in}. When V_{in} is positive, T_4 is turned on continuously; in addition, T_1 is turned on during the "on" phase and T_2 is turned on during the "off" phase. When V_{in} is negative, T_2 is turned on continuously and T_3 and T_4 are switched alternately. Thus, when the input voltage is positive, the motor voltage, V_m, becomes

$$V_m = \begin{cases} V_s & 0 \leqslant t < t_1 \\ 0 & t_1 \leqslant t < t_f \end{cases} \tag{11-2}$$

A similar expression for V_m may be written for the case where V_{in} is negative.

The two modes described above share the same disadvantage: during a single instant, one of the transistors of the pair (T_1, T_2) or (T_3, T_4) must be turned off while the other is being turned on. Due to storage and fall time of the transistor, it is possible for both transistors to conduct simultaneously, forming a short circuit across the power supply. In order to avoid this, it is necessary to introduce a delay interval between turning one transistor off and turning the other one on. The delay interval must be longer than the storage time of the transistor and thus limits the switching frequency.

A third mode of operation, the limited unipolar PWM amplifier, eliminates the need for a delay interval. Here again, the switching depends on the polarity of the input voltage, V_{in}. When V_{in} is positive, T_4 is turned on continuously; T_1 is switched on during the "on" phase. Thus, during the "on" phase, both T_1 and T_4 are on, resulting in a motor voltage, V_m:

$$V_m = V_s \qquad 0 \leqslant t < t_1 \tag{11-3}$$

During the off phase, only T_4 is turned on; the resulting motor voltage depends on the motor current I_{AB}. As long as $I_{AB} > 0$, which is a normal condition when $V_{AB} > 0$, the current I_{AB} will flow through D_2 and T_4, resulting in $V_A = 0$ and

$$V_m = V_{AB} = 0 \quad \text{for} \begin{cases} t_1 \leqslant t < t_f \\ I_{AB} > 0 \end{cases} \tag{11-4}$$

On the other hand, if I_{AB} is negative, the current will flow through D_1 and D_4, resulting in $V_A = V_s$ and

$$V_m = V_{AB} = V_s \quad \text{for} \begin{cases} t_1 \leqslant t < t_f \\ I_{AB} < 0 \end{cases} \tag{11-5}$$

This condition may occur after a change in the polarity of V_{in}.

Finally, there is a possibility that $I_{AB} = 0$. Here neither D_1 nor D_4 conducts, and the voltage V_m may be anywhere between zero and V_s.

$$0 < V_m < V_s \quad \text{for} \begin{cases} t_1 \leqslant t < t_f \\ I_{AB} = 0 \end{cases} \tag{11-6}$$

However, since $I_{AB} > 0$ is the common condition when $V_{in} > 0$, the unipolar and the limited unipolar modes are very similar. A summary of the operation modes and the resulting voltages is given in Table I.

We will now discuss each of the three operational modes in greater detail.

Table I. The Operation Modes of PWM Amplifiers.

Operation Mode and Input Voltage V_{in}	Transistors Condition and Motor Voltage	
	"On" Phase	"Off" Phase
Bipolar mode	T_1, T_4 on T_2, T_3 off $V_m = V_s$	T_2, T_3 on T_1, T_4 off $V_m = -V_s$
Unipolar $V_{in} > 0$	T_1, T_4 on T_2, T_3 off $V_m = V_s$	T_2, T_4 on T_1, T_3 off $V_m = 0$
Unipolar $V_{in} < 0$	T_2, T_3 on T_1, T_4 off $V_m = -V_s$	T_2, T_4 on T_1, T_3 off $V_m = 0$
Limited unipolar $V_{in} > 0$	T_1, T_4 on T_2, T_3 off $V_m = V_s$	T_4 on T_1, T_2, T_3 off $V_m = 0$ if $I_{AB} > 0$ $V_m = V_s$ if $I_{AB} < 0$ $0 < V_m < V_s$ if $I_{AB} = 0$
Limited unipolar $V_{in} < 0$	T_2, T_3 on T_1, T_4 off $V_m = -V_s$	T_2 on T_1, T_3, T_4 off $V_m = 0$ if $I_{AB} < 0$ $V_m = -V_s$ if $I_{AB} > 0$ $-V_s < V_m < 0$ if $I_{AB} = 0$

11.3 BIPOLAR PWM AMPLIFIERS

The input voltage of the amplifier, V_{in}, is assumed to change slowly, and therefore may be considered a constant over a switching period. The input voltage determines the load factor, ρ, as follows:

$$\rho = \frac{V_{in}}{V_{max}} \tag{11-7}$$

where V_{max} is the maximum of $|V_{in}|$. It then follows that

$$-1 \leqslant \rho \leqslant 1 \tag{11-8}$$

The length of the "on" phase, t_1, is selected so that when $\rho = 1$, $t_1 = t_f$; the amplifier is thus continuously on. Similarly, when $\rho = -1$, t_1 is selected to be zero, resulting in a continuously negative voltage. Such a relation between ρ and t_1 is given by

$$t_1 = \frac{1+\rho}{2} t_f \tag{11-9}$$

It then follows that $t_1 = t_f/2$ if $\rho = 0$.

If one ignores the delay intervals (assuming that they are relatively short), the motor voltage V_m will be as given in Fig. 11-2. This voltage can be expressed by a Fourier series of the form:

$$V_m = a_0 + \sum_{n=1}^{\infty} a_n \cos(2\pi n f_s t + \phi_n) \tag{11-10}$$

where

$$a_0 = \rho V_s = \frac{V_s V_{in}}{V_{max}} \tag{11-11}$$

and

$$a_n = \frac{4V_s}{n\pi} \sin \frac{n\pi(1+\rho)}{2} \tag{11-12}$$

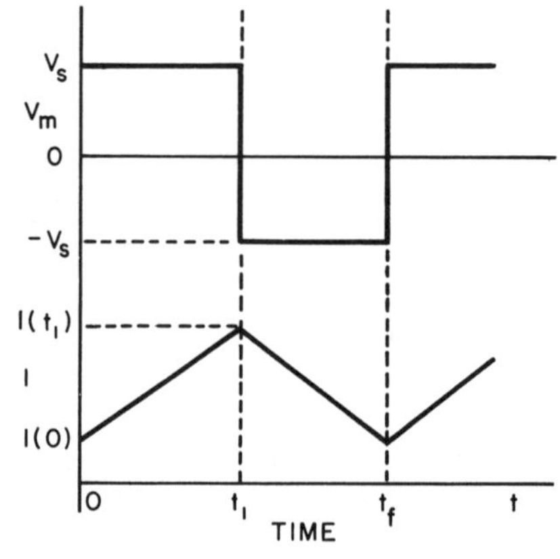

Fig. 11-2. The voltage and current of a bipolar PWM amplifier.

Thus, the motor voltage includes the slowly varying component, ρV_s, as well as high-frequency components. Only the slow component of the voltage needs to be considered because the switching frequency, f_s, is selected to be high enough above the motor bandwidth and resonance modes; therefore, all of the high-frequency components will be attenuated by the motor. Under these assumptions, the bipolar PWM amplifier can be represented by the block diagram of Fig. 11-3. The equivalent voltage gain of the amplifier is given by

$$A_v = \frac{V_m}{V_{in}} = \frac{V_s}{V_{max}} \tag{11-13}$$

The model of Eq. (11-13) is sufficient for the analysis of the amplifier behavior and its effect on the feedback system.

In order to analyze the current waveform, one must examine the electric equation of the motor:

$$V_m = L\frac{dI}{dt} + RI + K_e\omega \tag{11-14}$$

where

V_m is the motor voltage

I is the motor current

ω is the motor velocity

L is the armature inductance

R is the armature resistance

K_e is the motor voltage constant

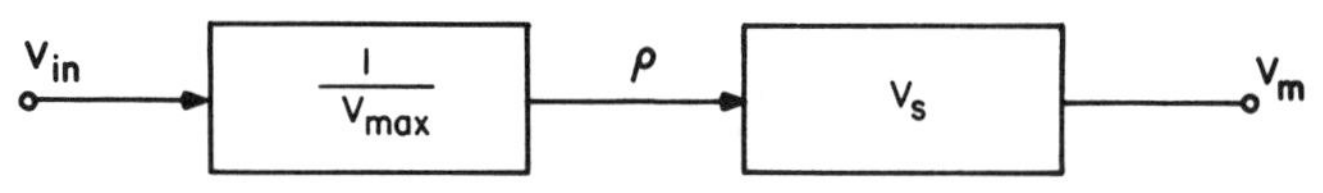

Fig. 11-3. Block diagram representation of a bipolar PWM amplifier.

The solution of Eq. (11-14) is exponential. However, the equation can be greatly simplified if we consider ω to be constant over a switching period and if we assume the variations in RI to be small. Define I_a and V_a as

$$I_a = \frac{1}{t_f}\int_0^{t_f} I(t)dt \tag{11-15}$$

and

$$V_a = RI_a + K_e\omega \tag{11-16}$$

Then, if we approximate RI by its average value RI_a, and if we assume V_a to be constant over a switching period, Eq. (11-14) can be simplified to the form

$$L \frac{dI}{dt} = V_m - V_a \tag{11-17}$$

In the case of a bipolar PWM amplifier, Eq. (11-17) becomes

$$\left.\begin{aligned} L \frac{dI}{dt} &= V_s - V_a && 0 \leqslant t < t_1 \\ L \frac{dI}{dt} &= -V_s - V_a && t_1 \leqslant t < t_f \end{aligned}\right\} \tag{11-18}$$

The solution of Eq. (11-18) is a triangular waveform as given by the following equation and is illustrated in Fig. 11-2.

$$\left.\begin{aligned} I(t) &= I(0) + \frac{V_s - V_a}{L} t && 0 \leqslant t < t_1 \\ I(t) &= I(t_1) - \frac{V_s + V_a}{L} (t - t_1) && t_1 \leqslant t < t_f \end{aligned}\right\} \tag{11-19}$$

During a steady-state operation, the current is periodic and

$$I(t_f) = I(0) \tag{11-20}$$

This requires that

$$\frac{V_s - V_a}{L} t_1 - \frac{V_s + V_a}{L} (t_f - t_1) = 0 \tag{11-21}$$

Combining Eq. (11-21) with Eq. (11-9) results in

$$\frac{V_a}{V_s} = \rho \tag{11-22}$$

Thus, the load factor, ρ, is proportional to V_a. The total current variation, ΔI, is given by combining Eq. (11-19) with Eqs. (11-9) and (11-22). We have

$$\Delta I = I(t_1) - I(0) = \frac{V_s t_f}{2L} (1 - \rho^2) \tag{11-23}$$

The largest current variation occurs when the load factor, ρ, is zero. This is equal to

$$\Delta I_{max} = \frac{V_s t_f}{2L} \tag{11-24}$$

Equation (11-24) shows the dependence of the current variation on the motor and amplifier variables, V_s, t_f, and L.

11.4 UNIPOLAR PWM AMPLIFIERS

Let us denote the input voltage by V_{in}, and let the load factor be defined as

$$\rho = \frac{V_{in}}{V_{max}} \tag{11-25}$$

where V_{max} is the same as in Eq. (11-7). Therefore, Eq. (11-8) holds here as well. The "on" phase, t_1, is selected to be $t_1 = t_f$ for $\rho = 1$, resulting in a fully-on amplifier. For $\rho = 0$, it is desired to have $t_1 = 0$ and a fully-off amplifier. The third possibility, when $\rho = -1$, should turn the amplifier fully on with a negative voltage, thus requiring $t_1 = t_f$. Such a relationship between t_1 and ρ is given by

$$t_1 = |\rho| t_f \tag{11-26}$$

The transistors to be switched are selected according to the following procedure:

Case 1 ($V_{in} > 0$)

T_4 turned on continuously
T_1 turned on during $0 \leqslant t < t_1$
T_2 turned on during $t_1 \leqslant t < t_f$

When the input voltage is negative, the following sequence is used.

Case 2 ($V_{in} < 0$)

T_2 turned on continuously
T_3 turned on during $0 \leqslant t < t_1$
T_4 turned on during $t_1 \leqslant t < t_f$

The resulting motor voltages for the two cases are illustrated in Fig. 11-4 and are given by

$$V_m = \begin{cases} V_s & 0 \leqslant t < t_1 \quad \text{and} \quad V_{in} > 0 \\ -V_s & 0 \leqslant t < t_1 \quad \text{and} \quad V_{in} < 0 \\ 0 & t_1 \leqslant t < t_f \end{cases} \tag{11-27}$$

The voltage V_m as given above may be expressed by a Fourier series.

$$V_m = a_0 + \sum_{n=1}^{\infty} a_n \cos(2\pi n f_s t + \phi_n) \tag{11-28}$$

where

$$a_0 = \rho V_s \tag{11-29}$$

and

$$a_n = \frac{2V_s}{n\pi} \sin n\pi|\rho| \tag{11-30}$$

Here again, the voltage components with high frequencies will be attenuated by the motor, and only a_0 will be considered. The transfer function of the simplified amplifier is thus the same as for the bipolar amplifier, and is given by

$$A_v = \frac{V_m}{V_{in}} = \frac{V_s}{V_{max}} \tag{11-31}$$

The current analysis is similar to that of the bipolar PWM amplifier. Define I_a and V_a as in Eqs. (11-15) and (11-16), respectively; then Eq. (11-17) holds and its solution for positive input voltage is

$$I(t) = \begin{cases} I(0) + \dfrac{V_s - V_a}{L} t & \text{for } \rho > 0, \quad 0 \leqslant t < t_1 \\ I(t_1) - \dfrac{V_a}{L}(t - t_1) & \text{for } \rho > 0, \quad t_1 \leqslant t < t_f \end{cases} \tag{11-32}$$

and for negative input voltage, it is

$$I(t) = \begin{cases} I(0) - \dfrac{V_s + V_a}{L} t & \text{for } \rho < 0, \quad 0 \leqslant t < t_1 \\ I(t_1) - \dfrac{V_a}{L}(t - t_1) & \text{for } \rho < 0, \quad t_1 \leqslant t < t_f \end{cases} \tag{11-33}$$

The current I(t) is illustrated in Fig. 11-4.

During a steady-state operation, the current is periodic, and therefore

$$I(t_f) = I(0) \tag{11-34}$$

The application of Eq. (11-34) to Eqs. (11-32) and (11-33) results in

$$\frac{V_a}{V_s} = \rho \tag{11-35}$$

for both ρ positive and negative. Note that Eq. (11-35) holds for the bipolar amplifier, as given by Eq. (11-22), as well.

The current variation during a switching period, ΔI, can be found from Eqs. (11-32), (11-33), (11-35), and (11-26). We get

$$I = |I(t_1) - I(0)| = \frac{V_s t_f}{L}(|\rho| - \rho^2) \tag{11-36}$$

The maximum current variation occurs when $|\rho| = 1/2$ and is given by

$$\Delta I_{max} = \frac{V_s t_f}{4L} \tag{11-37}$$

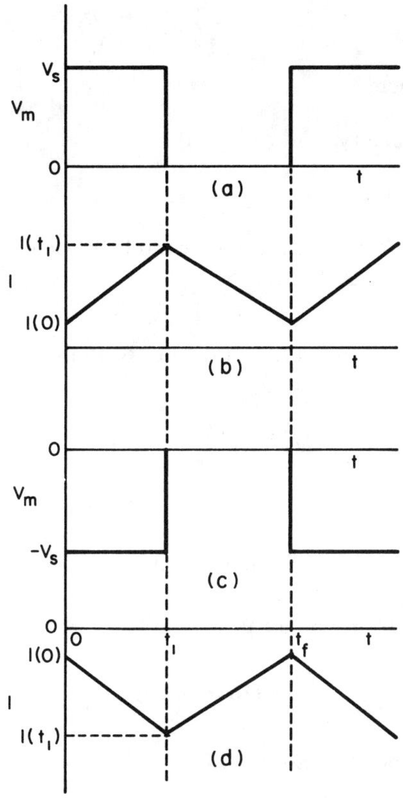

Fig. 11-4. The voltage and current of a unipolar PWM amplifier. (a) and (b) are the voltage and current for $\rho > 0$, whereas (c) and (d) illustrate the voltage and current for $\rho < 0$.

11.5 LIMITED UNIPOLAR PWM AMPLIFIERS

The operation of a limited unipolar PWM amplifier is identical to that of a unipolar amplifier if the polarities of the input voltage, V_{in}, and the polarity of the motor current, I_{AB}, are the same. This condition may be written as

$$V_{in} I_{AB} > 0 \tag{11-38}$$

When Eq. (11-38) is not satisfied, the operation in the two modes is different. This is illustrated by the following example.

Example 1

A dc motor is to be driven at a constant velocity by a PWM amplifier. The parameters of the motor and the amplifier are as follows:

$$V_s = 40 \text{ volts}$$

$$V_a = 30 \text{ volts}$$

$$I_a = 1 \text{ ampere}$$

$$L = 1 \text{ millihenry}$$

$$f_s = 2000 \text{ hertz}$$

Determine the current waveform under the three modes of operation.

Case 1 Bipolar Amplifier

The load factor, ρ, can be found from Eq. (11-16) to be

$$\rho = \frac{V_a}{V_s} = 0.75$$

The switching period equals

$$t_f = 0.5 \text{ millisecond}$$

and the "on" time is found from Eq. (11-9)

$$t_1 = \frac{1+\rho}{2} t_f = 0.4375 \text{ millisecond}$$

The current variation is found from Eq. (11-23)

$$\Delta I = \frac{V_s t_f}{2L}(1-\rho^2) = 4.375 \text{ amperes}$$

Thus, in order to have an average of $I_a = 1$ ampere, the current will be triangular with a maximum of 3.19 amperes and a minimum of −1.19 amperes, as shown in Fig. 11-5(a).

Case 2 Unipolar Amplifier

The load factor is found from Eq. (11-35) to be

$$\rho = 0.75$$

and therefore

$$t_1 = \rho t_f = 0.375 \text{ millisecond}$$

The current variation is found from Eq. (11-36)

$$\Delta I = \frac{V_s t_f}{L}(|\rho| - \rho^2) = 3.75 \text{ amperes}$$

The current waveform which has an average value of 1 ampere and peak variation of 3.75 amperes is a triangular waveform with a maximum of 2.875 amperes and a minimum of -0.875 amperes, as shown in Fig. 11-5(b).

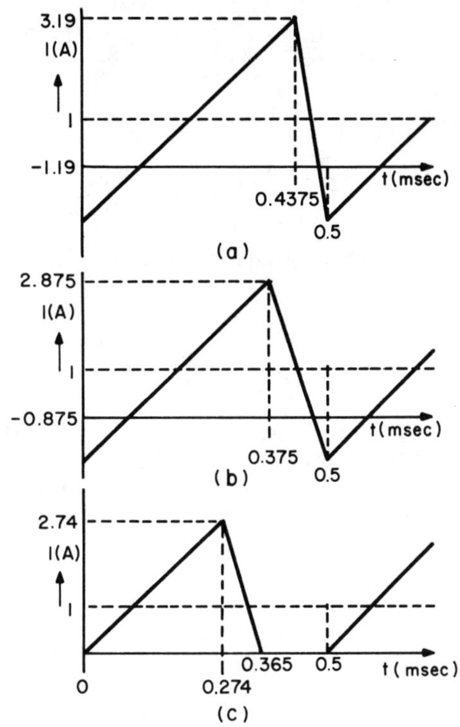

Fig. 11-5. The current waveforms for Example 1. (a) The current of a bipolar PWM amplifier, (b) the current of a unipolar PWM amplifier, (c) the current of a limited unipolar amplifier.

Case 3 Limited Unipolar Amplifier

The first step in the analysis of this mode is to determine whether it is identical to the unipolar mode. However, since ρ is positive, and since the current of the unipolar amplifier may be negative [as can be seen in Fig. 11-5(b)], Eq. (11-38) will not always hold. Therefore, the operation in this mode must be investigated separately.

The current equation during the "on" phase can be approximated by

$$L\frac{dI}{dt} = V_s - V_a \tag{11-39}$$

Consequently, the current will be

$$I(t) = I(0) + \frac{V_s - V_a}{L} t \qquad 0 \leqslant t < t_1 \tag{11-40}$$

At $t = t_1$, the transistor T_1 is turned off and the current will flow through D_2, resulting in $V_m = 0$ and

$$L \frac{dI}{dt} = -V_a \tag{11-41}$$

Then the current becomes

$$I(t) = I(t_1) - \frac{V_a}{L}(t - t_1) \tag{11-42}$$

Note that Eqs. (11-41) and (11-42) are valid as long as $I(t) > 0$. Suppose that at $t = t_2$, the current of Eq. (11-42) equals zero; i.e.,

$$I(t_2) = I(t_1) - \frac{V_a}{L}(t_2 - t_1) = 0 \tag{11-43}$$

Then Eqs. (11-41) and (11-42) are valid for $t_1 \leqslant t < t_2$. During the time $t_2 \leqslant t < t_f$, the current cannot be negative because this would require conduction through D_1, and this would occur only if $V_a > V_s$. Consequently, the current value is forced to remain at zero during this interval, and V_m becomes equal to V_a to satisfy the motor equation.

$$\left.\begin{aligned} V_m &= V_a \\ I(t) &= 0 \end{aligned}\right\} \text{ for } t_2 \leqslant t < t_f \tag{11-44}$$

The current waveform for this case is illustrated in Fig. 11-5(c).

The above example has shown that the limited unipolar amplifier may operate differently from the unipolar one, and hence a separate analysis is required. However, because a complete analysis of the limited unipolar mode is very complex, we will limit this discussion to an examination of the steady-state conditions.

During the steady-state condition, the motor runs at a constant velocity ω and generates a torque T_g. The torque T_g is used to overcome the load torque, T_L, and the internal friction torque. Consequently, every steady-state condition may be represented by a point in the speed-torque plane (ω, T_g). Every steady-state condition may also be characterized by the average current I_a required to generate the torque, T_g, and the voltage, V_a, as given by

$$V_a = K_e \omega + I_a R \tag{11-45}$$

This enables us to represent every steady-state condition as a point in the V_a versus I_a plane. In the following analysis, it is assumed that the input voltage V_{in} is positive; therefore, both I_a and V_a are positive as well. (In the case of negative input voltage, both I_a and V_a are negative, and the analysis is the same.)

In the steady-state condition, the limited unipolar amplifier will operate as a unipolar amplifier as long as the current I(t) is positive. The operation is different, however, if the current is nonpositive. In order to analyze the operation, the V_a versus I_a plane is divided into two regions: Region I, where the current is always positive, and therefore the limited unipolar and the unipolar mode operate identically; and Region II, where the current I(t) is sometimes nonpositive, and therefore the two modes operate differently. Typical currents in the two regions are illustrated in Fig. 11-6.

The boundary between the two regions can be found by noting that, in the limiting case, the current I(t) equals zero at one point; or equivalently,

$$I_a = \frac{1}{2} \Delta I \tag{11-46}$$

Applying Eq. (11-36) to Eq. (11-46), and noting that $\rho > 0$, results in

$$I_a = \frac{V_s t_f}{2L} (\rho - \rho^2) = \frac{t_f}{2LV_s} V_a(V_s - V_a) \tag{11-47}$$

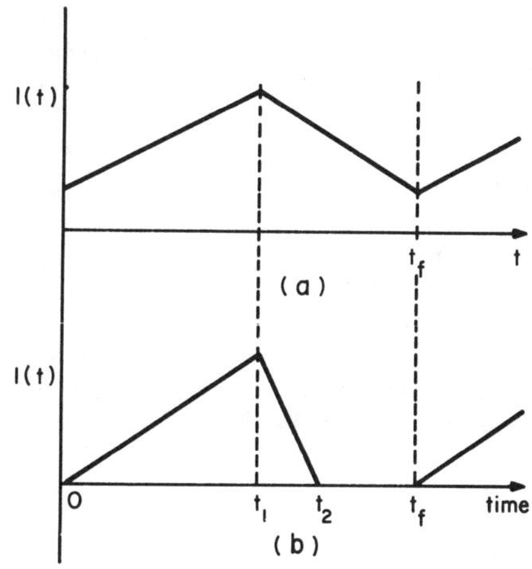

Fig. 11-6. Steady-state current where $I(t) > 0$ in Region I (a), and $I(t) \geqslant 0$ in Region II (b).

Equation (11-47) describes the boundary between the two regions. As can be seen, Eq. (11-47) represents a parabola in the V_a versus I_a plane. Notice also that the boundary depends on the inductance, L, and the switching period, t_f. As t_f is decreased, or as L is increased, Region II diminishes; the difference between the two modes is therefore reduced.

Figure 11-7 shows the two regions in the V_a versus I_a plane. Note that the maximum value of V_a is V_s and it occurs when $\rho = 1$. Similarly, the maximum of I_a occurs when $\rho = 1$ and when $\omega = 0$ and is equal to V_s/R. The two regions shown in Fig. 11-7 are for an amplifier and a motor with the following parameters:

$V_s = 40$ volts

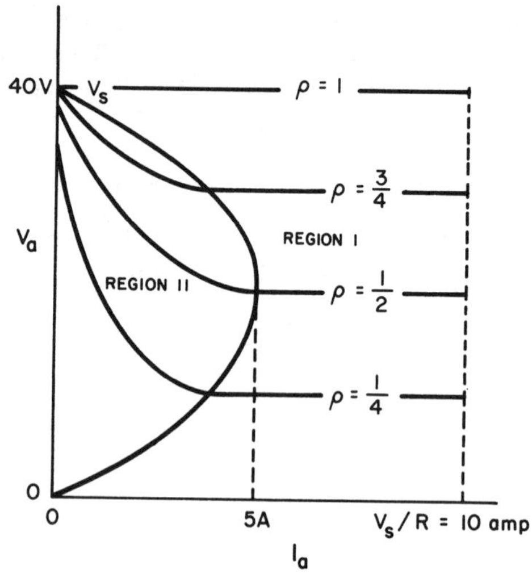

Fig. 11-7. The two regions of operation of a limited unipolar PWM amplifier and the load curves for constant ρ and variable load torque.

$R = 4$ ohms

$L = 1$ millihenry

$t_f = 1$ millisecond

Once the boundaries have been determined, the amplifier equations must be derived for each of the two regions. This is done below.

Region I

In this region, the limited unipolar amplifier operates as a unipolar amplifier; therefore, Eqs. (11-25) through (11-37) hold. Thus, the "on" time is given by

$$t_1 = |\rho| t_f \tag{11-48}$$

and V_a is equal to

$$V_a = \rho V_s \tag{11-49}$$

Finally, the current variation, ΔI, is given by

$$\Delta I = \frac{V_s t_f}{L} (|\rho| - \rho^2) \tag{11-50}$$

Note that if the load factor, ρ, is constant, then V_a is also constant, as given by Eq. (11-49). This results in horizontal lines in the V_a versus I_a plane, as illustrated in Fig. 11-7.

Region II

The objective of the following analysis is to determine the relationship between ρ, I_a, and V_a in this region. Note that t_1 is given by

$$t_1 = |\rho| t_f \tag{11-51}$$

If we approximate the term $K_e\omega + RI(t)$ by the constant V_a, the resulting current variations will be linear, having a positive slope $(V_s - V_a)/L$ during $0 \leqslant t < t_1$, and a negative slope $-V_a/L$ during $t_1 \leqslant t < t_2$; t_2 is the time when the current $I(t)$ becomes zero. The motor current is illustrated in Fig. 11-8.

Using the above approximation, the maximum value of the current, I_{pk}, is given by

$$I_{pk} = \frac{V_s - V_a}{L} t_1 \tag{11-52}$$

or

$$I_{pk} = \frac{V_a}{L}(t_2 - t_1) \tag{11-53}$$

The average current, I_a, is given by

$$I_a = I_{pk}\frac{t_2}{2t_f} \tag{11-54}$$

Combining Eq. (11-52) with Eq. (11-53) results in

$$V_s t_1 = V_a t_2 \tag{11-55}$$

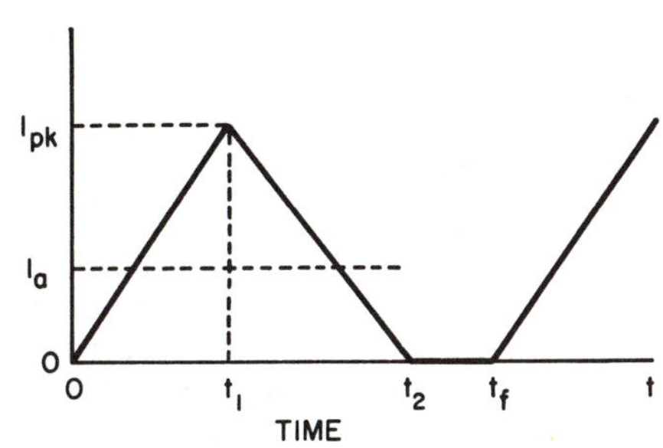

Fig. 11-8. The motor current in Region II.

Now we can substitute Eqs. (11-52) and (11-55) in Eq. (11-54) to yield

$$I_a = \frac{(V_s - V_a)t_1 t_2}{2Lt_f} = \frac{V_s t_1^2}{2Lt_f}\left[\frac{V_s}{V_a} - 1\right] \tag{11-56}$$

or, in view of Eq. (11-51),

$$I_a = \frac{V_s t_f \rho^2}{2L}\left[\frac{V_s}{V_a} - 1\right] \tag{11-57}$$

Equation (11-57) indicates the relationship between I_a, ρ, and V_a. Note that if ρ is held constant, Eq. (11-57) forms a hyperbola in the V_a versus I_a plane; also, whenever $I_a = 0$, V_a equals V_s independently of ρ. This is illustrated by the curves in Fig. 11-7.

A physical understanding of the operation may be facilitated by representing the working conditions as points in the speed-torque plane. Note that the transformation from V_a, I_a to ω, T_g is accomplished by the following equations:

$$T_g = K_t I_a \tag{11-58}$$

$$\omega = \frac{1}{K_e}(V_a - I_a R) \tag{11-59}$$

Upon application of the transformation in Eq. (11-58), the boundary equation, Eq. (11-47), becomes

$$\frac{T_g}{K_t} = \frac{t_f}{2LV_s}\left[K_e\omega + \frac{T_g R}{K_t}\right]\left[V_s - K_e\omega - \frac{T_g R}{K_t}\right] \tag{11-60}$$

The boundary in the speed-torque plane is given by a complex equation of Eq. (11-60) which resembles a parabola, as can be seen in Fig. 11-9. The horizontal lines in Region I (Fig. 11-7) are transformed into straight lines of the form:

$$\omega = \frac{\rho V_s}{K_e} - \frac{T_g R}{K_e K_t} \tag{11-61}$$

The transformation of the curves in Region II (Fig. 11-7) results in a more complex form:

$$\frac{2LT_g}{K_t V_s t_f \rho^2} = \frac{V_s K_t}{K_e K_t \omega + T_g R} - 1 \tag{11-62}$$

Equation (11-62) is represented by several curves in Region II of Fig. 11-9.

Note that if ρ is held constant and if T_g is reduced gradually, the motor velocity will reach the limit of V_s/K_e when $T_g = 0$ for all of the values of ρ. However, this condition cannot be seen in actual measurements, since the motor will always have some internal friction torque, T_f; therefore, it is impossible to reduce T_g below T_f or, equivalently, to measure anything to the left of the line $T_g = T_f$.

11.6 SPECIAL FEATURES

Because of the nature of the PWM amplifier, the dc servo system employing this kind of amplifier possesses some unique characteristics. The most important features are discussed below.

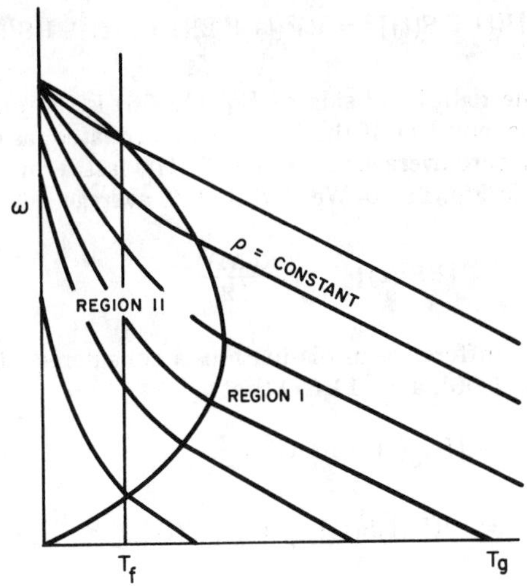

Fig. 11-9. A family of speed-torque curves.

Power Dissipation In The Motor

The first question to be considered deals with the effect of the amplifier on power dissipation in the motor. This can be illustrated by comparing the PWM amplifier with a linear amplifier. The power dissipation in a motor with armature resistance, R, internal friction torque, T_f, and viscous damping factor, D, is derived in [2].

$$P = I^2R + \frac{K_e T_f \omega}{K_t} + \frac{K_e D \omega^2}{K_t} \tag{11-63}$$

Let us suppose that the motor is to be driven at a velocity ω. This can be done with a linear amplifier having an armature current I(t) and power dissipation P_L. If a PWM amplifier is used to drive the motor at the same velocity, the required current is $I_p(t)$, and the power dissipation is P_{Lp}.

The only difference between the currents I(t) and $I_p(t)$ is that I_p is the sum of I(t) and the triangular wave S(t), with a total variation peak-to-peak of ΔI. The waveform of S(t) is shown in Fig. 11-10. Therefore,

$$I_p(t) = I(t) + S(t) \tag{11-64}$$

The power dissipation corresponding to the linear amplifier, P_L, is found from Eq. (11-63),

$$P_L = I^2(t)R + \frac{K_e T_f \omega}{K_t} + \frac{K_e D \omega^2}{K_t} \tag{11-65}$$

In contrast, P_{Lp} can be found by substituting Eq. (11-64) into Eq. (11-63). Note that only the first term depends on $I_p(t)$. This term equals

$$RI_p^2(t) = R[I(t) + S(t)]^2 = RI^2(t) + 2RI(t)S(t) + RS^2(t) \qquad (11\text{-}66)$$

The second term on the right-hand side of Eq. (11-66) is the product of S(t), which has a zero average value, and I(t). If the two are uncorrelated, as occurs in most cases, the product will have a zero average value as well. The last term of Eq. (11-66) can be found for S(t) as given in Fig. 11-10. We find that its average value is

$$[RS^2(t)]_{ave} = \frac{R(\Delta I)^2}{12} \qquad (11\text{-}67)$$

Next, when the average difference in dissipation is considered, it can be found from Eqs. (11-63), (11-64), (11-66), and (11-67) that:

$$[P_{Lp} - P_L]_{ave} = [RI_p^2(t) - RI^2(t)]_{ave}$$

$$= 2R[I(t)S(t)]_{ave} + [RS^2(t)]_{ave} = \frac{R(\Delta I)^2}{12} \qquad (11\text{-}68)$$

Typical values for ΔI and R are 2 amperes and 2 ohms, respectively. These results translate to power dissipation differences of 2/3 watts, which is negligible. Thus, a PWM amplifier does not affect power dissipation in the motor.

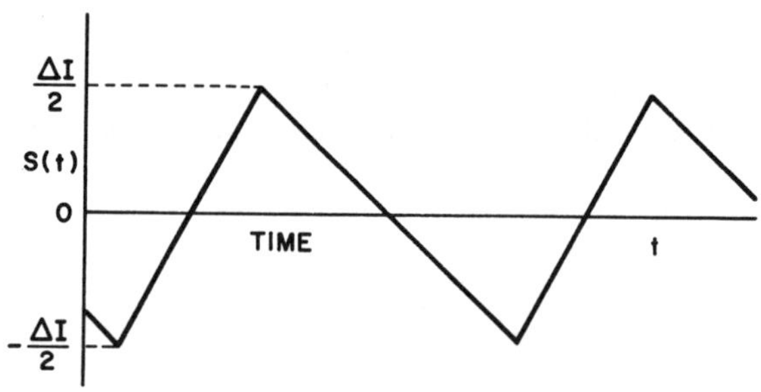

Fig. 11-10. The current triangular waveform.

Selection Of The Switching Frequency

The system designer has some freedom in selecting the switching frequency, f_s. The main considerations in choosing f_s are the following:

a. The switching frequency must be high enough that the inductance of the motor has a large impedance at the chosen frequency. Thus,

$$2\pi f_s L \gg R \qquad (11\text{-}69)$$

This can be stated equivalently as a limit on ΔI, the current variation which was derived above.

b. The switching rate must be high enough that the servo system does not respond to it. Thus, if the motion requirements are such that a bandwidth f_{BW} is needed, one should select

$$f_s > 10 f_{BW} \tag{11-70}$$

c. The switching frequency f_s must be higher than all the significant resonance frequencies of the loop.

$$f_s > (f_{res})_{max} \tag{11-71}$$

The above conditions might suggest that it is always desirable to increase f_s. However, increasing f_s has some drawbacks:

d. An increase in the switching rate results in proportionally increased power dissipation in the output transistors due to the finite switching time; therefore, it is desirable to maintain f_s as low as possible, subject to the other conditions listed above.

e. The delay phase required in bipolar and unipolar amplifiers to prevent short circuiting during transistor switching (see Sec. 11.2) limits the switching frequency by setting an upper boundary for it. (The delay phase usually lasts between 5 microseconds and 20 microseconds.)

Thus, there are reasons to increase f_s, and also reasons to decrease it. The final selection of a switching frequency will always reflect some degree of compromise between these conflicting conditions.

11.7 SUMMARY

In this chapter we described and analyzed the three modes of operation for PWM amplifiers: the bipolar, the unipolar, and the limited unipolar modes. We saw that the limited unipolar amplifier operates differently under varying current inputs; we then divided the operating conditions for this amplifier into two regions to account for its operation under both large and small currents.

The main results of the analysis of PWM amplifiers include information on the load factor, ρ, the "on" time, t_1, the current variation, ΔI, and the amplifier gain. These results are summarized in Table II.

Table II. Summary of Analysis Results.

Amplifier Mode / Variable	Bipolar	Unipolar and Limited Unipolar in Region I	Limited Unipolar in Region II
Load factor ρ	$\frac{V_{in}}{V_{max}}$	$\frac{V_{in}}{V_{max}}$	$\frac{V_{in}}{V_{max}}$
"On" time t_1	$\frac{1+\rho}{2} t_f$	$\|\rho\| t_f$	$\|\rho\| t_f$
Amplifier gain A_v	$\frac{V_s}{V_{max}}$	$\frac{V_s}{V_{max}}$	Depend on the velocity ω and on the current I_a
Current Variation ΔI	$\frac{V_s t_f}{2L}(1-\rho^2)$	$\frac{V_s t_f}{2}(\|\rho\|-\rho^2)$	
ΔI_{max}	$\frac{V_s t_f}{2L}$	$\frac{V_s t_f}{4L}$	

REFERENCES

1. Tal, J. "Design and Analysis of Pulsewidth-Modulated Amplifiers for DC Servo Systems," *IEEE Trans. on IECI*, Vol. IECI-23, No. 1, pp. 47-55, Feb. 1976.
2. *DC Motors, Speed Controls, Servo Systems*, Engineering Handbook, 3rd ed., Electro-Craft Corporation, 1975.
3. Kuo, B. C., *Automatic Control Systems*, 3rd ed., Prentice-Hall, Inc., Englewood Cliffs, N.J., 1975.
4. Elgerd, O. I., *Control Systems Theory*, McGraw-Hill Book Company, New York, 1967.

12 Velocity Control Systems

J. TAL

12.1 INTRODUCTION

The design requirements for incremental motion servo systems often call for the generation of a specific velocity profile. The achievement of this profile by the motor and the load can be accomplished most easily by employing a velocity control system.

Velocity control systems are capable of generating both constant and variable velocity profiles. They make use of the principle of feedback to achieve a high degree of velocity accuracy. Briefly, this is how the systems work: the motor speed, measured by a velocity sensor, is fed into the operational amplifier and compared to the desired velocity level. The difference between these two values, termed the velocity error, is then amplified and applied to the motor in order to modify the velocity. This modification eliminates velocity deviation, allowing a close correspondence between the motor velocity and the command signal. Once in operation, velocity control systems are capable of responding to either a constant or variable command signal, provided that the variations of the command are within the system bandwidth.

This chapter will deal with the operation of the most common of the velocity control systems, those in which the velocity feedback is a continuous analog signal. Systems in which the feedback is derived from either a modulated or discrete-time signal (e.g., phase-locked servo systems) do not fall in this category, and the following discussion will not necessarily apply to these systems. In such cases, one must analyze the individual system and apply the appropriate design. An analysis of the phase-locked servo system, a velocity control system in which the feedback is transmitted in a modulated form, is given in Chapter 14.

We will start the discussion of analog velocity control systems by describing these systems in terms of a block diagram. Later, we will distinguish between systems employing current amplifiers and those employing voltage amplifiers; the two systems will then be analyzed separately. In the system analysis we will derive two transfer functions: the first is the function between the input and the velocity and the second is between the load and the velocity.

12.2 SYSTEM BLOCK DIAGRAM

The block diagram describing velocity control systems is shown in Fig. 12-1. The input command, v_{in}, is compared with the feedback signal, v_g; their difference, e, is applied through an amplifier to the motor, resulting in a correction of the velocity error.

The first element in the loop, the amplifier, may be linear or pulsewidth modulated. Furthermore, the amplifier may be a current source or a voltage source. In either case, we will assume in this discussion that the amplifier may be represented by the transfer function A(s). The second element in the loop, the motor, is represented by the transfer function M(s) (as discussed in Chapter 6). The third element in the velocity loop is the velocity sensor. It generates a continuous analog signal whose magnitude is proportional to the motor velocity. In most cases, this sensor is a tachometer (as discussed in Chapter 8). Where cost or size considerations prohibit the use of a tachometer, however, an equivalent velocity feedback can be generated by combining the current and voltage signals. As an illustration of this method, consider a simplified model for the motor voltage as given by Eq. (12-1).

$$v_i(t) = R_a i_a(t) + K_e \omega(t) \tag{12-1}$$

where K_e is the back emf constant.

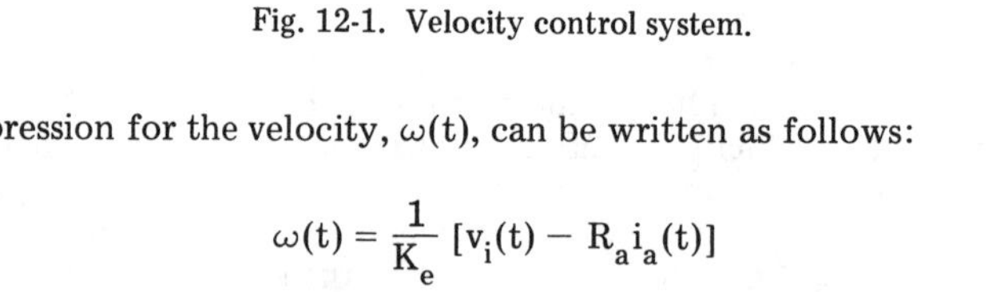

Fig. 12-1. Velocity control system.

An expression for the velocity, $\omega(t)$, can be written as follows:

$$\omega(t) = \frac{1}{K_e} [v_i(t) - R_a i_a(t)] \tag{12-2}$$

The velocity feedback relation of Eq. (12-2) can be implemented by a circuit such as the one shown in Fig. 12-2. Assuming that the output operational amplifier has an infinite gain, the voltage v_g is given by

$$v_g(t) = -\frac{R_f}{R_1} v(t) + \frac{R_f}{R_2} v_s(t) \tag{12-3}$$

However, since

$$v_s(t) = R_s i(t) \tag{12-4}$$

Equation (12-3) becomes

$$v_g(t) = -\frac{R_f}{R_1} v(t) + \frac{R_f R_s}{R_2} i(t) \tag{12-5}$$

In order for v_g to be proportional to the velocity, ω, the ratio of the two terms in Eq. (12-2) must be the same as the ratio of the terms in Eq. (12-5). Therefore, we require

$$\frac{R_f R_s}{R_2} = R_a \tag{12-6}$$

This leads to the relationship

$$v_g(t) = -\frac{K_e R_f}{R_1} \omega(t) \tag{12-7}$$

The method described here illustrates the principle of generating a velocity feedback without a tachometer. The circuit in Fig. 12-2 may be simplified by using a single amplifier with differential inputs.

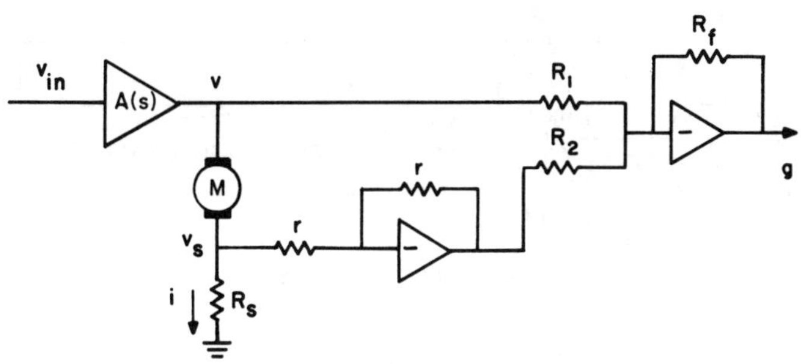

Fig. 12-2. Velocity sensing circuit.

The main problem with using this method for generating velocity feedback is that motor resistance changes with temperature, and, therefore, Eq. (12-6) will not exactly hold. Consequently, one must also consider a more complex model for the motor, one which includes the effect of the armature inductance, in order to develop a practical design. More details on this velocity feedback generation method and on temperature effects may be found in [2].

An alternative method for sensing velocity without a tachometer is to generate the velocity signal from the output of an incremental encoder, which may already be present in the loop for the purpose of position sensing. An incremental encoder generates a signal whose frequency is proportional to the motor angular velocity. In other words, the encoder output is frequency modulated by the velocity. In order to reconstruct the velocity signal, however, we must frequency demodulate the encoder output. This can be accomplished by using an FM demodulator, as shown in Fig. 12-3, or by incorporating a special circuit, as suggested by [3]. Further details on the applications of incremental encoder utilization as tachometer replacements are given in Chapter 18.

12.3 SYSTEM DESIGN AND ANALYSIS

All velocity control systems, regardless of their motion requirements, must satisfy the following two design conditions.

1. Speed Of Response.

In general, the velocity control system should be capable of following the command signal rapidly. This requirement may be expressed either in terms of the step response of the system or in terms of the system bandwidth.

2. Velocity Variations.

The variations in velocity due to load changes and other disturbances must be as small as possible. Furthermore, the transient behavior due to load change should decay as soon as possible.

In the following discussion we will analyze the system behavior with a special emphasis on the above requirements. The analysis will start with systems employing current amplifiers; later, systems containing voltage amplifiers will be discussed.

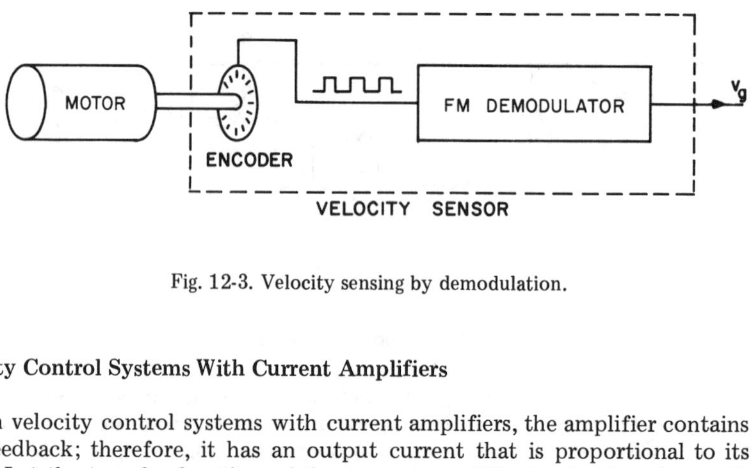

Fig. 12-3. Velocity sensing by demodulation.

Velocity Control Systems With Current Amplifiers

In velocity control systems with current amplifiers, the amplifier contains a current feedback; therefore, it has an output current that is proportional to its input signal. Let the transfer function of the current amplifier be $A_c(s)$ and note that the generated torque is proportional to the current

$$T_g(t) = K_t i(t) \tag{12-8}$$

Assuming the viscous friction to be negligible, the motor acceleration is proportional to the net torque $T_g - T_L$.

$$J\frac{d\omega(t)}{dt} = T_g(t) - T_L(t) \qquad (12\text{-}9)$$

where J is the combined moment of inertia of both the motor armature and the load, and T_L is the load torque.

Next, assume that the system includes a tachometer with a voltage gain of K_g. The system may be represented by the block diagram of Fig. 12-4. The closed-loop transfer function between the input command, v_{in}, and the output velocity, ω, is given by

$$\frac{\Omega(s)}{V_{in}(s)} = \frac{A_c(s)K_t/Js}{1 + A_c(s)K_tK_g/Js} \qquad (12\text{-}10)$$

The frequency range we are interested in is the low range, and low frequencies are well within the loop bandwidth. For such frequencies, the open-loop gain is much greater than one, i.e.,

$$\frac{A_c(s)K_tK_g}{Js} \gg 1 \qquad (12\text{-}11)$$

where $s = j\omega$; ω is the frequency in radians/second.

Therefore, under the assumption in Eq. (12-11), we may ignore the 1 term in the denominator of Eq. (12-10). This results in the simplified expression:

$$\frac{\Omega(s)}{V_{in}(s)} \cong \frac{A_c(s)K_t/Js}{A_c(s)K_tK_g/Js} = \frac{1}{K_g} \qquad (12\text{-}12)$$

Equation (12-12) indicates that the velocity $\omega(t)$ is proportional to the input command. This is the fundamental property of velocity control systems.

In order to determine the dynamic behavior of the system, we need to solve the characteristic equation of the loop, which is obtained by setting the denominator of Eq. (12-10) to zero;

$$1 + \frac{A_c(s)K_tK_g}{Js} = 0 \qquad (12\text{-}13)$$

The roots of Eq. (12-13) are the closed-loop system poles. These poles determine the bandwidth and the dynamic behavior of the velocity control system.

The system velocity is influenced by the input command, v_{in}, and by the load

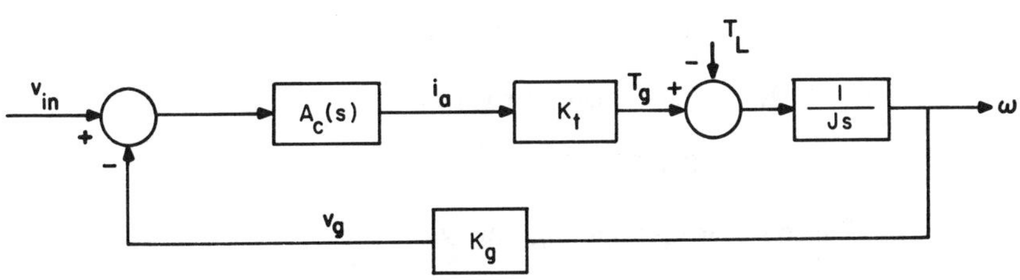

Fig. 12-4. Velocity control system with a current amplifier.

torque, T_L. The effect of the command on the velocity is described by Eqs. (12-10) and (12-12). This is the transfer function between the command signal and the velocity. We may derive the transfer function between the load torque and the velocity in a similar manner, as described below.

Torque-Velocity Transfer Function

The transfer function between the load torque, T_L, and the velocity, ω, can be found from the block diagram of Fig. 12-4. The transfer function turns out to be

$$\frac{\Omega(s)}{T_L(s)} = \frac{-1/Js}{1 + A_c(s)K_tK_g/Js} \tag{12-14}$$

Under the condition of Eq. (12-11), Eq. (12-14) becomes

$$\frac{\Omega(s)}{T_L(s)} \cong \frac{-1/Js}{A_c(s)K_tK_g/Js} = \frac{-1}{A_c(s)K_tK_g} \tag{12-15}$$

We may conclude from Eq. (12-15) that, in order to reduce the effect of T_L on the velocity, we should keep the open-loop gain, $A_c(s)K_tK_g$, as high as possible. An interesting point to be observed here is that all the zeros of the amplifier transfer function, $A_c(s)$, become the poles of the closed-loop transfer function. This has a significant effect on the system behavior in cases where the amplifier has a lag-lead compensation. In such cases the amplifier transfer function becomes

$$A_c(s) = \frac{K(s + b)}{s + a} \qquad b > a \tag{12-16}$$

The zero of the amplifier transfer function at $s = -b$ becomes a pole of the transfer function of Eq. (12-15), and since this typically occurs at low frequencies, it may result in a slow transient response to load variations.

It should be noted that the amplifier with the transfer function in Eq. (12-16) does not slow the velocity response to an input command, but only to load change. Therefore, *a step response to an input command does not indicate the system response to a load change.*

This effect is illustrated by the following example.

Example 1

Consider a velocity control system, as illustrated by Fig. 12-4, with the following parameters:

$K_t = 0.2$ Newton-meter/ampere

$K_g = 0.05$ volt/(radian/second)

$J = 2 \times 10^{-4}$ kilogram-meter^2

The amplifier transfer function is given by

$$A_c(s) = \frac{20(s + 10)}{(s + 1)} \tag{12-17}$$

The resulting transfer function between the input command and the velocity, as given by Eq. (12-12) becomes

$$\frac{\Omega(s)}{V_{in}(s)} = \frac{2 \times 10^4(s + 10)}{(s + 10.1)(s + 991)} \tag{12-18}$$

The zero at s = −10 nearly cancels the pole at s = −10.1, and the transfer function may be written as

$$\frac{\Omega(s)}{V_{in}(s)} = \frac{20}{1 + \frac{s}{991}} \tag{12-19}$$

In order to determine the step response to a step input command, let the input transform (a unit step function) be:

$$V_{in}(s) = \frac{1}{s} \tag{12-20}$$

This results in velocity of the form:

$$\Omega(s) = \frac{20}{s(1 + \frac{s}{991})} \tag{12-21}$$

The time response, $\omega(t)$, which corresponds to Eq. (12-21), is

$$\omega(t) = 20(1 - e^{-991t}) \qquad t \geqslant 0 \tag{12-22}$$

The function $\omega(t)$ is represented by curve a in Fig. 12-5.

In order to determine the step response to load changes, note that Eq. (12-15) becomes

$$\frac{\Omega(s)}{T_L(s)} = -\frac{5 \times 10^3(s + 1)}{(s + 10.1)(s + 991)} \tag{12-23}$$

Let the load change be a negative step of magnitude two,

$$T_L(s) = -\frac{2}{s} \tag{12-24}$$

Then the resulting velocity response is

$$\Omega(s) = \frac{10^4(s + 1)}{s(s + 10.1)(s + 991)} \tag{12-25}$$

Equation (12-25) may be written as the sum of partial fractions:

$$\Omega(s) = \frac{1}{s} + \frac{9.18}{s + 10.1} - \frac{10.18}{s + 991} \tag{12-26}$$

and the inverse Laplace transform of $\Omega(s)$ is

$$\omega(t) = 1 + 9.18e^{-10.1t} - 10.18e^{-991t} \qquad t \geqslant 0 \tag{12-27}$$

which is illustrated by curve b in Fig. 12-5.

A comparison between curve a and curve b of Fig. 12-5 shows that the response to the input command has a time constant of approximately 1 millisecond. On the other hand, it would take several hundreds of milliseconds for the response to a load change to decay.

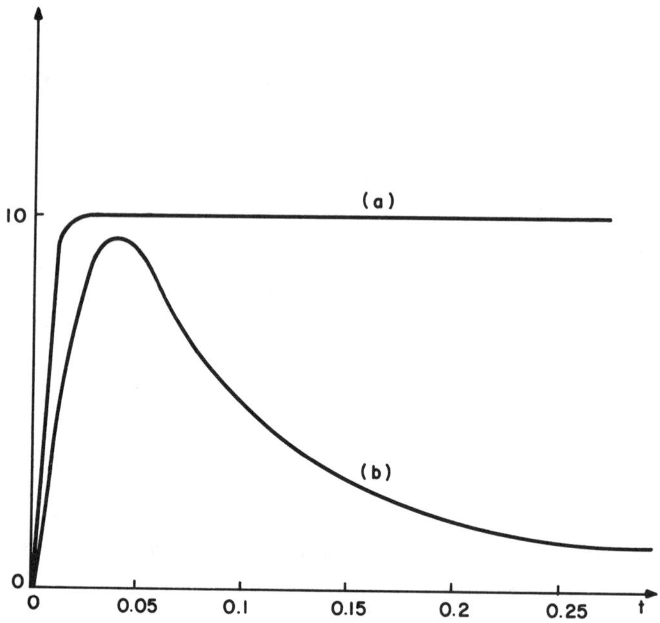

Fig. 12-5. Response of the control system. Curve a shows the response to a change in command; curve b shows the response of the same system to a load change.

Velocity Control Systems With Voltage Amplifiers

In the previous section we analyzed velocity control systems in which current amplifiers were used. Now we will consider the alternative of using voltage amplifiers.

Consider the block diagram of the system shown in Fig. 12-6. Let the transfer function of the amplifier be $A_v(s)$. Note that the motor representation in this case is more complex due to the effects of the armature inductance and the back emf voltage.

Let the transfer function of the motor be denoted by M(s). For the given block diagram, we obtain

$$M(s) = \frac{\Omega(s)}{V_{in}(s)} = \frac{\dfrac{K_t}{(sL+R)Js}}{1 + \dfrac{K_tK_e}{(sL+R)Js}} = \frac{K_t}{Js(sL+R) + K_tK_e} \tag{12-28}$$

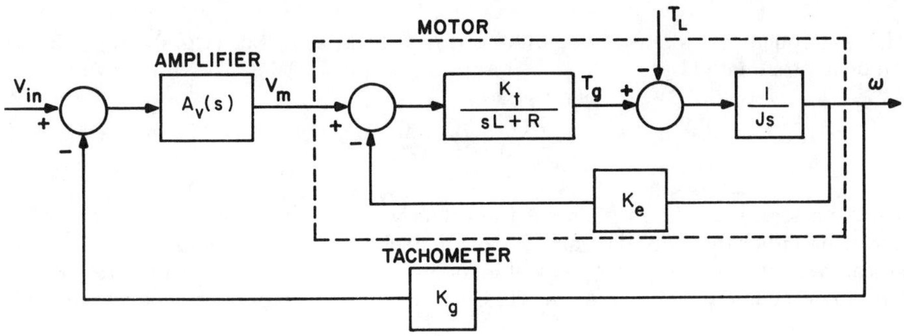

Fig. 12-6. Block diagram of velocity control system with voltage amplifier.

The transfer function between the command signal, v_{in}, and the velocity, ω, is equal to

$$\frac{\Omega(s)}{V_{in}(s)} = \frac{A_v(s)M(s)}{1 + K_gA_v(s)M(s)} \tag{12-29}$$

At low frequencies, where the open-loop gain is very high,

$$|K_gA_v(s)M(s)| \gg 1 \tag{12-30}$$

where $s = j\omega$.

Therefore, the 1 term in the denominator of Eq. (12-29) may be ignored and the system transfer function of Eq. (12-29) can be approximated by

$$\frac{\Omega(s)}{V_{in}(s)} \cong \frac{1}{K_g} \tag{12-31}$$

which is identical to the transfer function of systems with current amplifiers in Eq. (12-12).

The transfer function between the load torque, T_L, and the velocity, ω, is given by

$$\frac{\Omega(s)}{T_L(s)} = \frac{-\dfrac{1}{Js}}{1 + \dfrac{K_tK_e}{Js(sL+R)} + \dfrac{A_v(s)K_tK_g}{Js(sL+R)}} \tag{12-32}$$

Equation (12-32) may be written as

$$\frac{\Omega(s)}{T_L(s)} = \frac{-\dfrac{sL + R}{Js(sL + R) + K_eK_t}}{1 + \dfrac{A_v(s)K_gK_t}{Js(sL + R) + K_eK_t}} = \frac{-\dfrac{sL + R}{K_t}M(s)}{1 + A_v(s)K_gM(s)} \tag{12-33}$$

At low frequencies, where the open-loop gain is high, as in Eq. (12-30), Eq. (12-33) can be approximated by

$$\frac{\Omega(s)}{T_L(s)} \cong -\frac{(sL + R)}{A_v(s)K_tK_g} \tag{12-34}$$

It can be seen that the response of the velocity to load changes can be reduced by increasing either the amplifier gain, A_v, or the tachometer feedback gain, K_g. Further, we can see that the zeros of $A_v(s)$ will be poles of the system transfer function; therefore, low-frequency zeros of $A_v(s)$ will slow the velocity response to changes in load.

12.4 VELOCITY VARIATIONS

There are several factors which cause the motor velocity to diverge from the desired profile. These velocity variations are divided into two groups: the slow variations, called long-term speed variations (LTSV); and the faster changes, called instantaneous speed variations (ISV). The threshold between the two groups depends on the specific application, but a typical value for the time boundary is one minute; changes taking longer than one minute to occur are termed LTSV, and faster variations are considered ISV.

LTSV occur as a result of changes in parameter values, which are in turn caused by temperature changes, voltage drive, and component wear. In applications where long-term stability is required, it is essential that all of the parameter changes fall within the acceptable tolerance levels. If this is not possible, one must adopt an approach which is less sensitive to parameter change, such as the phase-locked servo systems described in Chapter 14.

The causes of ISV are more complex and depend largely on the specific application. In general, the main reasons for ISV are the following.

1. Response Time.

As the command changes, the system goes through a transient mode before reaching the desired velocity. This transient response is predictable and depends on the loop bandwidth and on the system stability.

2. Changes In Load Torque.

When the load torque changes, it causes a change in the motor velocity. This is an undesirable effect since we want the motor velocity to be independent of the load. The relationship between the load change and the corresponding speed variations is not well-known because of the difficulty in measuring the effect of load change. If a fast response to load changes is required, the system designer should avoid the use of lag-lead compensation networks with low-frequency zeros, because they slow the system response to load changes. This effect was discussed in Sec. 12.3.

3. Torque Ripples.

The generated torque is not exactly proportional to the current; it varies slightly as a function of the shaft angular position. These torque variations are called torque ripples, and they cause some ISV. The effect of torque ripples is more significant at low speeds because the frequency of the ripples is low. Torque ripples and their frequencies can be measured, and the resulting velocity changes can be predicted.

4. Internal Friction.

Variable friction in the motor will change the velocity in much the same way that a change in load will. The internal friction in the motor can be measured, and is specified by most motor manufacturers.

5. Noise.

Velocity changes may occur as a response to noise, disturbances, and other undesirable signals. This is especially significant in systems employing pulsewidth-modulated amplifiers or phase detectors. These undesirable signals can be measured at various points in the system; if their amplitude cannot be reduced, their frequency should be increased in order to minimize their effects.

6. Tachometer Disturbances.

The worst type of speed variations are those caused by disturbances in the tachometer, because the system is not protected against them. These disturbances are interpreted by the system as velocity error, and the loop tries to "correct" them. It is, therefore, extremely important to have a low-noise tachometer in applications where the velocity level is determined by the velocity feedback.

In the above discussion, the main causes for speed variations in velocity control systems were described. The relative significance of these and any additional factors is determined by the particular design requirements in each individual application.

REFERENCES

1. *DC Motors, Speed Controls, Servo Systems*, 3rd ed., Engineering Handbook, Electro-Craft Corporation, 1975.

2. D. T. Phan and V. H. Malakian, "The Design of a Velocity Control System Without a Tachometer—The EMF Feedback Approach," *Proceedings, Sixth Annual Symposium on Incremental Motion Control Systems and Devices*, ed., B. C. Kuo, University of Illinois, pp. 243-259, 1977.

3. M. A. Lewis, "Design Strategies for High-Performance Incremental Servos," *Proceedings, Sixth Annual Symposium on Incremental Motion Control Systems and Devices*, ed., B. C. Kuo, University of Illinois, pp. 141-151, 1977.

4. J. Tal, "Phase-Locked Servo Systems—A New Approach to Speed Control," *Proceedings, Fourth Annual Symposium on Incremental Motion Control Systems and Devices*, ed., B. C. Kuo, University of Illinois, pp. AA-1—AA-5, 1975.

5. J. Tal, "Speed Control by Phase-Locked Servo Systems—New Possibilities and Limitations," *IEEE Trans. IECI*, Vol. IECI-24, No. 1, pp. 118-125, Feb. 1977.

6. B. C. Kuo, *Automatic Control Systems*, 3rd ed., Prentice-Hall, Inc., Englewood Cliffs, N.J., 1975.

13 Position Control Systems

J. TAL

13.1 INTRODUCTION

Incremental motion systems are often required to move a given load, to stop it at a specified position, and to hold it there until a subsequent motion command is initiated. Alternatively, these systems are sometimes required to move a load along some assigned position trajectory. Both of these motion requirements can be accomplished by position control systems.

In position control systems, the angular position of the motor and, therefore, of the load which is coupled to it are both synchronized with the input command. When the command signal is a constant, the motor shaft locks to the desired position; when the command signal varies continuously, the motor position follows the command, as long as the signal variations are within the loop bandwidth.

The position sensors may be continuous elements, such as potentiometers, in which case the resulting control system is a continuous, or analog, control system. Continuous position control systems have infinite resolution, a very favorable characteristic. However, the continuous position signal is limited to a certain position range, and this limits the motion.

Another kind of position sensor is the digital type. The incremental encoders fall into this category. These devices provide an infinite range of angular position. However, they have a disadvantage in that the position resolution is finite. Also, since the position is transmitted in digital form, the encoder pulses must be counted and the count must be applied to a digital-to-analog converter before an analog position signal can be derived.

The main problem in designing position control systems is maintaining stability under adequate gain and sufficient loop bandwidth. Therefore, the stability problem and the design approach will be emphasized in the following discussion. Furthermore, since the stability of the system depends largely on whether the system uses a tachometer for velocity feedback, the following discussion will be divided into two parts: the first will deal with systems incorporating tachometers, and the second will consider tach-free systems.

13.2 POSITION CONTROL SYSTEMS WITH TACHOMETER FEEDBACK

The system under consideration is represented by the block diagram in Fig. 13-1. All of the system components are assumed to be linear and are, therefore, represented by their transfer functions. The amplifier is represented by A(s); the motor's transfer function is M(s). The tachometer gain is K_g, and the tachometer compensation is F(s). These four components form the inner loop, which is also a velocity loop. The block with the transfer function 1/s represents the physical relationship between the velocity, ω, and the position, θ. The position sensor has a gain of K_p, and the position compensation transfer function is G(s).

In order to analyze the loop, the transfer function of the inner loop is denoted by H(s). Thus,

$$H(s) = \frac{\Omega(s)}{E(s)} = \frac{A(s)M(s)}{1 + A(s)M(s)K_gF(s)} \tag{13-1}$$

The closed-loop transfer function becomes

$$\frac{\Theta(s)}{V_{in}(s)} = \frac{H(s)\,\frac{1}{s}}{1 + \frac{1}{s}\,K_pH(s)G(s)} \tag{13-2}$$

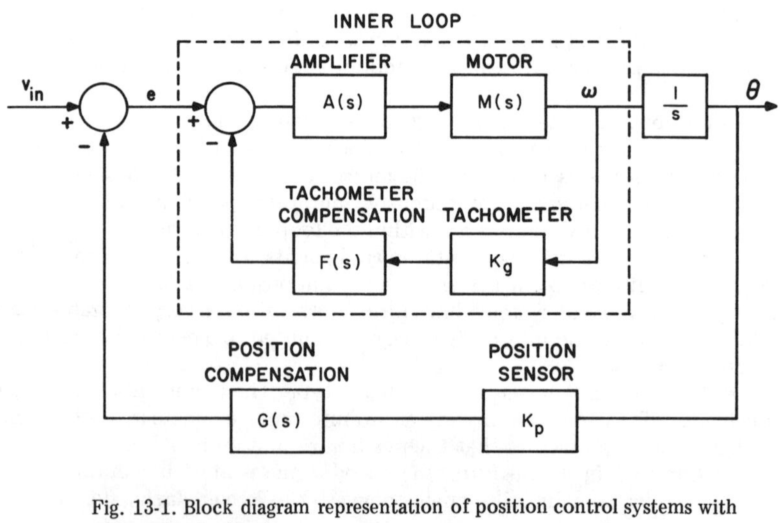

Fig. 13-1. Block diagram representation of position control systems with tachometer feedback.

The open-loop transfer function, L(s), is given by

$$L(s) = \frac{K_p H(s)G(s)}{s} \tag{13-3}$$

Upon examination of Eq. (13-3), we find that there is a pole at the origin, causing a phase lag of 90 degrees. This phase lag is the reason why it is more difficult to stabilize position control loops than velocity loops. In order to increase the phase margin and thereby improve the stability situation, we must reduce the phase lag of Eq. (13-3) as much as possible. One way to achieve this objective is to use a lead-lag compensation for G(s). Thus, let

$$G(s) = \frac{1 + s/a}{1 + s/b} \qquad a < b \tag{13-4}$$

The parameters a and b are selected to provide the desired phase lead at the critical frequency range. (For more details on this method, see [3] or [4].)

Another possible approach to the problem of increasing the phase margin is to reduce the phase lag of H(s). This can be done by increasing the gain of the inner loop. The effect of such an increase can be enhanced by the addition of a lag network in F(s). Thus,

$$F(s) = \frac{1}{1 + s/c} \tag{13-5}$$

where c is a positive constant.

When the gain of the inner loop is large, we may ignore the 1 term in the denominator of Eq. (13-1). This results in the approximation:

$$H(s) \cong \frac{1}{K_g F(s)} \tag{13-6}$$

Combining Eq. (13-6) with Eq. (13-5) yields

$$H(s) = \frac{1 + s/c}{K_g} \tag{13-7}$$

Thus, H(s) as given above provides phase lead, which increases the phase margin.

If the system is to operate as a velocity control system in a multimode capacity, the lag network in the inner loop and the increased gain could have an adverse effect on the response of the velocity loop. Therefore, we could use different compensation networks for the velocity loops in the two operation modes. Such an arrangement is illustrated by the block diagram in Fig. 13-2. The compensation $F_1(s)$ is designed for the velocity control mode, whereas the sum $F_1(s) + F_2(s)$ constitutes the effective compensation for the inner loop in the position control mode.

13.3 POSITION CONTROL SYSTEMS WITHOUT TACHOMETER FEEDBACK

Considerations of system stability make the incorporation of a tachometer a desirable feature of any position control system. However, some control systems do not include a tachometer due to economic reasons or size limitations. In these cases, the

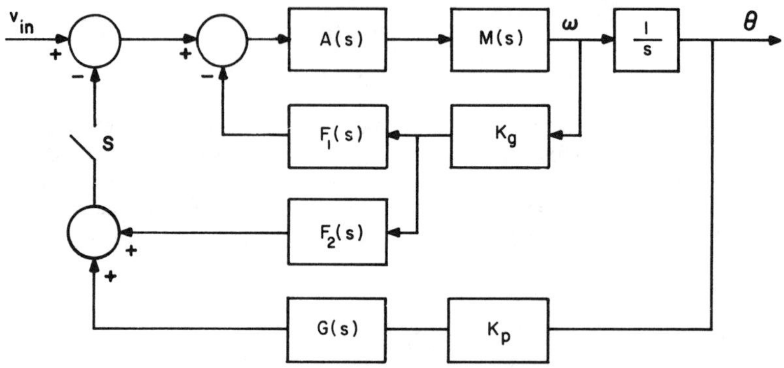

Fig. 13-2. Multimode control system when the switch S is closed, the system operates as a position control system. Opening the switch transfers the system to the velocity control mode.

stability problem is more severe, and the resulting systems have, as a general rule, a lower bandwidth.

The tachometer-free system under consideration is represented by the block diagram in Fig. 13-3. This is the same as its counterpart in Fig. 13-1, with the exception that the velocity feedback is now removed. The open-loop transfer function in this case equals

$$L(s) = \frac{A(s)M(s)K_p G(s)}{s} \tag{13-8}$$

In order to improve the loop stability, we must reduce the phase lag of L(s) as much as possible. This can be done in several ways.

1. The Use Of Voltage Amplifiers.

It is preferable to use a voltage amplifier rather than a current amplifier in position control systems because the combined transfer function of the current amplifier and the motor has a pole at the origin. This pole, when added to the pole at the origin due to the velocity integration effect, tends to make the system unstable.

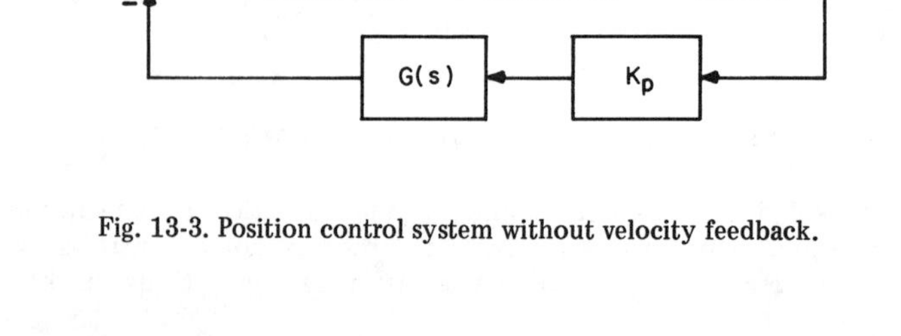

Fig. 13-3. Position control system without velocity feedback.

2. The Use Of Positive Current Feedback.

This is recommended because positive current feedback may have the same effect as that of velocity feedback, as discussed in Chapter 12. However, this approach requires a careful analysis; deviations from the desired values have adverse effects on the system stability.

3. The Use Of Compensation Networks.

A compensation network in the loop can improve the stability greatly. The most common types of networks to be used here are the lag-lead and the lead-lag compensations. While the lag-lead compensation is easier to apply, it usually results in a slower system response. The lead-lag compensation increases the system bandwidth, but also requires a careful system analysis to determine the location of the zero and the pole.

REFERENCES

1. *DC Motors, Speed Controls, Servo Systems*, 3rd ed., Engineering Handbook, Electro-Craft Corporation, 1975.

2. B. C. Kuo, *Automatic Control Systems*, 3rd ed., Prentice-Hall, Inc., Englewood Cliffs, N.J., 1975.

3. V. W. Eveleigh, *Control Systems Design*, McGraw-Hill Book Company, New York, 1972.

4. I. M. Horowitz, *Synthesis of Feedback Systems*, Academic Press, London, 1963.

14 Phase-Locked Servo Systems

J. TAL and C. K. RUSHFORTH

14.1 INTRODUCTION

Phase-locked loops (PLL) are widely used in communication systems for frequency demodulation and bit synchronization. Another popular application of the PLL is the velocity control of servo systems. These phase-locked servo systems (PLS) are used to achieve a high degree of speed regulation and accuracy. Although PLS may be designed to follow a time-varying command, we consider only constant-velocity systems in order to simplify the discussion.

A PLL in its basic form is represented by the block diagram of Fig. 14-1. The input signal and the feedback signal are sinusoidal and, upon locking, both signals will have the same frequency and a constant phase difference. Any deviation from the desired phase difference is detected by the phase detector and is transmitted to the voltage-controlled oscillator (VCO) to correct the error. Clearly, since the phase

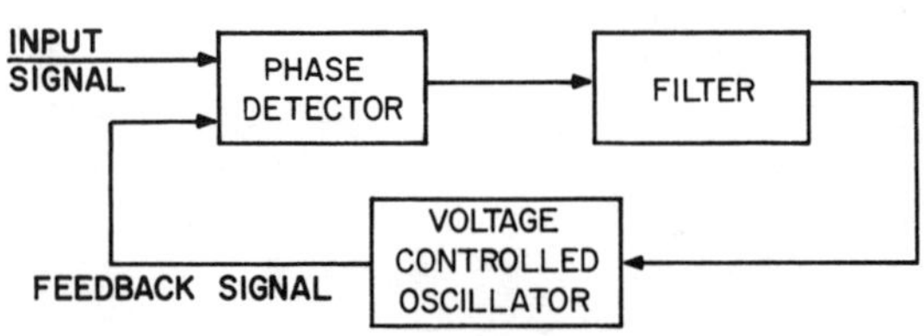

Fig. 14-1. Phase-locked loop.

difference is constant, the frequencies of the input and the feedback signals are the same. This principle can be applied to servo systems in order to regulate the motor velocity. The VCO is replaced by an amplifier, a motor, and an encoder, which produces a sinusoidal signal at a frequency proportional to the motor angular velocity. The resulting system is the phase-locked servo system (PLS) shown in Fig. 14-2.

When the system in Fig. 14-2 is locked, the reference and feedback signals have identical frequencies and a constant phase difference. Any deviation from this condition is detected, filtered, amplified, and fed back to the motor for velocity correction.

While the fundamentals of PLL and PLS are the same, there are several important differences arising out of different design requirements and system components.

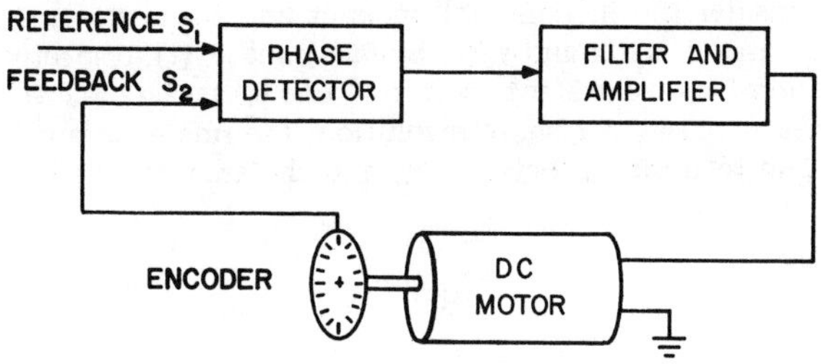

Fig. 14-2. Phase-locked servo system.

The main differences between PLL and PLS are:

1. **Frequency Range.**

The frequency of the input signal in PLL is usually high and its range is narrow. In contrast, the input frequency of PLS is much lower and its range is wide, up to 1000:1.

2. **Loop Components.**

The feedback signal is generated in the PLL by the VCO, which responds instantly to changes in command. In PLS, the generation of the feedback signal is done by the motor and an amplifier, which have considerably longer response time, thereby introducing additional lag in the loop.

3. **Noise**

The main source for noise in PLL is the input signal which is usually corrupted. In contrast, the noise in PLS is generated primarily by the loop components.

In this chapter, we discuss the analysis and design of PLS systems. We begin with a development of mathematical models for two different types of systems, one using an analog phase detector and the other using a discrete-time phase-frequency detector. These models are then used to analyze system performance, with an emphasis on stability considerations. Finally, we discuss possible solutions to some common design problems.

14.2 SYSTEM MODEL

In this section we develop two models for the PLS depicted in Fig. 14-2. As a first step, we combine all the linear continuous components, namely the filter, the amplifier, and the motor. The transfer function of this combination of components is the ratio of the Laplace transform of the motor velocity ω to the Laplace transform of the phase comparator output x:

$$F(s) = \frac{\Omega(s)}{X(s)} \tag{14-1}$$

Next we consider the incremental encoder and the waveforms $s_1(t)$ and $s_2(t)$. We denote the phases of $s_1(t)$ and $s_2(t)$ by $\theta_1(t)$ and $\theta_2(t)$, respectively. Since s_2 is a nonlinear function of θ_2, we prefer to regard θ_1 and θ_2 as the basic variables. Since the encoder produces n pulses per shaft revolution, the phase θ_2 is n times the angular shaft position. The relationship between θ_2 and the angular velocity ω is, therefore,

$$\frac{\Theta_2(s)}{\Omega(s)} = \frac{n}{s} \tag{14-2}$$

Finally we consider the phase detector (PD). We will examine two types of PD, the continuous-multipler type and the digital phase-frequency detector. We start with the continuous detector.

The continuous phase detector simply multiplies the two input signals. Let the input $s_1(t)$ and the feedback signal $s_2(t)$ be sinusoidal:

$$s_1 = V_1 \sin(\theta_1 - \tfrac{\pi}{2}) \tag{14-3}$$

$$s_2 = V_2 \sin\theta_2 \tag{14-4}$$

Then the multiplier output is

$$V_d = K_d s_1 s_2 = \frac{1}{2} K_d V_1 V_2[\sin(\theta_1 - \theta_2) + \sin(\theta_1 + \theta_2)] \tag{14-5}$$

The frequency of the second term of V_d is the sum of the frequencies of s_1 and s_2, and, therefore, is relatively high. Consequently, this component will be attenuated by the filter and the motor, and it may be ignored. The voltage V_d may, therefore, be approximated by the first term:

$$V_d = \frac{1}{2} K_d V_1 V_2 \sin(\theta_1 - \theta_2) \tag{14-6}$$

In other words, the output of the phase detector is proportional to the sine of the phase difference.

The model described by Eqs. (14-1) through (14-6) can be represented by the block diagram of Fig. 14-3. The model for the PD is nonlinear, but may be linearized for small values of $\theta_1 - \theta_2$, resulting in a linear continuous model for the PLS [6, 9].

One of the most serious drawbacks of the PLS system using a continuous phase detector is its limited acquisition range. During startup, when the encoder output frequency is significantly lower than the input frequency, the discrepancy may be so great that the system cannot respond properly in order to acquire the desired velocity. This problem is often solved by adding an acquisition circuit or by programming the command signal to increase gradually in frequency.

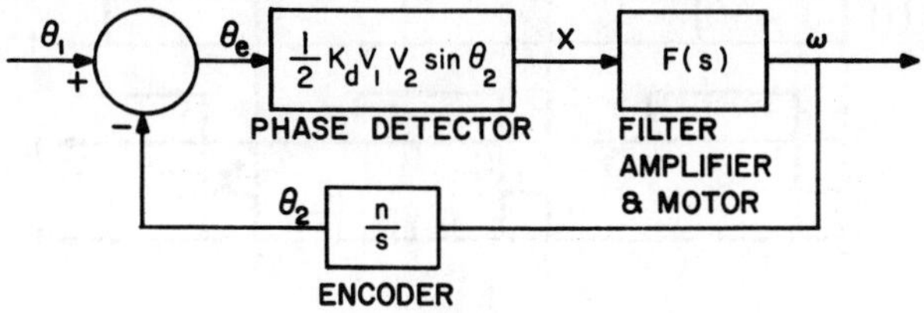

Fig. 14-3. PLS with a continuous phase detector.

An alternative solution to the acquisition problem has been provided by the development in recent years of the digital phase-frequency detector (PFD). This device is a digital circuit which changes states on the leading edges of the incoming pulses. Although several different PFDs have been developed, most of them can be described using the three-state model shown in Fig. 14-4. As seen in that figure, state transitions occur upon positive transitions in $s_1(t)$ or $s_2(t)$. When the frequencies f_1 and f_2 of $s_1(t)$ and $s_2(t)$ are equal and s_1 leads s_2, the output switches between 0 and V_s, producing a positive average output. Conversely, if s_1 lags s_2, the output switches between 0 and $-V_s$ and the average output is negative.

When $f_1 > f_2$, the steady-state output will switch between 0 and V_s, again producing a positive average value. Similarly, the average output is negative if $f_1 < f_2$. The ability of this device to sense both phase differences and frequency differences suggests the name phase-frequency detector.

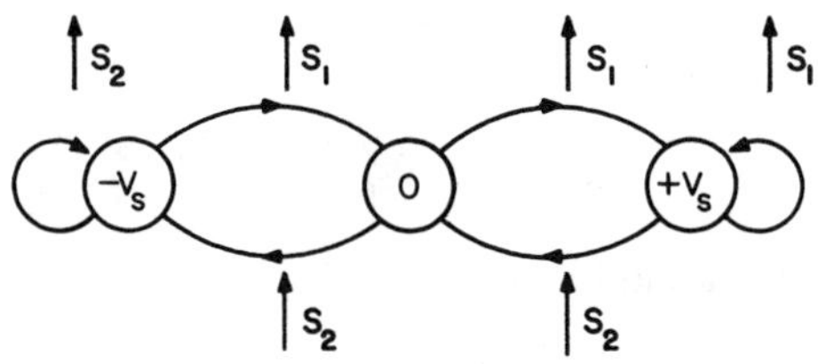

Fig. 14-4. Three-state model of PFD.

Although the state diagram of Fig. 14-4 is quite simple, it does not lead directly to a simple input-output model which can be used for stability analysis. Previous work on the development of such input-output models by Moore [3] and Sharpe[5] was limited to relating the average value of the output to the inputs. These models do not describe the dynamic behavior of the system and are, therefore, not adequate for stability analysis. We now develop a linearized input-output model of the PFD which removes this limitation.

The following model was developed by the authors [1] to describe the dynamic behavior of the PLS. Since the main objective is to examine loop stability, the model is linearized around the locking condition in which the frequencies of $s_1(t)$ and $s_2(t)$ are the same and constant, and the phase difference is constant as well.

Assume with no loss of generality that the signal $s_1(t)$ leads $s_2(t)$. The resulting output, x(t), will switch between 0 and V_s according to the state diagram. The output signal is illustrated in Fig. 14-5.

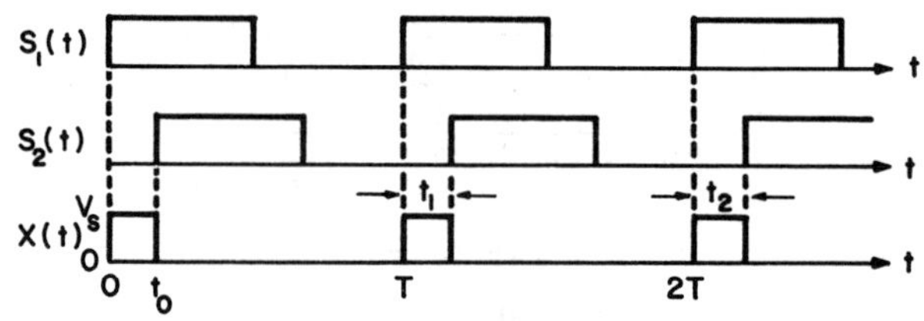

Fig. 14-5. The inputs $s_1(t)$ and $s_2(t)$ and the output x(t).

We wish to consider the effects on the output waveform x(t) of small perturbations in the relative phase of the input waveforms. As the model is a linearized one, we consider first the nominal waveforms of $s_1(t)$ and $s_2(t)$, which are denoted by $s_{1n}(t)$ and $s_{2n}(t)$, respectively. The phases of these signals are denoted by $\theta_{1n}(t)$ and $\theta_{2n}(t)$, respectively. Since the transitions in x(t) occur upon positive changes in s_1 and s_2, the phases $\theta_1(t)$ and $\theta_2(t)$ at the transition times must be multiples of 2π:

$$\theta_1(kT) = 2\pi k \tag{14-7}$$

$$\theta_2(kT + t_k) = 2\pi k \tag{14-8}$$

Define the phase difference θ_e as

$$\theta_e = \theta_1 - \theta_2 \tag{14-9}$$

and denote its nominal value by

$$\theta_{en} = \theta_{1n} - \theta_{2n} \tag{14-10}$$

The nominal value of the output is

$$x_n(t) = \begin{cases} V_s & \text{for } kT < t \leqslant kT + t_{kn} \\ 0 & \text{elsewhere} \end{cases} \tag{14-11}$$

where t_{kn} is the nominal value of the pulse duration t_k. Next assume that the value of θ_2 is varied slightly about the nominal value:

$$\theta_2 = \theta_{2n} - \delta\theta \tag{14-12}$$

This results in a change in θ_e,

$$\theta_e = \theta_{en} + \delta\theta \tag{14-13}$$

The variation also affects the time t_k and the output signal x(t):

$$t_k = t_{kn} + \delta t_k \tag{14-14}$$

and

$$x(t) = x_n(t) + \delta x(t) \tag{14-15}$$

The objective is to develop a linearized relationship between $\delta x(t)$ and $\delta\theta$.

Note that Eq. (14-8) must be satisfied for θ_{2n}:

$$\theta_{2n}(kT + t_{kn}) = 2\pi k \tag{14-16}$$

The same relation must hold for θ_2 as given by Eq. (14-12). Combining Eq. (14-8) with Eqs. (14-12) and (14-14), we obtain

$$\theta_{2n}(kT + t_{kn} + \delta t_k) - \delta\theta(kT + t_{kn} + \delta t_k) = 2\pi k \tag{14-17}$$

Since $\delta\theta$ and δt_k are small, we may expand Eq. (14-17) in a Taylor series and retain only the first-order terms to yield

$$\theta_{2n}(kT + t_{kn}) + \frac{d\theta_{2n}(kT + t_{kn})}{dt}\,\delta t_k - \delta\theta(kT + t_{kn}) = 2\pi k \tag{14-18}$$

Denoting the angular frequency of the nominal signals by

$$\frac{d\theta_{2n}(t)}{dt} = \omega(t) \tag{14-19}$$

and combining Eqs. (14-16), (14-18), and (14-19) yields

$$\omega(kT + t_{kn})\delta t_k - \delta\theta(kT + t_{kn}) = 0 \tag{14-20}$$

Since the nominal frequency is constant, we may write $\omega(kT + t_{kn}) = \omega$, and Eq. (14-20) leads to

$$\delta t_k = \frac{1}{\omega}\,\delta\theta(kT + t_{kn}) \tag{14-21}$$

The resulting change in $x(t)$ is a pulse of duration δt_k. When $\delta\theta$ is positive, $\delta x(t)$ is given by

$$\delta x(t) = \begin{cases} V_s & kT + t_{kn} < t \leqslant kT + t_{kn} + \delta t_k \\ 0 & \text{elsewhere} \end{cases} \tag{14-22}$$

Similarly, if $\delta\theta$ is negative, $\delta x(t)$ is a negative pulse.

Since δt_k is very small, we may approximate the series of pulses of Eq. (14-22) by impulses of the same area. The area A of each pulse is

$$A = V_s\delta t_k = \frac{V_s}{\omega}\,\delta\theta(kT + t_{kn}) \tag{14-23}$$

The same area will be obtained by multiplying an impulse of weight V_s/ω by $\delta\theta$. Thus $\delta x(t)$ may be represented as a product of an impulse train and $\delta\theta$:

$$\delta x(t) = \frac{V_s}{\omega} \sum_{k=0}^{\infty} \delta\theta(kT + t_{kn})\delta(t - kT - t_{kn}) \tag{14-24}$$

In other words, $\delta x(t)$ is the result of an impulse sampling of $\delta\theta$ with impulses of weight V_s/ω.

When the frequency is constant, t_{kn} is constant. By shifting the origin of time, we may eliminate t_{kn}. Furthermore, the signal frequency is related to the period by $\omega = 2\pi/T$. This leads to a simplified expression for $\delta x(t)$:

$$\delta x(t) = \frac{V_s T}{2\pi} \sum_{k=0}^{\infty} \delta\theta(kT)\delta(t - kT) \tag{14-25}$$

Equation (14-25) is the desired linearized input-output model for the digital PFD. This equation can be combined with Eqs. (14-1) and (14-2) to yield the linearized hybrid model for the phase-locked servo system shown in Fig. 14-6. We use this model in the next section to analyze the performance of the PLS.

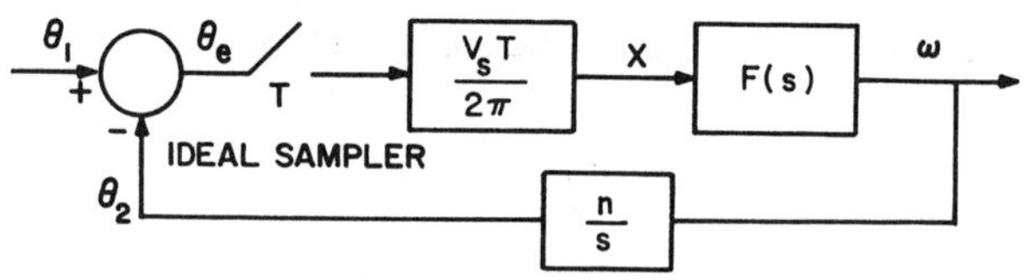

Fig. 14-6. Linearized model for the PLS.

14.3 SYSTEM ANALYSIS

In this section we use the models depicted in Fig. 14-3 and in Fig. 14-6 to analyze the stability of the PLS. Starting with the continuous system of Fig. 14-3, we note that the system is nonlinear. However, for the purpose of stability analysis, we may linearize the model by replacing the sine function with argument

$$\sin\theta \cong \theta \tag{14-26}$$

The resulting linear system can be analyzed using standard control theory (Chapter 4). The characteristic equation of the loop becomes

$$1 + \frac{nK_d V_1 V_2 F(s)}{2s} = 0 \tag{14-27}$$

The condition for stability is that all the roots of Eq. (14-27) must have negative real parts. This results in restrictions on the form of F(s) and on the values of the loop parameters.

The analysis of the discrete system is more complex as it involves determining the characteristic roots of the loop using techniques of discrete-time system analysis [10, 11]. We first combine all the continuous elements in Fig. 14-6 to yield

$$D(s) = \frac{nV_sTF(s)}{2\pi s} \tag{14-28}$$

We next denote by D(z) the discrete-time function corresponding to D(s). Specifically, D(z) is the z-transform of the sampled inverse Laplace transform of D(s). The characteristic equation of the loop is

$$1 + D(z) = 0 \tag{14-29}$$

The loop is stable if and only if all the roots of Eq. (14-29) are inside the unit circle $|z| = 1$. Furthermore, roots close to the origin result in rapid response while roots close to the unit circle result in slow response.

As an illustration of the analysis, consider a system with F(s) of the form

$$F(s) = \frac{k}{s\tau + 1} \tag{14-30}$$

Although the combined transfer function of the amplifier, filter, and motor is typically of third or fourth order, Eq. (14-30) is a valid approximation for the combination, as the mechanical time constant of the motor is often the dominant factor.

The resulting open-loop transfer function is

$$D(s) = \frac{C}{s(s\tau + 1)} \tag{14-31}$$

where

$$C = \frac{nkTV_s}{2\pi} \tag{14-32}$$

The corresponding discrete transfer function is [11]

$$D(z) = \frac{Cz}{z - 1} - \frac{Cz}{z - e^{-T/\tau}} \tag{14-33}$$

Define

$$B = e^{-T/\tau} \tag{14-34}$$

and combine Eqs. (14-33) and (14-34) with Eq. (14-29). This leads to the characteristic equation

$$z^2 + z(-1 - B - BC + C) + B = 0 \tag{14-35}$$

Equation (14-35) allows us to determine the loop behavior for different loop parameters. Loop stability may be analyzed using standard methods of stability tests for linear discrete-data systems [11]. The resulting condition for stability is

$$C < 2\frac{1 + B}{1 - B} \tag{14-36}$$

Combine Eq. (14-36) with Eq. (14-32) to obtain the following explicit condition:

$$k < \frac{4\pi(e^{T/\tau} + 1)}{nV_s T(e^{T/\tau} - 1)} \tag{14-37}$$

This may be approximated [2] by

$$k < k_{max} = \frac{8\pi\tau}{nV_s T^2} \tag{14-38}$$

Equation (14-38) sets an upper limit for T, the period of the reference signal:

$$T_{max} = \left[\frac{8\pi\tau}{nkV_s}\right]^{1/2} \quad \text{(sec)} \tag{14-39}$$

Similarly, we may describe the limit in terms of the frequency of the reference signal

$$f_{min} = \frac{1}{T_{max}} = \frac{nkV_s}{8\pi\tau} \quad \text{(Hz)} \tag{14-40}$$

This can be translated to an equivalent motor velocity

$$\Omega_{min} = \frac{f_{min}}{n} = \left[\frac{kV_s}{8\pi n\tau}\right]^{1/2} \quad \text{(rad/sec)} \tag{14-41}$$

Equation (14-41) indicates that the PLS has a lower velocity limit. If the velocity drops below this threshold, the system becomes unstable. This result is confirmed by experimental results. Note that analog velocity feedback systems, or PLS with continuous PD, do not have a lower limit for the motor velocity, and use of a discrete-time PFD adds a limitation to the design.

14.4 SYSTEM DESIGN

In this section, we review several important problems which arise in connection with the design of PLS systems. Although the design of such a system depends in detail on the problem at hand, there are certain basic considerations common to all such systems. The first of these is the system bandwidth, ω_B, which determines the speed of the response to changes in the input. The value of ω_B typically lies between 0.1 and 300 radians/second.

A second important system parameter is the range of velocities (Ω_L, Ω_H) which the system can follow, commonly referred to as the lock range. The upper and lower limits of the lock range, Ω_H and Ω_L, will be discussed separately below.

A final and most important design consideration is the speed variation which can occur during system operation. Unfortunately there is little that can be said about this problem in general, aside from the obvious fact that we want to keep speed variations as small as possible. Detailed design and performance considerations depend heavily on the particular application.

We now turn to a somewhat more detailed discussion of each of the above design considerations. This discussion applies to PLS systems containing analog multiplier phase detectors as well as to those employing the digital PFD, but since the latter is currently more common, we use the nomenclature and notation appropriate for the hybrid system.

PLS Bandwidth

In order to study the bandwidth of the loop, we need an equivalent continuous model of it. It is shown in Chapter 4 that if the sampling period T is much smaller than the time constant τ, the transfer function of the ideal sampler can be approximated by a gain of 1/T. Under these conditions, the equivalent system model is that shown in Fig. 14-7.

The characteristic equation of the system of Fig. 14-7 is

$$s(\tau s + 1) + a = 0 \tag{14-42}$$

where a, the equivalent gain, is

$$a = \frac{nV_sK}{2\pi} \tag{14-43}$$

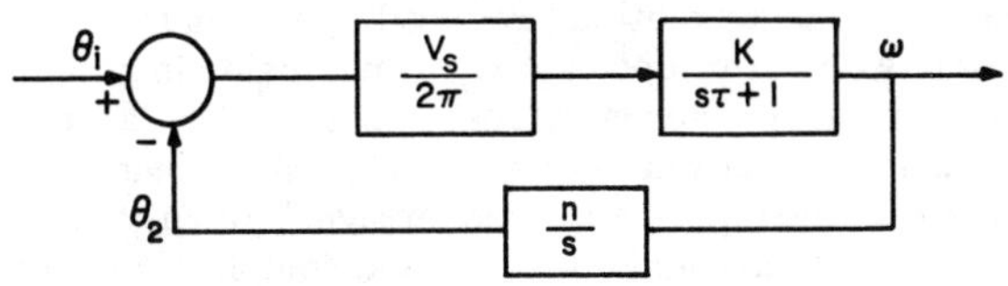

Fig. 14-7. A continuous model for PLS under the assumption $T \ll \tau$.

The roots of the characteristic equation, as obtained from Eq. (14-42), are

$$P_{1,2} = \frac{-1 \pm j\sqrt{4a\tau - 1}}{2\tau} \tag{14-44}$$

For illustrative purposes, suppose we set

$$4a\tau = 2 \tag{14-45}$$

This leads to roots with a damping ratio of 0.707

$$P_{1,2} = -\frac{1}{2\tau} \pm j\frac{1}{2\tau} \tag{14-46}$$

The corresponding bandwidth of the system is

$$\omega_B = \frac{1}{2\tau} \tag{14-47}$$

Equation (14-47) indicates the first difficulty in design. If one requires a certain bandwidth, ω_B, one must start with a motor which has the right time constant. If the motor time constant is too small, it is easy to create a larger one using a filter. Unfortunately, τ is usually too large and the problem is to reduce it. This can be done, for example, by adding velocity feedback or current feedback.

Lower Speed Limit Ω_L

It was previously shown that a PLS system will be unstable at low speeds. The design problem is to choose the parameters of the system such that it will be stable at the desired lower limit Ω_L.

Combining Eq. (14-41) with Eq. (14-43) yields the lower speed limit

$$\Omega_L = \frac{1}{n}\left[\frac{a}{4\tau}\right]^{1/2} \tag{14-48}$$

For the illustrative case of Eqs. (14-45), (14-46), and (14-47), Eq. (14-48) becomes

$$\Omega_L = \frac{\omega_B}{n\sqrt{2}} \tag{14-49}$$

This expression relates the lower speed limit to the system bandwidth and to n, the number of pulses per shaft revolution produced by the shaft encoder. It is clear from Eq. (14-49) that if we wish to decrease Ω_L while maintaining a fixed system bandwidth, we must increase n. However, cost and size place a practical limit on the achievable value of n. Typical values of n are between 500 and 1000. One can double n by sensing the zero crossings of the encoder signal, or further increase it by using separate sensors, but this also adds to the cost and complexity of the system.

Upper Speed Limit Ω_H

In order for the loop to follow speeds in the lock range, the gain must be such that when the PFD output varies between extremes, the motor speed will vary over the lock range. Specifically,

$$\Delta\Omega = \frac{V_s K}{\pi} \geqslant \Omega_H - \Omega_L \tag{14-50}$$

On the other hand, we can combine Eq. (14-50) with Eqs. (14-43), (14-45), and (14-47) to obtain

$$\Delta\Omega = \frac{2a}{n} = \frac{2\omega_B}{n} \tag{14-51}$$

Thus, once we have chosen ω_B and n, $\Delta\Omega$ is determined; it may or may not satisfy Eq. (14-50). Note that we need consider only the values of $\Delta\Omega$ and $(\Omega_H - \Omega_L)$, and not the actual velocities. The reason for this is that the speed can always be shifted by adding a bias voltage to the amplifier.

When $\Delta\Omega$ from Eq. (14-51) does not satisfy Eq. (14-50), the resulting problem of insufficient lock range can be solved in one of two ways. The first method is to divide the desired lock range into several bands, each narrower than the value of $\Delta\Omega$ given by Eq. (14-51). A band selector switch is then used to provide the necessary bias for a particular band, and the PFD can then vary the speed within that band. While this solution is relatively simple, it is not acceptable when a continuous speed variation is required. In that case, one can use the second method, which was developed independently by several authors [3, 4, 6]. This method is based on the idea of leaving F(s) as in Eq. (14-30) in the range of frequencies which affect stability, but increasing the low-frequency gain by the addition of lag-lead network G(s) of the form

$$G(s) = \frac{s + ma}{s + a} \qquad (m > 1) \tag{14-52}$$

The pole frequency, a, is selected to be much smaller than ω_B. Consequently, $G(s) \approx 1$ for frequencies near or above ω_B. The actual filter transfer function, $F_1(s)$, equals

$$F_1(s) = G(s)F(s) = \frac{s + ma}{s + a}\frac{K}{s\tau + 1} \tag{14-53}$$

Bode plots of $F(j\omega)$ and $F_1(j\omega)$ are shown in Fig. 14-8. Note that at higher frequencies, the two functions are actually the same. On the other hand, the dc gain of $F_1(s)$ is $F_1(0) = mK$, whereas $F(0) = K$. Thus, the lag-lead network allows a gain increase by a factor of m, causing a proportional increase in $\Delta\Omega$. This allows us to extend $\Delta\Omega$ as much as necessary, so the lock range is limited on the lower end by Ω_{min}, and on the higher end by the maximum allowed velocity of the motor, or by the amplifier saturation voltage.

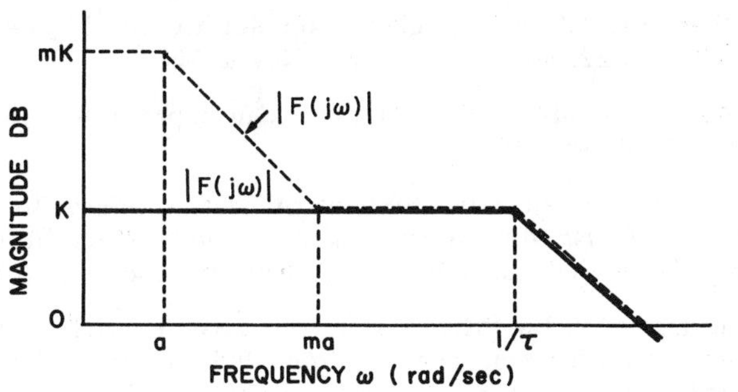

Fig. 14-8. Bode plots of the original function $F(j\omega)$, and the compensated function, $F_1(j\omega)$.

Speed Variations

It is useful to divide speed variations into two categories: long-term speed variations (LTSV), which are slow changes caused by such factors as temperature changes or component wear; and instantaneous speed variations (ISV), caused by load variations, external disturbances, or other short-term perturbations. Most of the low values of speed variation quoted in the literature (e.g., 0.1 percent) refer only to LTSV and not to ISV. It is quite possible for a system to exhibit negligible long-term variation while undergoing ISV of even 10 percent. Thus, care must be exercised in interpreting speed-variation data.

It is instructive to compare the speed variation of a PLS system with that of an analog velocity feedback system. Because of the locking behavior of the PLS, its performance is generally far superior to the analog velocity feedback system with regard to long-term variations. This favorable comparison does not extend to ISV, however. The substantially greater bandwidth of the velocity feedback system makes it more immune to ISV than is the PLS.

The ISV performance of a PLS system can be improved in one of two ways. The first and most obvious method is to eliminate insofar as possible the causes of ISV. This can be done by removing friction from the load, using a low-friction or brushless motor, and carefully designing the amplifier to attenuate noise. A second approach, which can be taken when the response time is long (several seconds), is to load the motor with a heavy mass which acts as a mechanical filter to attenuate load disturbances. The first approach is often used in tape drives, the second in turntables.

REFERENCES

1. J. Tal and C. K. Rushforth, "Speed Control Improvement by Phase-Locked Loops—Modeling Analysis and Design," *Proceedings, Joint Automatic Control Conference*, 1977.
2. J. Tal, "Speed Control by Phase-Locked Servo Systems—New Possibilities and Limitations," *IEEE Trans. IECI*, Vol. IECI-24, No. 1, pp. 118-125, Feb. 1977.
3. A. W. Moore, "Phase-Locked Loops for Motor-Speed Control," *IEEE Spectrum*, pp. 61-67, Apr. 1973.
4. D. H. Smithgall, "A Phase-Locked Loop Motor Control System," *IEEE Trans. IECI*, Vol. IECI-22, No. 4, pp. 487-490, Nov. 1975.
5. C. A. Sharpe, "A 3-State Phase Detector Can Improve Your Next PLL Design," *EDN*, pp. 55-59, Sept. 1976.
6. J. Tal, "Phase-Locked Servo Systems—A New Approach to Speed Control," *Proceedings, Fourth Annual Symposium on Incremental Motion Control Systems and Devices*, ed., B. C. Kuo, University of Illinois, pp. AA-1—AA-5, 1975.
7. D. C. Schlick and J. Tal, "High-Performance Phase-Locked Incremental Drive for Microfilming Moving Documents," *Proceedings, Sixth Annual Symposium on Incremental Motion Control Systems and Devices*, ed., B. C. Kuo, University of Illinois, pp. 237-242, 1977.
8. *DC Motors, Speed Controls, Servo Systems*, 3rd ed., Engineering Handbook, Electro-Craft Corporation, 1975.
9. A. J. Viterbi, *Principles of Coherent Communication*, McGraw-Hill Book Company, New York, 1966.
10. J. P. Cadzow and H. R. Martens, *Discrete-Time and Computer Control Systems*, Prentice-Hall, Inc., Englewood Cliffs, N.J., 1970.
11. B. C. Kuo, *Analysis and Synthesis of Sampled-Data Control Systems*, Prentice-Hall, Inc., Englewood Cliffs, N.J., 1963.

15 Optimal Design Of Incremental Motion Servo Systems

J. TAL

15.1 INTRODUCTION

In recent years there has been a growing recognition of the need for improved performance in incremental motion systems. Devices capable of faster and better performance are built every year, and the trend toward optimal performance continues. As a consequence, merely "adequate" systems are no longer satisfactory. The system designer is continually pressed to obtain the maximum performance from given components in his search for the best system possible. Unfortunately, the best—or the optimal—design cannot always be implemented due to various practical considerations; but even so, the study of the optimal design can still prove valuable for two main reasons: first, such a study indicates how much can be achieved, and, therefore, provides a basis against which nonoptimal solutions can be compared; it is important for the designer to know whether the optimal design will improve the performance by 40 percent or by 2 percent. Secondly, an analysis of the optimal design points out the direction for improvement. Even if this direction cannot be followed all the way to the optimal design, it can still be followed part of the way, providing at least a degree of increased performance.

It is assumed in the following discussion that a dc motor is used as an actuator, and that our goal is to achieve the best design. There is no general agreement on what the "best" performance is: one designer may want to maximize the repetition rate, f, while another may want to minimize the motion time, τ, or reduce the power dissipation. However, since the major limitation on the design is due to temperature rise in the motor resulting from power dissipation, all three design objectives mentioned above are equivalent to the objective of performing the required motion while minimizing the motor temperature. Therefore, the minimization of the motor temperature has been chosen as our design objective. In addition to the temperature minimization, the designer may be required to conform to some additional constraints such as limits on current and voltage. In these cases, the optimization solution must be modified accordingly.

In order to keep the discussion as general as possible, we will consider systems with both linear and rotary motion. We will also discuss geared drive and direct drive, with a special emphasis on capstan drives.

We will begin our analysis by stating the design objective as an optimization problem. This is put forward in Sec. 15.2, where it is shown that minimizing the motor temperature is equivalent to minimizing the power dissipation. The optimization problem itself is divided into two parts: the selection of components, and the control selection problem. Since, as it turns out, it is easier to apprehend the analysis if we deal with the control selection first and the component selection second, that is the sequence we shall follow. Therefore, in Sec. 15.3 we will discuss the problem of optimal and suboptimal controls; the component selection problem will be considered in Sec. 15.4.

15.2 DESIGN OBJECTIVES

The requirements of angular motion are to rotate a load, which consists of some inertia J_L and a constant opposing torque T_L, by an angle θ_L, and to come to a halt at time τ. The stepping may be repeated f times every second. The analogous requirements for linear motion are to move a load of mass M and a constant opposing force F along a distance d, and to stop at time τ. Here, also, the repetition rate is f. The design objective in each case is to perform the motion while minimizing the motor temperature.

The objective includes the selection of the system components as well as the selection of the control method minimizing the motor temperature. The system components to be selected are the motor and the coupling which connects it to the load. In the case of rotary motion, the coupling is accomplished by a gear train, and the problem is to determine the gear ratio. If the required motion is linear, the coupling can be achieved with a capstan, a belt-pulley drive, or a lead screw. The various couplings are shown in Figs. 15-1, 15-2, 15-3, and 15-4.

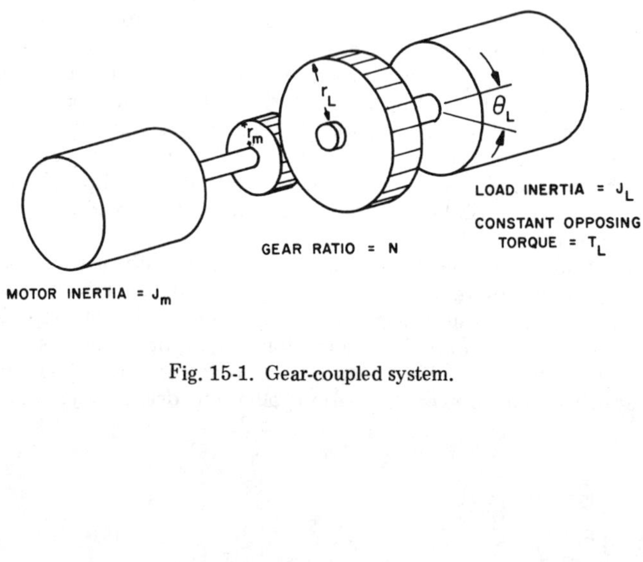

Fig. 15-1. Gear-coupled system.

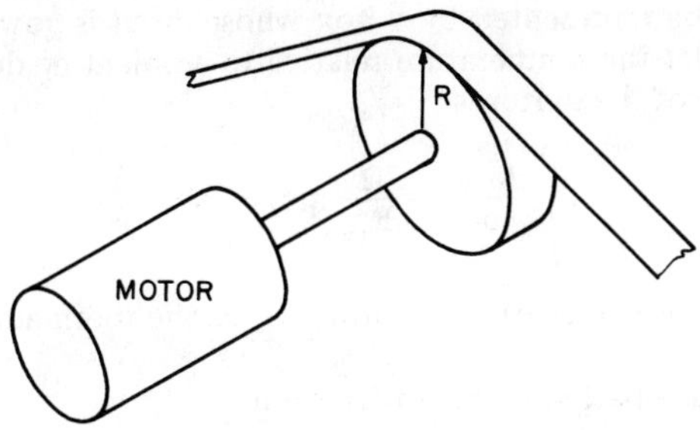

Fig. 15-2. Capstan drive system.

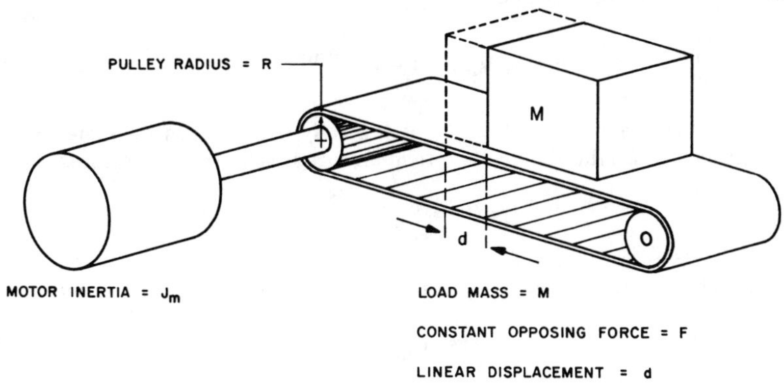

Fig. 15-3. Belt-pulley drive system.

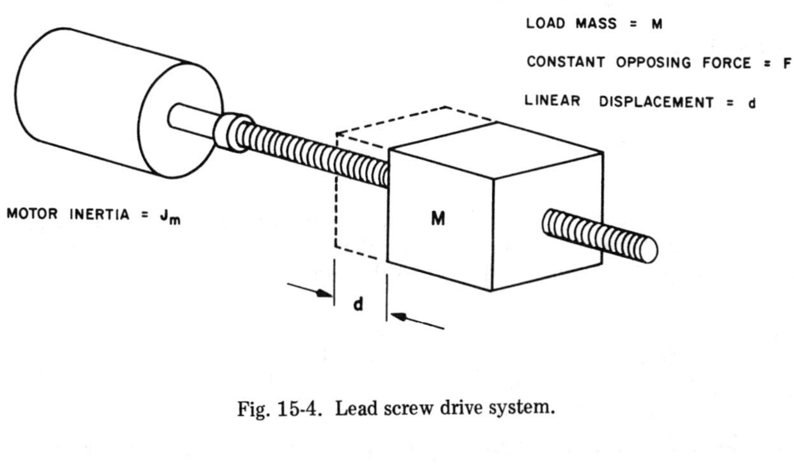

Fig. 15-4. Lead screw drive system.

Let the motor be represented by a box whose input is power and whose output is temperature, and let the temperature relative to ambient be denoted by q(t). Then the thermal equation of the motor is

$$C_h \frac{dq(t)}{dt} + \frac{1}{R_{th}} q(t) = P(t) \tag{15-1}$$

where C_h is the heat capacity of the motor, R_{th} is the thermal resistance, and P(t) is the power dissipation.

Let the power dissipation be a step function:

$$P(t) = \begin{cases} P_0 & t \geqslant 0 \\ 0 & t < 0 \end{cases} \tag{15-2}$$

Then the solution of Eq. (15-1) is

$$q(t) = R_{th} P_0 (1 - e^{-t/R_{th} C_h}) \tag{15-3}$$

Since the thermal time constant $R_{th}C_h$ is typically shorter than the system run time, it can be assumed that the steady-state temperature is achieved. Therefore,

$$q(t) = R_{th} P_0 \tag{15-4}$$

If we denote the energy dissipation per step by E_n, then Eq. (15-4) becomes

$$q(t) = R_{th} E_n f \tag{15-5}$$

where f is the repetition rate.

Having described the design objective, we turn now to the control selection problem.

15.3 CONTROL SELECTION

System Equations

Let the motor current be i(t). Then the motor torque $T_m(t)$ is given by

$$T_m(t) = K_t i(t) \tag{15-6}$$

where K_t is the torque constant. Let the motor angular position and velocity be denoted by $x_1(t)$ and $x_2(t)$, respectively. Also, let the total moment of inertia and the constant opposing torque, reflected to the motor, be J and T, respectively. Then the equation of motion is

$$T_m(t) = J \frac{dx_2(t)}{dt} + T \tag{15-7}$$

Equation (15-7) can be combined with Eq. (15-6) and rewritten in the form

$$\frac{dx_1(t)}{dt} = x_2(t)$$
$$\frac{dx_2(t)}{dt} = \frac{1}{J}[K_t i(t) - T] \tag{15-8}$$

The motion requirements are:

$$x_1(0) = 0 \qquad x_1(\tau) = \theta$$
$$x_2(0) = 0 \qquad x_2(\tau) = 0 \tag{15-9}$$

where τ is the time at the end of one step.

We now need to study the effect of the constant torque, T, on the energy dissipation. The current i can be found from Eq. (15-8)

$$i(t) = \frac{1}{K_t}[J\frac{dx_2(t)}{dt} + T] \tag{15-10}$$

If we define

$$u(t) = \frac{J}{K_t}\frac{dx_2(t)}{dt} = i(t) - \frac{T}{K_t} \tag{15-11}$$

then the energy dissipation per step, E_n, becomes

$$E_n = \int_0^\tau R_a i^2(t)dt = R_a\int_0^\tau u^2(t)dt + \frac{2R_aT}{K_t}\int_0^\tau u(t)dt + \frac{R_aT^2\tau}{K_t^2} \tag{15-12}$$

Note that u(t) is proportional to the motor acceleration, as seen from Eq. (15-11). Therefore, the integral of u(t) is proportional to the velocity change $[x_2(\tau) - x_2(0)]$. However, the velocity change is zero, as given by Eq. (15-9); therefore, the second integral in Eq. (15-12) is zero. As a consequence, the expression for E_n is simplified to the form:

$$E_n = R_a\int_0^\tau u^2(t)dt + \frac{R_aT^2\tau}{K_t^2} \tag{15-13}$$

It is clear from Eq. (15-13) that the effect of the constant torque is to add to E_n some fixed amount which is independent of the velocity trajectory. Therefore, the minimization of E_n is equivalent to minimizing the following integral:

$$C(u) = R_a\int_0^\tau u^2(t)dt \tag{15-14}$$

In the following discussion, the optimal control to minimize C(u) is shown. Later, the use of nonoptimal control will be discussed.

Optimal Control

Our objective is to find a piecewise-continuous control function u(t) which satisfies Eqs. (15-8) and (15-9) while minimizing the cost functional C(u). Such a control function is called "optimal" and is denoted by u*(t). This optimization problem was first solved by E. A. Rozenman [2]. It may be solved using the Minimum Principle [1] which is formulated as follows.

Define the Hamiltonian as

$$\mathcal{H}(x,\lambda,u) = \lambda_1(t)x_2(t) + \lambda_2(t)\frac{K_t}{J}u(t) + R_a u^2(t) \tag{15-15}$$

Then the necessary conditions for the optimal control, u*(t), and the trajectory, x*(t), as stated in reference [1], are:

$$\frac{dx^*(t)}{dt} = \left.\frac{\partial \mathcal{H}}{\partial \lambda}\right|_* \tag{15-16}$$

$$\frac{d\lambda^*(t)}{dt} = -\left.\frac{\partial \mathcal{H}}{\partial x}\right|_* \tag{15-17}$$

$$\mathcal{H}(x^*,\lambda^*,u^*) \leqslant \mathcal{H}(x,\lambda,u) \tag{15-18}$$

where x and λ denote the state vector and the costate vector, respectively. For the present problem, x and λ are 2 × 1 vectors. The boundary conditions of Eqs. (15-16) and (15-17) are given in Eq. (15-19).

Substitution of the Hamiltonian $\mathcal{H}$ into Eq. (15-16) yields

$$\frac{dx_1^*(t)}{dt} = x_2^*(t) \tag{15-19}$$

$$\frac{dx_2^*(t)}{dt} = \frac{K_t}{J}u^*(t) \tag{15-20}$$

which are the state equations in Eq. (15-8) under the optimal condition.

Equation (15-17) leads to

$$\frac{d\lambda_1^*(t)}{dt} = 0 \tag{15-21}$$

$$\frac{d\lambda_2^*(t)}{dt} = -\lambda_1^*(t) \tag{15-22}$$

which are the costate equations.

Equation (15-18) implies that the Hamiltonian is minimum along the optimal trajectory. Thus, $\mathcal{H}$ must satisfy:

$$\frac{\partial \mathcal{H}(x^*,\lambda^*,u^*)}{\partial u^*(t)} = 0 \tag{15-23}$$

which leads to

$$u^*(t) = -\frac{K_t}{2R_aJ}\lambda_2^*(t) \tag{15-24}$$

which is the optimal control.

Solution of Eqs. (15-21) and (15-22) yields

$$\lambda_1^*(t) = \lambda_1^*(0) \tag{15-25}$$

$$\lambda_2^*(t) = -\lambda_1^*(0)t + \lambda_2^*(0) \tag{15-26}$$

Thus,

$$u^*(t) = -\frac{K_t}{2R_aJ}[-\lambda_1^*(0)t + \lambda_2^*(0)] \tag{15-27}$$

In order to determine $\lambda_1^*(0)$ and $\lambda_2^*(0)$, we substitute Eq. (15-27) into Eq. (15-20). Now Eqs. (15-19) and (15-20) are solved using the boundary conditions in Eq. (15-9). The results are:

$$x_1^*(t) = \frac{K_t^2}{2J^2R_a}[\lambda_1^*(0)\frac{t^3}{6} - \lambda_2^*(0)\frac{t^2}{2}] \tag{15-28}$$

$$x_2^*(t) = \frac{K_t^2}{2J^2R_a}[\lambda_1^*(0)\frac{t^2}{2} - \lambda_2^*(0)t] \tag{15-29}$$

Thus,

$$\lambda_1^*(0) = \frac{-24\theta J^2R_a}{K_t^2\tau^3} \tag{15-30}$$

$$\lambda_2^*(0) = \frac{-12\theta J^2R_a}{K_t^2\tau^2} \tag{15-31}$$

Substituting the last two equations into Eq. (15-27), we get

$$u^*(t) = \frac{6\theta J}{K_t\tau^3}(\tau - 2t) \tag{15-32}$$

The optimal trajectories of the system are described by

$$x_1^*(t) = \frac{\theta}{\tau^3}(3\tau - 2t)t^2 \tag{15-33}$$

$$x_2^*(t) = \frac{6\theta}{\tau^3}(\tau - t)t \tag{15-34}$$

The resulting cost function $C(u^*)$ is

$$C(u^*) = R_a \int_0^\tau [u^*(t)]^2 dt = \frac{12R_a\theta^2J^2}{K_t^2\tau^3} \tag{15-35}$$

and the total energy dissipation per step, E_n, given by Eq. (15-13) becomes

$$E_n = \frac{12R_a\theta^2J^2}{K_t^2\tau^3} + \frac{R_aT^2\tau}{K_t^2} \tag{15-36}$$

The optimal current and velocity are illustrated graphically in Fig. 15-5.

The optimal control proposed by Eq. (15-32) is not a very practical one because it is difficult to implement. As a consequence, the system designer may want to use some suboptimal control functions which are easier to implement. However, such utilization may result in an increase in power dissipation. Suboptimal control functions are described below, as are problems associated with the resultant increase in power dissipation.

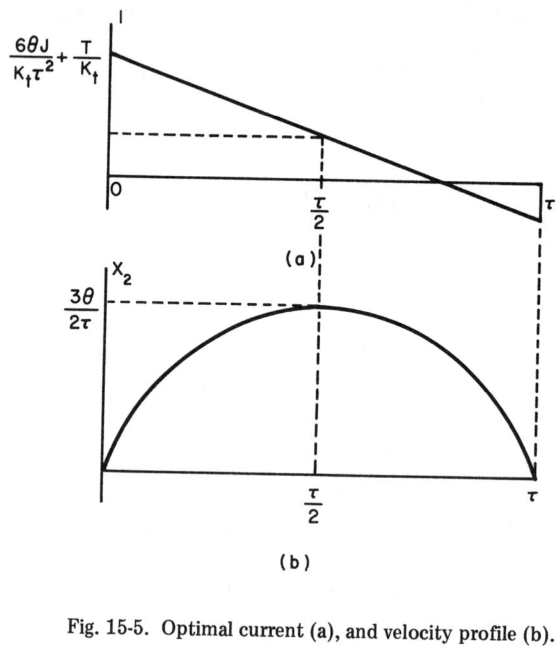

Fig. 15-5. Optimal current (a), and velocity profile (b).

Nonoptimal Controls

Let the nonoptimal velocity trajectory be $\omega(t)$ and let its symmetric and asymmetric parts be defined as:

$$\omega_s = \frac{1}{2}[\omega(t) + \omega(\tau - t)] \tag{15-37}$$

$$\omega_a = \frac{1}{2}[\omega(t) - \omega(\tau - t)] \tag{15-38}$$

It is now clear that

$$\omega = \omega_s + \omega_a \tag{15-39}$$

It also can be seen that

$$\int_0^{\tau} \omega_a(t)dt = 0 \tag{15-40}$$

Therefore,

$$\int_0^{\tau} \omega(t)dt = \int_0^{\tau} \omega_s(t)dt \tag{15-41}$$

Thus, the total displacement is not altered if the velocity is substituted by its symmetric part. The effect of such a substitution on the cost functional will now be examined.

Let the cost functional which corresponds to ω_s be C_s. Then, from Eqs. (15-11) and (15-14),

$$C_s(u) = R_a \int_0^{\tau} \left[\frac{J}{K_t} \dot{\omega}_s(t)\right]^2 dt$$

$$= C(u) - \frac{R_a J^2}{K_t^2} \int_0^{\tau} \dot{\omega}_a^2(t)dt \leqslant C(u) \tag{15-42}$$

Since $C_s(u)$ is always less than or equal to $C(u)$, we are led to the conclusion that substituting a velocity by its symmetric part reduces the energy dissipation. Therefore, a symmetric velocity profile should be used whenever possible.

We will now turn our attention to the selection of specific velocity trajectories.

Triangular Velocity: The velocity trajectory used to drive the load is often triangular. Such a velocity is shown in Fig. 15-6.

Two questions are to be considered here: what is the value of ξ, which minimizes the cost function; and what is the resulting cost?

From the principle of symmetry given above, it is obvious that the cost is minimized if the velocity is symmetric and if $\xi = 1/2$. For that value, the velocity is

$$\omega = \begin{cases} \dfrac{4\theta}{\tau^2}\, t & 0 \leqslant t \leqslant \dfrac{\tau}{2} \\[2ex] \dfrac{4\theta}{\tau^2}\, (\tau - t) & \dfrac{\tau}{2} \leqslant t \leqslant \tau \end{cases} \tag{15-43}$$

and the control is

$$u = \begin{cases} \dfrac{4\theta J}{K_t \tau^2} & 0 \leqslant t \leqslant \dfrac{\tau}{2} \\[2ex] \dfrac{-4\theta J}{K_t \tau^2} & \dfrac{\tau}{2} < t < \tau \end{cases} \tag{15-44}$$

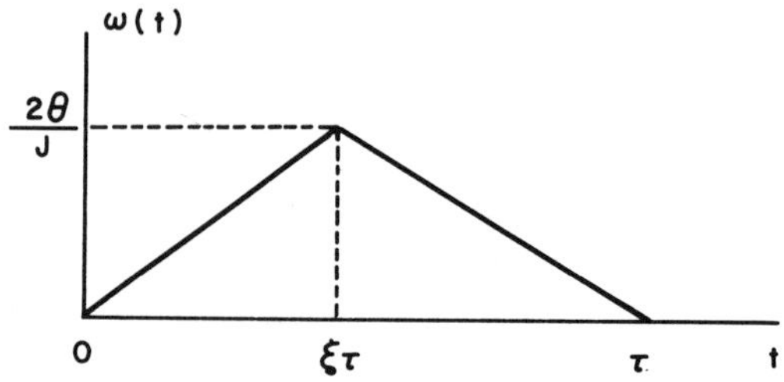

Fig. 15-6. Triangular velocity profile.

The corresponding cost functional becomes

$$C(u) = R_a \left(\frac{4\theta J}{K_t \tau^2}\right)^2 = \frac{16 R_a \theta^2 J^2}{K_t^2 \tau^3} \tag{15-45}$$

The efficiency η of a nonoptimal control is defined by

$$\eta = \frac{C(u^*)}{C(u)} = \frac{12 R_a \theta^2 J^2}{K_t^2 \tau^3 C(u)} \tag{15-46}$$

Then, for triangular velocity, the efficiency is

$$\eta = \frac{3}{4} \tag{15-47}$$

Trapezoidal Velocity: In some systems, a trapezoidal velocity profile, as shown in Fig. 15-7, is preferred. Here we must consider two questions: what are the best values of β and δ in order to minimize C(u); and what is the corresponding efficiency?

It follows from the principle of symmetry that the acceleration and the deceleration times must be equal; hence,

$$\delta = 1 - 2\beta \tag{15-48}$$

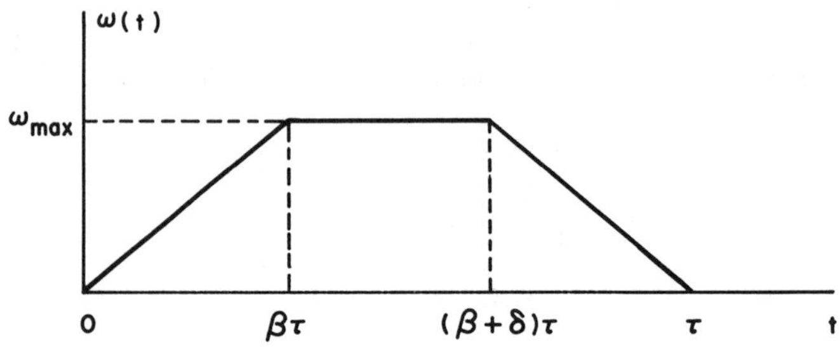

Fig. 15-7. Trapezoidal velocity profile.

This reduces the problem to the selection of β. Since the total displacement should be θ, the maximum velocity is

$$\omega_{max} = \frac{\theta}{\tau(1-\beta)} \tag{15-49}$$

The velocity is given by

$$\omega = \begin{cases} \dfrac{\theta}{\beta(1-\beta)\tau^2}\, t & 0 \leqslant t \leqslant \beta\tau \\ \dfrac{\theta}{(1-\beta)\tau} & \beta\tau \leqslant t \leqslant (1-\beta)\tau \\ \dfrac{\theta}{\beta(1-\beta)\tau^2}\,(\tau - t) & (1-\beta)\tau \leqslant t \leqslant \tau \end{cases} \tag{15-50}$$

and the corresponding dissipation is

$$C(u) = R_a \int_0^\tau u^2 dt = \frac{R_a J^2}{K_t^2} \int_0^\tau \dot{\omega}^2 dt$$

$$= \frac{2R_a J^2}{K_t^2} \left[\frac{\theta}{\beta(1-\beta)\tau^2} \right]^2 \beta\tau$$

$$= \frac{2R_a \theta^2 J^2}{K_t^2 \tau^3} \frac{1}{\beta(1-\beta)^2} \tag{15-51}$$

Minimizing C(u) in Eq. (15-51) is equivalent to maximizing the term:

$$D(\beta) = \beta(1-\beta)^2 \tag{15-52}$$

In order to accomplish this, we require

$$\frac{dD}{d\beta} = (1-\beta)^2 - 2\beta(1-\beta)$$

$$= (1-\beta)(1-3\beta) = 0 \tag{15-53}$$

$D(\beta)$ attains its maximum at $\beta = 1/3$ where $D = 4/27$, and the cost functional becomes

$$C(u) = \frac{13.5 R_a \theta^2 J^2}{K_t^2 \tau^3} \tag{15-54}$$

The corresponding efficiency of the trapezoidal velocity is given by

$$\eta = \frac{8}{9} \tag{15-55}$$

It is evident that if a trapezoidal velocity profile is utilized instead of a triangular one, the efficiency is increased considerably.

The energy dissipation per step, when a nonoptimal control is used, can be found by combining Eqs. (15-13) and (15-14) with Eq. (15-46). Thus,

$$E_n = \frac{12R_a\theta^2J^2}{K_t^2\tau^3\eta} + \frac{R_aT^2\tau}{K_t^2} \tag{15-56}$$

This concludes the discussion of control selection. The selection of components for the system will now be considered.

15.4 COMPONENT SELECTION

Geared Drive

Once the control function has been determined, the system components can be selected. In the following discussion, the gear ratio and the motor are selected for an indirect drive system. It is assumed that the gear to be selected is an ideal one; i.e., that there is no backlash or friction, and its moment of inertia is zero. This is obviously an unrealistic assumption, but it can serve as a first-order approximation. Such a gear-coupled system is illustrated in Fig. 15-1.

Define the angular position and velocity of the load by y_1 and y_2, respectively. Also, let the tangential force between the two disks at the point of contact be F. Then the torque generated in the motor, T_m, equals

$$T_m = J_m\dot{x}_2 + Fr_m \tag{15-57}$$

and the dynamic equation of the load is

$$Fr_L = J_L\dot{y}_2 + T_L \tag{15-58}$$

where J_L and T_L represent the load inertia and the constant opposing torque, respectively. The linear displacement of the two disks at the point of contact must be the same; therefore,

$$r_m\dot{x}_2 = r_L\dot{y}_2 \tag{15-59}$$

If we define the gear ratio, N, as

$$N = \frac{r_L}{r_m} \tag{15-60}$$

Then Eq. (15-59) becomes

$$\dot{x}_2 = N\dot{y}_2 \tag{15-61}$$

We can then combine Eq. (15-57) with Eq. (15-58) to obtain

$$T_m = \left(J_m + \frac{1}{N^2}J_L\right)\dot{x}_2 + \frac{1}{N}T_L \tag{15-62}$$

Equation (15-62) is equivalent to Eq. (15-7) if the following relation is defined:

$$J = J_m + \frac{1}{N^2}J_L$$

$$T = \frac{1}{N} T_L \tag{15-63}$$

If we now suppose that the load is to be rotated by an angle θ_L in time τ, then the corresponding rotation of the motor is

$$\theta = \int_0^\tau x_2 dt = \int_0^\tau N y_2 dt = N\theta_L \tag{15-64}$$

Equations (15-63) and (15-64) establish the equivalence between the real problem, in which the requirements are given in terms of the load, and the reflected problem, in which all the variables are viewed from the motor end. The control selection for the reflected system was discussed in the previous section. When Eq. (15-13) is combined with Eq. (15-14) and Eq. (15-46), it follows that the total energy dissipation per step is

$$E_n = \frac{12R_a\theta^2J^2}{K_t^2\tau^3\eta} + \frac{R_aT^2\tau}{K_t^2} \tag{15-65}$$

The corresponding motor temperature, q, is given by Eq. (15-5):

$$q = fE_nR_{th} = \frac{fR_aR_{th}}{K_t^2}\left(\frac{12\theta^2J^2}{\tau^3\eta} + T^2\tau\right) \tag{15-66}$$

In order to express the temperature in terms of the load requirements, we substitute Eqs. (15-63) and (15-64) into Eq. (15-66). This results in

$$q = \frac{fR_aR_{th}}{K_t^2}\left[\frac{12N^2\theta_L^2\left(J_m + \frac{1}{N^2}J_L\right)^2}{\tau^3\eta} + \frac{T_L^2\tau}{N^2}\right] \tag{15-67}$$

The problem to be discussed now is the selection of the gear ratio N and the motor in order to minimize q. Define the opposing torque factor, a, as

$$a = \frac{T_L^2\tau^4\eta}{12\theta_L^2J_L^2} \tag{15-68}$$

Also define:

$$D = (1 + a)^{1/2} \tag{15-69}$$

Then Eq. (15-67) becomes

$$q = \frac{12fR_a\theta_L^2R_{th}}{K_t^2\tau^3\eta}\left[N^2\left(J_m + \frac{J_L}{N^2}\right)^2 + \frac{aJ_L^2}{N^2}\right] \tag{15-70}$$

The optimal gear ratio, N*, which minimizes the temperature, can be determined from

$$N^{*2} = \frac{J_L}{J_m}(1 + a)^{1/2} = \frac{J_L D}{J_m} \tag{15-71}$$

Note that if $T_L = 0$, then $a = 0$, and the gear ratio becomes

$$N^{*2}\Big|_{T_L=0} = \frac{J_L}{J_m} \tag{15-72}$$

This condition is known as inertial match because the load inertia being reflected to the motor equals the motor inertia.

In order to select the motor, substitute Eq. (15-71) in Eq. (15-70). This results in

$$q\Big|_{N=N*} = \frac{24fR_a\theta_L^2 J_L J_m(1 + D)R_{th}}{K_t^2\tau^3\eta} \tag{15-73}$$

The motor dependent parameters in Eq. (15-73) can be combined into one term. This term, which is given by

$$\frac{R_a J_m R_{th}}{K_t^2} \rightarrow \min \tag{15-74}$$

must be minimized by the motor selection. Thus, the quantity in Eq. (15-74) should be examined for all available motors to determine the best motor selection.

Capstan Drive

The motion requirements for a capstan drive are to advance a tape along a linear distance d in time τ and to come to a halt. This motion may then be repeated at a frequency f. The tape is coupled to the motor by a capstan as shown in Fig. 15-2.

The capstan inertia, J_c, depends on the specific design. Generally, it can be described by

$$J_c = K_c R^p \tag{15-75}$$

where R is the capstan radius, and p is a constant in the range between 3 and 4.

The total inertia reflected to the motor in this case becomes

$$J = J_m + J_c = J_m + K_c R^p \tag{15-76}$$

and the motor rotation, θ, is

$$\theta = \frac{d}{R} \tag{15-77}$$

The friction level in capstan drives is usually low, and one can assume

$$T = 0 \tag{15-78}$$

The problem to be discussed here is the selection of the correct capstan radius and the best motor in order to minimize the motor temperature.

When Eqs. (15-76) and (15-77) are substituted in Eq. (15-66), the motor temperature becomes

$$q = \frac{12fR_a(\frac{d}{R})^2 (J_m + K_c R^p)^2 R_{th}}{K_t^2 \tau^3 \eta} \tag{15-79}$$

If we define

$$h = \frac{J_m}{R} + K_c R^{p-1} \tag{15-80}$$

then

$$q = \frac{12fR_a d^2 R_{th}}{K_t^2 \tau^3 \eta} h^2 \tag{15-81}$$

The selection of R affects q through the term h^2. Since h is positive, the minimization of q subject to the radius R is equivalent to minimizing h. To accomplish this, let us require that

$$\frac{dh}{dR} = \frac{-J_m}{R^2} + (p-1)K_c R^{p-2} = 0 \tag{15-82}$$

Then

$$J_m = (p-1)K_c R^p = (p-1)J_c \tag{15-83}$$

Equation (15-83) gives the relation between the inertia of the motor and that of the capstan. Clearly this is not an inertial match as the motor inertia is approximately three times larger than the capstan inertia. The optimal radius, R*, can be found from Eqs. (15-75) and (15-83):

$$R^* = \left[\frac{J_m}{K_c(p-1)}\right]^{1/p} \tag{15-84}$$

When the optimal radius is used, the motor temperature, Eq. (15-79), becomes

$$\begin{aligned} q(R^*) &= \frac{12fR_a d^2 J_m^2 R_{th}}{K_t^2 \tau^3 \eta R^{*2}} \left(1 + \frac{1}{p-1}\right)^2 \\ &= \frac{12fd^2}{\tau^3 \eta} \frac{R_{th} R_a J_m^{2-2/p}}{K_t^2} \left(\frac{p}{p-1}\right)^2 [K_c(p-1)]^{2/p} \end{aligned} \tag{15-85}$$

The selection criterion for the motor follows immediately from Eq. (15-85). The motor dependent term to be minimized is:

$$\frac{R_a J_m^{2-2/p} R_{th}}{K_t^2} \to \min \tag{15-86}$$

Note that the power of J_m is larger than one, and for $p = 4$ it becomes 1.5. This indicates that the motor inertia plays a more important role in capstan drives than in geared drives. This is one of the reasons why the motors used for capstan drive are usually smaller than those used for geared drive.

Belt-Pulley Drive

The motion requirement for a belt-pulley drive is to advance a given load, which consists of a mass M and a constant opposing force F, along a distance d, and to stop at time τ. The motion is then repeated f times every second. A typical drive system is shown in Fig. 15-3. Here again, the coupling is assumed to be ideal with no inertia or friction.

Since the pulley radius is R, the required rotation of the motor is

$$\theta = \frac{d}{R} \tag{15-87}$$

and the reflected load is given by

$$J = J_m + MR^2 \tag{15-88}$$

$$T = FR \tag{15-89}$$

The objective is to select the pulley radius R and a motor which most effectively minimizes q. Substitute Eqs. (15-87) and (15-88) into Eqs. (15-56) and (15-5). This yields

$$q = \frac{12R_a f R_{th} d^2}{K_t^2 \tau^3 \eta R^2}(J_m + MR^2)^2 + \frac{f R_a R_{th} F^2 R^2 \tau}{K_t^2} \tag{15-90}$$

Define the opposing force factor, σ, as

$$\sigma = \frac{F^2 \tau^4 \eta}{12 d^2 M^2} \tag{15-91}$$

and define

$$G = (1 + \sigma)^{1/2} \tag{15-92}$$

Then Eq. (15-90) becomes

$$q = \frac{12R_a f R_{th} d^2 M^2}{K_t^2 \tau^3 \eta}\left[\frac{1}{R^2}\left(\frac{J_m}{M} + R^2\right)^2 + \sigma R^2\right] \tag{15-93}$$

Note that Eq. (15-93) has the same structure as Eq. (15-70). The optimal pulley radius, R*, which minimizes the temperature is given by

$$R^{*2} = \frac{J_m}{M(1 + \sigma)^{1/2}} = \frac{J_m}{MG} \tag{15-94}$$

When the opposing force is zero, $\sigma = 0$, and the optimal radius becomes

$$R^{*2}\Big|_{F=0} = \frac{J_m}{M} \tag{15-95}$$

which corresponds to inertial match.

The temperature rise under optimal radius can be found by substituting Eq. (15-94) into Eq. (15-93). This results in

$$q\Big|_{R=R^*} = \frac{24R_a fR_{th} d^2 MJ_m (G + 1)}{K_t^2 \tau^3 \eta} \tag{15-96}$$

The motor dependent variables in Eq. (15-96) are the same as in Eq. (15-73), and therefore the motor selection for this case is identical to that of geared drive.

Lead Screw Drive

Lead screw is another form of translating rotary motion into linear motion. The following discussion assumes the screw to be free of inertia and friction. A diagram of a lead screw drive is shown in Fig. 15-4.

The motion requirement in this case is identical to that for the belt drive. The objective is to select the optimal screw pitch, p (which is the number of rotations per unit length), and the best motor in order to minimize the motor temperature.

Note that the linear velocity of the load, V, is proportional to the angular velocity of the motor, ω:

$$\omega = 2\pi pV \tag{15-97}$$

Thus, the lead screw is equivalent to a pulley-belt drive of radius

$$R = \frac{1}{2\pi p} \tag{15-98}$$

and therefore the results of the previous section on the belt-pulley drive can be applied here.

Thus, the optimal pitch is given by

$$p^* = \frac{1}{2\pi}\left(\frac{MG}{J_m}\right)^{1/2} \tag{15-99}$$

and the criterion for selecting the motor is the same.

15.5 CONCLUSIONS

In the analysis of the control selection, the optimal control for minimizing energy dissipation was determined. It was found that under optimal control, the resulting velocity is parabolic. However, such a control may not be practical because of the complexity of the controller; therefore, some nonoptimal controls were considered.

It was found that if the velocity profile is not symmetric with respect to the middle of the motion time, $\tau/2$, then replacing that velocity by its symmetric part will not affect the rotation angle, but will reduce the energy dissipation. The efficiency of nonoptimal controls was defined as the ratio between the energy dissipation under optimal controls and under suboptimal controls. As an example, it was shown how a triangular velocity profile (Bang-Bang Control) corresponds to an efficiency of 0.75, whereas a slight change to trapezoidal velocity increases the efficiency to 0.89.

Finally, the selection of components for the system was considered. This discussion covered cases where the coupling is accomplished by gear trains, capstans, belts, and lead screws. The optimal coupling was determined in each case, and selection criteria for the most suitable motor were developed. The results on component selection are summarized in Table I.

REFERENCES

1. M. Athen and P. L. Falb, *Optimal Control,* McGraw-Hill Book Company, New York, 1966.
2. E. A. Rozenman, "On Optimum Transient Response in Power-Saturating Systems" (in Russian), *Avtomatika, Telemekhanika,* Vol. 18, p. 6, 1957.
3. J. Tal and E. K. Persson, "Incremental Tape Transports," *Digital Design,* Vol. 2, No. 1, pp. 42-45, Jan. 1972.
4. J. Tal and S. R. Wilhelm, "Velocity Profile and Coupling Ratio Optimization in Incremental Motion Control Systems," Automation Control Equipment Conference, France, Feb. 1972.
5. J. Tal and S. J. Kahne, "Control and Component Selection for Incremental Motion Systems," *Preprints of the IFAC Fifth World Congress,* Paris, France, June 1972. Also in *Automatica,* Vol. 9, pp. 501-507, 1973.
6. H. Waagen, "Incremental Motor Selection," *Electromechanical Design,* Vol. 15, No. 3, pp. 10-13, 32-33, March 1971.
7. E. Masterson, "High Speed Incremental Motions Generated by Moving Coil Motors," ASME Design Engineering Conference Paper No. 67-DE-10, 1967.
8. E. P. Grenda, "Graphical Performance Criteria for Moving Coil Motors," ASME Paper No. 69-DE-31, 1969.
9. *DC Motors, Speed Controls, Servo Systems,* 3rd ed., Engineering Handbook, Electro-Craft Corporation, 1975.

Table I. Optimal Selection Of System Components.

System Type	Gear Drive	Pulley-Belt Drive	Lead Screw Drive	Capstan Drive
Step size	θ_L	d	d	d
Coupling parameter	Gear ratio N	Pulley radius R	Screw pitch p Define $R = \frac{1}{2\pi p}$	Capstan radius R
Load	Inertial load J_L Opposing torque T_L	Mass M Constant opposing force F		Capstan inertia $J_c = K_c R^p$ where $3 \leqslant p \leqslant 4$
Total reflected inertia and torque	$J = J_m + \frac{1}{N^2} J_L$ $T = T_L/N$	$J = J_m + MR^2$ $T = FR$		$J = J_m + K_c R^p$ $T = 0$
Reflected rotation	$\theta = N\theta_L$	$\theta = d/R$		$\theta = d/R$
Opposing torque or force factor	$a = \frac{T_L^2 \tau^4 \eta}{12\theta_L^2 J_L^2}$ $D = (1 + a)^{1/2}$	$\sigma = \frac{F^2 \tau^4 \eta}{12 d^2 M^2}$ $G = (1 + \sigma)^{1/2}$		None
Optimal coupling	$N^2 = \frac{J_L D}{J_m}$	$R^2 = \frac{J_m}{MG}$	$p^2 = \frac{1}{(2\pi)^2} \frac{MG}{J_m}$	$R^p = \frac{J_m}{K_c(p-1)}$
Motor selection criterion, minimize	$\frac{R_a R_{th} J_m}{K_t^2}$			$\frac{R_a R_{th} J_m^{2-2/p}}{K_t^2}$

16 Microprocessor Programming And Application To DC Motor Control

S. H. POLLACK

16.1 INTRODUCTION

In the early days of computer programming, a computer user had no choice but to learn a "machine code," the binary string of 1's and 0's that a computer understands. Not only was this tedious, but every time a new line of computers was developed the user had to learn a new set of instructions specific to that machine. In order to help alleviate this problem, higher-level machine-independent languages such as FORTRAN were developed. The user of languages like FORTRAN could concentrate his efforts on problem solving, without worrying about which machine his program would eventually run on. In this approach to computer programming, only a small number of people, the FORTRAN compiler writers and systems programmers, for example, had to learn the detailed architecture and machine code of a particular computer. As a result, many electrical engineers have not taken the time to learn the basic principles of assembly-language programming.

The advent of the microprocessor in the early 1970's has rekindled the desire of a large group of engineers to learn assembly language. While most engineers today are familiar with at least one higher-level language, they often find this knowledge does not readily enable them to learn assembly language. Assembly-language programming is so different from FORTRAN programming that the engineer trying to teach himself how to use a microcomputer can become quite frustrated.

Therefore, it is the purpose of this chapter to offer a complete self-teaching guide to assembly-language programming. Throughout this chapter it will be assumed that the reader has a knowledge of some higher-level programming language. Specific examples will be taken from FORTRAN. No other knowledge about computers or computer programming will be assumed.

In order to have a specific assembly language to talk about in the examples, the Intel 8080 language will be used throughout this chapter. However, every attempt will be made to keep the discussion general so that the reader will be easily able to learn any assembly language, after having learned the principles discussed in this section.

16.2 BASIC PROGRAMMING CONCEPTS

In many respects the earliest computers resembled the modern pocket calculator. Instructions were entered one at a time through a keyboard and only data were placed in the memory. To perform an "ADD" instruction, for example, the user would enter the two operands into some memory locations and push the "ADD" key on his keyboard. The computer would perform the operation at a speed governed by electronics, but would then wait idly until the user entered the next instruction. Obviously, the machine spends almost all of its time waiting for the next command, and very little time is actually spent on processing data.

The key idea behind all modern computer programming is the concept of the *stored program*. The idea is simply to put the computer program in the memory as well as putting the data in the memory. To see the advantage of this, suppose it is desired to add two numbers X and Y and then take the square of the result. Assuming X and Y have been entered into the memory previously, we must give several sequential commands: Add X and Y, save the result in some designated memory location, multiply the result by itself, and finally, save the answer. If each of these commands are activated by separate keys on a keyboard, it will take four keystrokes to carry out our program, with the computer waiting idly between operations while we search for the next key to push. If instead, the commands are somehow stored in the memory, then as soon as the machine finishes executing the first "ADD" instruction it can look up in the memory and determine the next instructions. It can then execute the "SAVE" instruction, then execute the "SQUARE" instruction, and then execute the command to save the result; that is, once the commands get into the memory, the entire program can be rapidly and automatically executed.

The stored program idea leads to the introduction of two other programming concepts. The first is the concept of *conventions*. If we are going to store instructions in the memory, and if the memory consists of a string of 1's and 0's, then we must make up numerical codes for each instruction and each piece of hardware and make use of these numerical conventions when writing the binary version of a program. For example, suppose we want to write some result on one of four teletypes. We can number the teletypes 0, 1, 2, 3. We can say that the binary pattern "11010011, $n_1 n_2$" means "write a result to device $n_1 n_2$." Then when the machine encounters the binary string "11010011, 10" the string can be decoded to mean "write a result to device #2" (since "10" is "2" in binary). We see that by use of prearranged conventions we can numerically code instructions and devices. Thus, the instructions can be placed in memory, and the machine can be wired to decode the instruction and perform the correct task.

The last basic notion of computer programming is the concept of *sequential processing*. The idea is simply that successive instructions will be stored next to each other, sequentially, in the memory. Thus, if we know what instruction was just executed, we know where to look for the next instruction: in the next memory location. This of course implies that all memory locations are numbered, so that we know where the "next" location is. So if we have just finished executing an "ADD" instruction located in numbered memory location 200, we should read the contents of location 201 to find out what to do next.

To summarize, the rapid processing of a computer program is possible because of the stored program; that is, the program as well as the data are placed in the memory. This implies conventions for numbering the instructions and numbering the hardware devices in some code. After finishing one instruction, the next numbered memory location must be inspected to find out what to do next, since sequential processing is used.

To actually implement the above schemes we will need certain pieces of electronics, the computer hardware. In order to understand assembly language, it is necessary to understand the building blocks of a computer that implement the stored program-sequential processing ideas. Thus, we must learn about computer architecture.

16.3 COMPUTER ARCHITECTURE

This section is by no means the definitive word on computer architecture. Every machine has its own special features which must be understood to program the machine most efficiently. However, certain pieces of hardware are common to most computers. The purpose of this section is to familiarize the reader with the hardware that most directly relates to the software itself. In learning about certain basic hardware devices we also learn precisely what the software instructions are controlling. Thus, studying computer architecture is a necessary prelude to studying assembly-language programming.

Memory

One hardware device common to all computers is the memory. The smallest unit of storage is the bit, which can be put in one of two states. Because it can be put into two different states, the bit can store a logical (binary) 1 or 0. The two "states" can be as simple as two different voltage outputs; i.e., if the memory is displaying 5 volts it is in the logical 1 state, and if it is tied to ground it is in the logical 0 state. The bit itself, however, is too small a number to be considered the basic unit of *information.* Bits are strung together to form larger, more useful units. Typically, 8 bits are placed together to form a "byte." Several of these 8-bit bytes may be placed together to form a "word." Words may be almost any bit length, from 4 bits to 128 bits or more. The distinction between a bit and a byte is somewhat like that between a letter of the English alphabet and a word in English. While a letter may be said to be the smallest unit that can be written down, it really has no meaning in itself until it is strung together with other letters to form the actual basic unit of information, the English word.

On the Intel 8080 microprocessor, the memory is organized in 8-bit bytes. Each byte is numbered to distinguish it from all other bytes. The 8080 is so designed that up to 64K of bytes can be attached to the machine. ("1K bytes" is actually 1024 or 2^{10} bytes, so 64K bytes represents 65,536 or 2^{16} bytes.)

The individual numbering of the memory locations is common to all machines. This is called *addressing.* If we imagine the bytes stacked up on top of each other, the address of a particular byte is the position of the byte in the stack. It is important to clearly distinguish between the numerical address and the numerical *contents* of that address. The address just locates a particular byte out of the many total bytes in the machine; the contents of that address can be any number that will fit in 8 bits. For example, in Fig. 16-1 memory address 2 contains the number 49.

It is also important to remember that the bit length of the byte (or word) is not necessarily related to the bit length of the address. The bit length of the address is

ADDRESS	CONTENTS OF ADDRESS
000	16
001	122
002	49
003	0028
004	14
005	106
⋮	⋮

Fig. 16-1. Schematic view of memory organized as a stack of individually numbered bytes.

solely determined by the maximum number of individual bytes the machine must be capable of distinguishing. Thus, if we desire a microprocessor to have a memory capacity of up to 65,536 bytes, we need a 16-bit address (since $2^{16} = 65{,}536$). If we had fewer than 16 bits in which to specify an address, we could not individually number all 65,536 bytes. (As a side point, it should be noted that the numbering goes from 0 to 65,535 and not from 1 to 65,536.) The actual byte (or word) length, on the other hand, could be any bit length at all; that is, each of the 65,536 bytes could be 8 bits long, or 64 bits long, or 1000 bits long, etc. Typically, however, the byte length is some integral multiple of the address length. Thus, a machine designed with a 64-bit address often has a byte length of 8, 16, 32, or 64 bits. We will shortly see why it is convenient to do this, although it is not necessary.

Central Processor Unit (CPU)

The central processor unit (CPU) is the piece of electronics in which data processing and control take place. For example, inside the CPU two operands may be added together, data may be transferred into the CPU from some input device, or the CPU may transmit a character to be printed on a terminal.

The CPU is responsible for centrally controlling almost the entire computer. It will generate most of the key timing pulses and control signals which the external devices will need to interact properly with the computer. The finite-state machine logic which will direct the computer to perform its tasks in an orderly sequential fashion are part of the CPU, as shown here. In short, the CPU directs the computer's activity. When one speaks of purchasing a microprocessor chip, that device is a CPU on a single chip, although now there are microprocessor chips that include memory as well.

As mentioned above, the computer control functions are derived from a *finite state machine* (FSM) driven by clock pulses. The FSM will direct the processor in accord with the stored program-sequential processing ideas discussed earlier. When the computer is first turned on, the FSM knows that the first instruction is stored in the memory, and so it generates the correct signals to retrieve the instruction. Next the FSM will place that instruction in some decoder so the particular instruction specified will become known to the processor, and so on. The complete sequence of events that take place in a step-by-step manner as the computer operates will be discussed in great detail after more hardware that is involved in the processing has been discussed.

Similarly, the data processing functions of the CPU are performed internally by hardware units which themselves have names. For example, it was said that two operands may be added together in the CPU. While this is correct, a computer scientist would probably prefer to say that the control logic of the CPU passes the two operands to an "adding" device (the ALU, to be discussed shortly) which performs the addition. Again, it is difficult to talk about the detailed data processing until certain other pieces of hardware have been discussed.

We, therefore, will begin presenting the individually named units that make up the CPU. As we build up a knowledge of what is inside the CPU, the overall operation of a computer will become clearer. Finally, in Sec. 16.4 the detailed events involved in the execution of a program by the CPU will be covered.

Arithmetic And Logic Unit (ALU)

The first hardware device that is a part of the CPU that will be discussed is the Arithmetic and Logic Unit (ALU). The ALU is the device which is wired to perform all the arithmetic and logical operations that the computer will be capable of performing using hardware (other operations may be software simulated). It is the job of the ALU to perform addition of two operands, subtract them, take the bit-for-bit logical "AND" of two operands*, etc. Data are handed to the ALU by the CPU control circuitry. The ALU performs the arithmetic or logic operation requested by the CPU controller and returns the answer to some prearranged place in the CPU.

Most microprocessor ALU's are fairly simple. They often cannot perform such common operations as multiplication or division. However, through the use of software, operations such as multiplication can be simulated (by repeated addition, for example). The 8080 ALU can add and subtract, and take logical operations such as OR, AND, NOT, and EXCLUSIVE-OR. The 8080 ALU cannot multiply or divide.

One final function performed by the ALU is to report interpretive information about results. For example, if the ALU subtracts B from A and the result is greater than zero, it may be useful to have the ALU report this fact. Similarly, it may be useful for the ALU to report if some result of an operation is less than zero, or exactly zero, or if the sum of two numbers overflowed the 8 bits of a byte and carried over to a ninth bit. Exactly what gets reported depends on the particular machine, but all machines require the ALU to be able to supply such interpretive data about results of past operations. This will be discussed in more detail in Sec. 16.4 when the operations of FLAGS are discussed.

Registers

Registers are special memory locations located inside the CPU. On the 8080, some registers are meant to hold a single byte of data and so are 8 bits long. Other registers are meant to store addresses and so must be 16 bits long. Because registers are physically a part of the CPU, they can be directly wired to the control logic, ALU, and other internal CPU devices. On some machines data must be first transferred from addressable memory to a register before it can be operated on. Even on machines which can operate directly on data in the core memory, it always is faster to execute an instruction in which all the operands are already in the registers. This is simply because in order to pull an operand out of the core memory the CPU must access the memory by first specifying which address the operand is in, wait for the memory to decode the address, and finally have the right memory location supply the operand to the CPU. On the other hand, an operand in a register is immediately available to the processor. Thus, register-using instructions are faster to execute.

For this reason, and more importantly for the sake of simplicity, we initially assume that each operand must be in a register before it can be further processed. For

*The following is meant by the "AND" of two N-bit operands: the first bit of operand 1 is ANDED with the first bit of operand 2, and the result is the first bit of the answer. Similarly, the jth bit of operand 1 is ANDED with the jth bit of operand 2 and the result is the jth bit of the answer.

example, to add the contents of two memory locations, we first execute instructions which bring the contents of these locations into two different registers. Now we can add the contents of these registers together and store the result in some other register for future use or transfer the result back to some memory location.

Many machines have a special register called the ACCUMULATOR, or "result" register. The ACCUMULATOR (AC) is special in two respects. On some machines, one of the operands to be processed in an instruction must be in the AC. On these machines we could not, for example, directly add the contents of register B and register C. We could only add the contents of the AC to one of the other registers. Thus, to perform the above addition, we must first move the contents of register C to the AC, then add the AC to register B. The second way in which the AC is a special register is that the results of most operations are placed automatically in the AC. Up to now we have not said where the result goes. Note that if we add the AC to register B we could, by convention, place the result in the AC, in register B, in a third register not yet specified, or place the result in different registers depending on the instruction specification. On machines with an ACCUMULATOR this problem is usually solved by the convention: the result will go into the AC. So when we add the content of the AC to that of register B, the sum is placed in the AC, thus erasing the operand that was in the AC with the new sum. Note that the content of register B is not erased if this convention is used. Again, for the sake of simplicity, we assume initially that this is how the 8080 works. The 8080 does indeed have an ACCUMULATOR, but the above rules are not always strictly followed by every instruction. We will relax the assumptions as we become more familiar with the 8080 and the assembly language.

The final assumption is that in order to transfer data between core memory and the CPU, the data must first go through the ACCUMULATOR. Thus, we assume that instructions are available to load the ACCUMULATOR with the contents of any memory location and to store the number in the ACCUMULATOR into any memory location. We assume that no other register is wired to do this. Once the information from the memory is in the AC, it can be freely moved about between the AC and any other register, or, in fact, between any two registers.

To see the format of some of these "data transfer" instructions, let us examine how certain "MOVE" instructions will be specifically written on the 8080. Suppose we wish to transfer the number in register B to register C. This transfer will erase the old contents in register C, but by convention will not erase the number in B. (Thus, the effect of MOVE is to make a copy of the content of register B in register C.) We must code the operation "move the content of register B to register C" in binary so the machine will understand it; that is, the CPU must ultimately read some memory byte which should be the binary equivalent of "move B to C." Just as every memory location had its own address to distinguish it from all other locations, each instruction the machine can perform must have a unique bit pattern associated with it so as to distinguish that instruction from all other instructions. In binary form, the agreed upon code for "move B to C" may look like "01001000," as it does in the 8080 code. Since it is cumbersome to write out the phrase, "move B to C," and it is confusing to us to write "01001000," we will abbreviate the instruction "move B to C" so that it can be written out quickly. To specify that the contents of register R1 should be moved to register R2, we will write "MOV R2,R1." Thus, the word "MOVE" gets shortened to "MOV," the *source* register (the register supplying the information to be transferred) will be denoted as the second operand, and the *destination* register (the register where the information is being moved to) will be denoted first. To say "move B to C" in our agreed upon shorthand, we would write "MOV C,B." Of course, the machine can only understand "01001000." Writing "MOV C,B" is a convenience only for the user.

We make up similar shorthand notation for all the other instructions. To move the contents of memory location 50 into the ACCUMULATOR, we write "LDA 50." The "LDA" stands for "load the ACCUMULATOR," while the 50 specifies which memory location's contents are to be loaded into the AC. Note that we are *not* loading the number 50 into the AC, but rather whatever is in location 50. This is the distinction between the address and the contents of the address. For our example, "50" is an address with unspecified contents. To store the number which is in the AC into location 225, we write "STA 225." The "STA" stands for "store the ACCUMULATOR" while the "225" specifies into which memory location the contents of the AC should be stored.

Suppose we wish to move the contents of location 8 into register B. Then by convention we must first move the contents of 8 into the AC, since it was assumed earlier that all data transfer between the memory and the CPU must take place through the ACCUMULATOR. We, therefore, have to execute two instructions consecutively, "LDA 8" and "MOV B,A." The first instruction loads the AC (also called "register A" on the 8080) with the contents of location 8. The final instruction transfers the data in the AC to register B, as desired. To store the contents of register C into location 36 we must execute the two instructions: "MOV A,C" and "STA 36." The reasoning behind the choice of these two instructions should be clear.

The 8080 has six general-purpose registers. They are denoted as registers B, C, D, E, H, and L. The 8080 also has the special ACCUMULATOR register, denoted as register A, which functions almost exactly as just described.

There are other special registers as well. We will discuss the program counter (PC) and stack pointer (SP) registers in separate sections. We will also discuss some special uses of the 8080 general-purpose registers H and L (standing for "HIGH" and "LOW," respectively) in other sections.

Buses

The block diagram of a computer system is shown in Fig. 16-2. From the figure we see that a computer can be thought of in block form as consisting of a memory which stores programs and data, input/output devices which enable the computer to interact and receive information from the outside world, and a CPU which controls all operations and can process data. The three blocks are interconnected by various *buses*. A bus is a set of wires, grouped together because of the similarity of their functions, which run to all the functional blocks of a computer system.

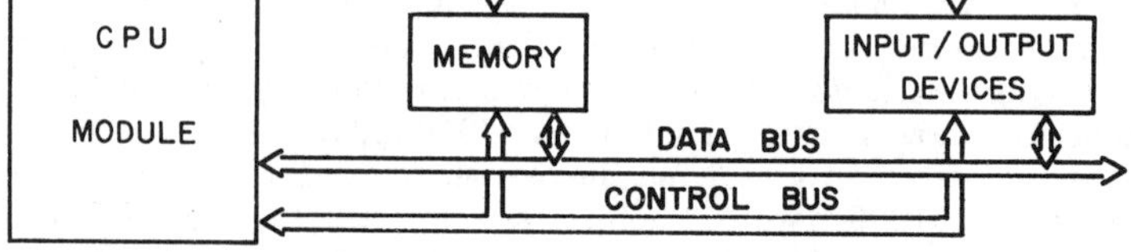

Fig. 16-2. Block diagram of a computer system.

To see how the bus structure works, consider the steps involved in retrieving a data byte from the memory and bringing it into the CPU. Suppose we are executing the instruction "LDA 50." The CPU will probably need to inform the memory and the I/O devices that it wishes to perform a memory read. It will send out certain control signals on wires that are part of the *Control Bus*. Since the Control Bus runs to the memory and all the I/O devices, the CPU can place the control signals on just this one set of wires. Each device separately decodes the information on the bus and independently determines what it is supposed to do. For our example, the I/O devices would decide to do nothing, while memory would be notified that it is soon to be accessed.

At this point the CPU would place the binary number "0000 0000 0011 0010" on the 16 wires that make up the "16 bits" of the Address Bus. This binary number is the decimal number 50, the location to be read. The Address Bus wires are numbered from 1 to 16 so the CPU can unambiguously make the correct wires carry a logical 1 voltage (say 5 volts) and force the rest of the wires to carry a logical 0 voltage (say ground potential). The memory will look at the Address Bus and find the number 50 on it. This number will be decoded, and the contents of memory location 50 will be placed on the 8 wires of the Data Bus. At this point the memory may signal the CPU, via the Control Bus, that the number on the Data Bus is now the contents of location 50. The CPU will now save the current value of the Data Bus by storing the logical 1's and 0's that are on the Data Bus in the CPU's ACCUMULATOR. As soon as this latching of data has been completed, the processor has successfully executed the "LDA 50" instruction with the help of all three buses.

The basic advantage of the bus structure is that it simplifies the communication between the hardware devices. Instead of having separate wires connecting each pair of devices, we have one set of wires running through the system. The bus effectively connects all the devices together by the fact that the devices are all connected to the bus. The disadvantage is that the system designer must make sure that two different devices do not try to put data on a bus at the same time. It would be disastrous, for example, if one device is trying to put a logical 1 (5 volts) on wire 3 of the Data Bus while another device is trying to ground wire 3. Thus, the system designer must make sure that there is no competition for the buses.

Instruction Register And Decoder

One-Byte Instructions And The IRD

The purpose of the Instruction Register and Instruction Decoder is to inform the CPU which instruction the programmer has requested, given the binary-coded version of the instruction. This is best illustrated by a simple example. Suppose the CPU knows that the next instruction has been coded in binary and stored in memory location 15. From the previous section we know that the CPU will issue signals to the memory on the Control Bus and place the number "0000 0000 0000 1111" on the Address Bus (decimal 15). The memory will respond by placing the contents of location 15 on the Data Bus. Let us assume that the binary number "01001000" has been placed on the Data Bus. If this number was intended to be a piece of data, it would be latched by the CPU into the ACCUMULATOR. But we know that this is not intended as data. The number "01001000" is meant to be the binary-coded version of one of the many instructions the CPU can perform. What happens to the number on the Data Bus in this case?

Since we assume that the CPU knows that the contents of location 15 is the next instruction, we know the CPU will not place the data in the ACCUMULATOR. Instead, the CPU will latch the data on the Data Bus into the Instruction Register and Decoder (IRD). The register part of the IRD is there so the IRD can store the bit pattern for future use. The decoder part of the IRD is just a large amount of logic, wired to detect every possible instruction code the machine could be sent. The decoder outputs are signals specifying which instruction is being requested. These signals go to the control circuitry within the CPU, and this circuitry then decides what to do next. In our example, the number "01001000" would be latched into the Instruction Register from the Data Bus. The decoder logic would indicate to the control circuitry that the instruction "move B to C" was being requested. At this point the processor would copy the contents of register B into register C and would then look to the memory to find out the next instruction to perform. So we see that through the use of conventions we could wire the IRD so that the binary string "01001000" would be decoded to mean "MOV C,B." We, as users, could then store the string "01001000" in the memory, tell the processor to retrieve the string, have the IRD decode it, and have the instruction executed. Thus, the IRD is a vital part of the stored program implementation.

Variable-Length Instructions And The IRD

Instructions on the 8080 are of variable length; that is, some instructions, such as "MOV C,B" can be completely specified in 8 bits (1 byte) but other instructions cannot. For example, consider the instruction "LDA 359." We know addresses are 16 bits long on the 8080, so just to store the "359" will take 16 bits (two full bytes)*. But now there is no room to specify the "LDA" part of the instruction. We must code the "LDA" information in a third byte. Thus, the "LDA 359" instruction will take 3 bytes to store. The first byte will contain the code for "load the ACCUMULATOR." This code is known as the "operation" code (OP-CODE). The next two successive bytes in the memory will contain the 16-bit address that specifies which location's contents to load. The upper 8 bits of the address (most significant part) are known as the high-order address byte. The last 8 bits (least significant part) are known as the low-order address byte†. The high- and the low-order address bytes are shown in Fig. 16-3.

16-BIT ADDRESS

0000 0001	0110 0111
HIGH-ORDER BYTE	LOW-ORDER BYTE

Fig. 16-3. High- and low-order address bytes. Address 359 is written as a 16-bit, 2-byte binary number. The high-order byte contains the number 1 while the low-order byte contains the number 103.

The 8080 format for storing the two address bytes is the opposite of what one would probably expect. The low-order byte is stored first, immediately following the OP-CODE byte. Then the high-order byte is stored last. This is shown in Fig. 16-4.

Because instructions on the 8080 are variable length (they may take 1, 2, or 3 bytes to store), the first thing the CPU must determine once the first byte of an

*Because the address bit length is twice the byte bit length, addresses can be conviently stored in exactly two memory locations. If the address length was such that storing an address required using a fractional part of another byte, the unused storage bits would be unnecessarily wasted. For example, if addresses were 17 bits long on the 8080, it would take two full bytes plus 1 bit of a third byte to store an address. The other 7 bits of the third byte would be wasted every time an address had to be stored. Because it is costly to waste storage, computer designers find it desirable to relate the bit length of the addresses and bytes so that addresses can be stored in some integral multiple of bytes. This is why the bit lengths of addresses and bytes are often chosen to be integral multiples of each other, although there is nothing that absolutely requires this choice.

†Note that if some number N is stored in the high-order byte, this number contributes 256 × N to the value of the 16-bit address. This is so because the high-order byte represents the upper 8 bits of the 16-bit address.

	0011 1010	} OP-CODE FOR LDA
LOW-ORDER BYTE	0110 0111	} METHOD OF STORING ADDRESS 359
HIGH-ORDER BYTE	0000 0001	

Fig. 16-4. Storing the instruction "LDA 359." The three bytes of the instruction are stored in successive memory locations. Thus, if the OP-CODE "LDA" is stored in location 25, the instruction "LDA 359" will be found in locations 25, 26, and 27.

an instruction is read is whether or not it has the entire instruction in the IRD. The first byte of each instruction will always be the OP-CODE so that the IRD can inform the control logic which instruction is being requested. The control logic knows the length of every instruction. If the control logic determines the instruction is only one byte long, it is immediately executed. If, however, the instruction is longer than one byte, the CPU will access the memory again, storing the remaining one or two bytes in the IRD until the entire instruction is stored in the IRD. At this time the 8080 has all the information it needs to process the instruction, and the instruction is executed.

The Program Counter

At the beginning of this chapter we mentioned the fact that data and instructions will both be stored in the memory. Data will be written as binary numbers, while instructions will first be translated to a prearranged binary code and then stored in the memory. We have noted that the bit pattern "01001000" means "move B to C" to the 8080. But "01001000" is also the decimal number 72. If the bit pattern "01001000" is stored in location 15, how does the CPU know whether the contents of 15 is the instruction "MOV C,B" or the number 72?

The answer centers around a piece of hardware called the Program Counter (PC). The PC is a register big enough to hold an address (the PC on the 8080, therefore, is 16 bits long). The CPU's convention for determining whether a byte contains an instruction or data is simple: whatever number is in the Program Counter is the address of the next instruction. We say that the PC "points" to the next instruction. After each instruction has been executed, the CPU will increment the PC automatically so that the PC once again is pointing to the next instruction. This cycle continues until the processor stops executing instructions. The PC keeps track of where the computer is in the program and enables the CPU to distinguish program from data. It is a vital part of the stored program implementation.

As an example of how the CPU uses the PC, suppose the bit patterns listed in Fig. 16-4 are stored in memory locations 25, 26, and 27. In addition, we assume that the number 25 is in the PC (that is, the PC contains "0000 0000 0001 1001"). We also assume that the last instruction has just been completed. The CPU finds itself at the start of what is called an "instruction cycle." The start of each instruction cycle is a "fetch cycle" in which the OP-CODE of the next instruction is fetched from the memory so that the machine will know what to do next. When the CPU finds itself at the beginning of an instruction cycle, it looks to its Program Counter to find the address of the next instruction. More specifically, the CPU signals the memory that a memory read will be initiated by sending out a signal on the Control Bus. Then the CPU places the contents of the PC on the Address Bus. Memory location 25 will respond by placing its contents, a "0011 1010," on the Data Bus. The CPU knows it is still at the beginning of an instruction cycle, so it places the data from the Data Bus into the Instruction Register and Decoder. The IRD signals the control logic that the OP-CODE for "load the ACCUMULATOR" has just been read in. "LDA" is a three-byte instruction, so the CPU adds 1 to the PC and places this number (26) on the

Address Bus. The next byte from location 26 ("0110 0111") comes in on the Data Bus and also gets placed in the IRD. Again the PC gets incremented by 1 and again the CPU repeats the memory access cycle. Now the contents of location 27 ("0000 0001") are placed on the Data Bus, and they too get latched into the IRD. The CPU now has all the information it needs to execute the instruction. It knows that the instruction is to load the ACCUMULATOR, and it knows it must load it with what is in location 359. The CPU then enters the execution cycle by placing the address stored in the Instruction Register ("0000 0001 0110 0111" = 359) on the Address Bus. Location 359 places its contents on the Data Bus, and this number is now latched into the ACCUMULATOR as required by the instruction. The last action taken by the CPU in the instruction cycle is to increment the PC to 28. Since we agreed very early that sequential processing would be used, the next instruction should, in fact, be in 28, the next available location after 27. The CPU once again finds itself at the start of a new instruction cycle, and again it places the contents of the PC on the Address Bus. Since the number in the PC has been previously incremented to 28, the CPU is correctly requesting the first byte of the next instruction. Thus, by simply correctly updating the PC after each instruction, the CPU can always keep track of where it is in the program.

Now suppose through some error the PC contained the number 26 at the start of a fetch cycle, but memory looked as it did in Fig. 16-4. That is, the PC at fetch time is pointing erroneously to the middle of the "LDA 359" instruction. Well, the CPU has no way of knowing an error is occurring. It will enter a fetch cycle, look at its PC, and fetch the contents of location 26 ("0110 0111"). Only now instead of interpreting "0110 0111" as the lower half of the 16-bit address 359, it will interpret the bit pattern as its next instruction. It must do this since it is in the fetch cycle, and the convention is that during the fetch cycle the number in the PC points to the next instruction. Every bit pattern is *some* instruction, even if an unintended one. In the example, the bit pattern "0110 0111" just happens to represent the instruction "MOV H,A." The CPU, will, therefore, copy the contents of the AC into register H, even though this was not intended by the programmer at all. The CPU would then increment the PC to 27 and interpret the contents of 27 as an instruction as well. We see that the user must insure that the program and the data are stored in appropriate locations and that the PC starts out with the correct value. From there, the sequential storage of instructions and the automatic incrementing of the PC will insure the rest of the program's instructions are executed in the proper order.

Summary

We have built up a rather complete picture of the parts of a computer. This is summarized in Fig. 16-5. A few hardware devices have not yet been discussed since the execution of simple programs can occur without them. As we require the computer to perform more complex tasks, we will complete the description of the devices that make digital computers work.

We next cover the execution of simple programs in detail. The programs will be written in the binary-coded conventions we have been discussing. Since the machine can directly understand such a binary code, programs written in this binary language are called "machine-language programs." Like the subject of computer architecture, it is necessary to study machine language before assembly language can be learned.

16.4 MACHINE-LANGUAGE PROGRAMMING

Overview Of Program Execution

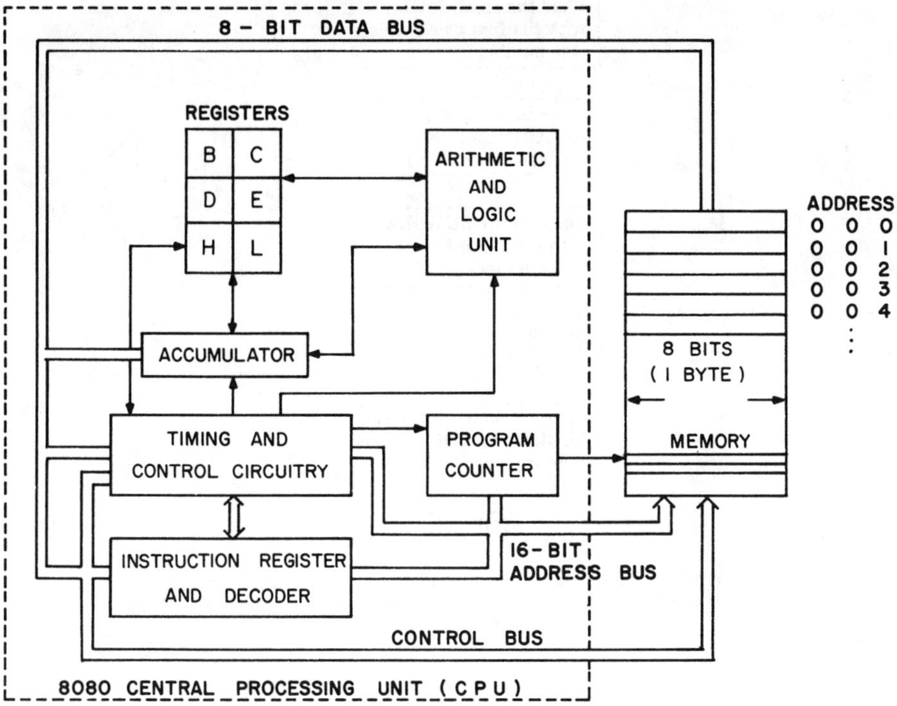

Fig. 16-5. Detailed diagram of typical (8080) CPU and bus structure.

We begin this section by carefully examining a simple machine-language program. Special attention will be paid to the flow of data through the machine and keeping track of each step the computer takes in executing the program.

The example will be the machine-language implementation of the FORTRAN program: X=3+4, STOP. The 3 and 4 must be data stored in designated data bytes in the memory. However, what is X? A moment's thought will show that X is intended to be a memory location which will contain the sum of 3 and 4; that is, X is actually an address, the address of the location where we choose to store the sum. Thus, the statement, X=3+4, actually means retrieve the number 3 from the memory, save it, retrieve the number 4 from the memory, add them together, and save the result in some memory location which will be called location X for future reference. A flowchart of the program is shown in Fig. 16-6.

Looking at the flowchart of Fig. 16-6, it appears that for this simple program each box in the flowchart will translate into one 8080 instruction. In addition to the six instructions that make up this program, we will need three more data bytes: one to store the number 3, one to store the number 4, and one to store the sum. If each instruction occupies only one byte, this would be a 9-byte program. However, certain instructions take more than one byte, and we actually end up with the 15-byte program shown in mnemonic form in Fig. 16-7 and in binary-coded form in Fig. 16-8.

Let us assume that the binary data bytes of Fig. 16-8 have been loaded into memory in locations 0-14, and that the Program Counter contains the number 0. We now trace through the execution of the program.

Since the CPU is at the start of an instruction cycle, the CPU will look to its PC to find out the location of its next instruction. More specifically, the CPU places the

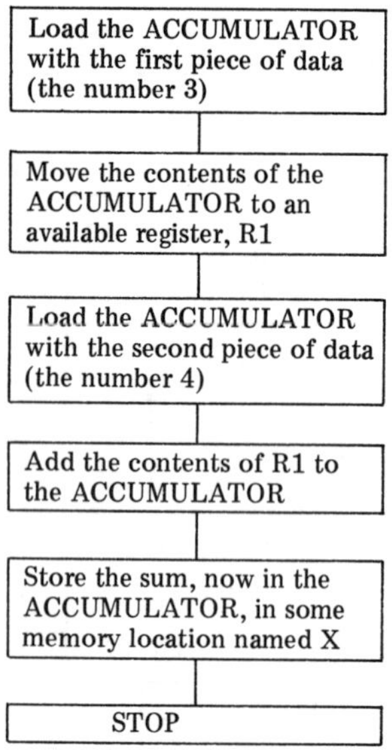

Fig. 16-6. Flowchart for implementing the FORTRAN program X=3+4, STOP.

ALGORITHM

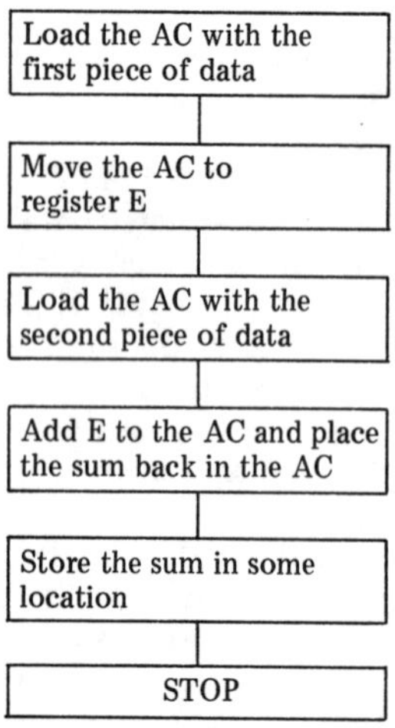

ADDRESS	INSTRUCTION OR DATA
000	LDA 12
003	MOV E,A
004	LDA 13
007	ADD E
008	STA 14
011	HLT
012	3
013	4
014	0

Fig. 16-7. Mnemonic coding of instructions to implement the flowchart of Fig. 16-6.

ADDRESS	CONTENTS OF ADDRESS (MNEMONIC)	CONTENTS OF ADDRESS (BINARY)
000	LDA	0011 1010
001	12	0000 1100
002	0	0000 0000
003	MOV E,A	0101 1111
004	LDA	0011 1010
005	13	0000 1101
006	0	0000 0000
007	ADD E	1000 0011
008	STA	0011 0010
009	14	0000 1110
010	0	0000 0000
011	HLT	0111 0110
012	3	0000 0011
013	4	0000 0100
014	0	0000 0000

Fig. 16-8. Mnemonic and binary-coded version of program X=3+4, STOP. Note that locations 0-11 contain instructions while locations 12-14 contain data. Also note that the instructions are of variable length.

number in the PC on the Address Bus after signaling the memory that a memory access is about to take place. Location 000 responds by placing the bit pattern "0011 1010" on the Data Bus. Since the processor is in the fetch cycle, this byte is placed into the IRD. The IRD signals the control logic that a "load the ACCUMULATOR" instruction is desired. This is a three-byte instruction, so the PC is incremented by 1, the address 1 is placed on the Address Bus, and location 1 places the byte "0000 1100" on the Data Bus. This byte is also clocked into the IRD. Again, the PC is incremented to 2, and the contents of location 2 (all zeros) is clocked into the IRD. At this point the CPU has all the information it needs. The 16-bit address "0000 0000 0000 1100" is placed on the Address Bus as the processor begins the execution cycle. Location 12 places the first operand, the number 3, on the Data Bus. This byte is clocked into the ACCUMULATOR as specified by the instruction. The PC is incremented to 3, and the first instruction has been completed.

Again the processor finds itself at the start of an instruction cycle. The same course of action as just described is followed again. The location specified by the PC is accessed, and the content of that location is placed in the IRD. The contents of location 3 is "0101 1111," the 8080 code for "MOV E,A." The execution phase starts immediately since this one byte contains all the information needed. The number that was in the ACCUMULATOR, 3, is copied into register E. The copy is made so that the content of the ACCUMULATOR can be replaced with new data, and a copy of the old data will still be on hand in some register. When the move is complete, the PC is incremented to 4.

Location 4 is now accessed. Upon determining that the three-byte instruction "LDA" is to be executed, the CPU brings the information stored in consecutive bytes 5 and 6 into the IRD. The PC now reads "6." To execute the instruction "LDA 13" the CPU accesses location 13 and latches the contents of 13 (the number 4) into the ACCUMULATOR, erasing the "3" that was previously stored there. At this point both operands, the 3 and the 4, are stored within the CPU in registers, and one of those registers is the ACCUMULATOR. The PC is incremented to 7.

The contents of location 7 ("1000 0011") is the code for having the CPU add the contents of register E to the AC, and store the sum back in the AC, erasing the old number in the AC. (The mnemonic for this instruction is "ADD E." It is not necessary to specify what to add E to since it is understood that we add E to the AC, and we do not specify where to place the sum since it is agreed to place the sum in the AC.) At the completion of this instruction, the 3 and 4 will have been added by the ALU, and the resulting sum of 7 will have been stored in the AC. The contents of register E are unchanged and remain at "3." The PC is then incremented to 8.

The contents of byte 8 informs the CPU that it must perform a "store the ACCUMULATOR" instruction, and the next two bytes of the instruction (bytes 9 and 10) tell the CPU to store the AC in location 14. The CPU then places the number 14 on the Address Bus, places the contents of the AC (the number 7) on the Data Bus, and uses the Control Bus to signal memory that it must write the number on the Data Bus into the specified location. Thus, the number 7 gets stored in location 14. The PC then gets incremented to 11.

When the bit pattern stored in location 11 gets placed in the IRD, the IRD signals the CPU to halt (the mnemonic for the halt instruction is "HLT"). The PC no longer gets incremented, and the processor takes no further actions.

The execution of the program "X=3+4, STOP" is complete. We note that "location X" (location 14) has the number 7 stored in it. We further note that placing the data, 3, 4, and the result of 7, in locations 12, 13, and 14 was completely arbitrary. That is, we could just have easily placed the data in locations 108 and 417 and placed the result in location 1912. The only thing that would change in the program if we did this, is the address part of the "LDA" and "STA" instructions. That is, if the number 3 had been stored in location 108, the first instruction would have had to be "LDA 108," etc.

Figure 16-9 pictorially shows the state of the machine as the various instructions are executed. It should be studied along with the explanatory text until the flow of data and the logic behind the choice of each instruction is clear to the reader.

Branching

The sample program just discussed provided an illustration of a sequentially-executed program. The instructions were performed one after the other with no possibility of any other instructions being executed. But all programmers know that often instructions are not executed in the order they appear in a program. FORTRAN statements such as "GO TO 50" and "IF (X+4) 10,20,10" force a transfer of control to a different part of the program. The FORTRAN do-loop forces the block of code to be performed over and over again. Since jumping and looping are such an important part of programming, we must investigate how a computer can be made to perform "branching," or nonsequential processing.

Unconditional Branching

The unconditional branch instruction causes the computer to begin executing instructions located anywhere in memory, as specified by the programmer. The next instruction executed need not physically be located in the byte following the unconditional branch instruction. Once the branch, or "jump," is made, the computer once again resumes sequential processing. In FORTRAN, the unconditional jump instruction is "GO TO." On the 8080, the instruction is "jump to location X," abbreviated "JMP X," where X is some integer between 0 and 65,535.

To understand how branching is accomplished, consider the program shown in Fig. 16-10. Assuming the PC has been set to 0, we know that the CPU will access location 0 to determine the next instruction. In location 0 it finds the "LDA" OP-

Before any instructions have been executed	REGISTERS	
	A	Random bit pattern
	E	Random bit pattern
	LOCATIONS	
	12*	0000 0011
	13	0000 0100
	14	0000 0000
After "LDA 12"	REGISTERS	
	A	0000 0011
	E	Random bit pattern
	LOCATIONS	
	12	0000 0011
	13	0000 0100
	14	0000 0000
After "MOV E,A"	REGISTERS	
	A	0000 0011
	E	0000 0011
	LOCATIONS	
	12	0000 0011
	13	0000 0100
	14	0000 0000
After "LDA 13"	REGISTERS	
	A	0000 0100
	E	0000 0011
	LOCATIONS	
	12	0000 0011
	13	0000 0100
	14	0000 0000
After "ADD E"	REGISTERS	
	A	0000 0111
	E	0000 0011
	LOCATIONS	
	12	0000 0011
	13	0000 0100
	14	0000 0000
After "STA 14"	REGISTERS	
	A	0000 0111
	E	0000 0011
	LOCATIONS	
	12	0000 0011
	13	0000 0100
	14	0000 0111

Fig. 16-9. Pictorial view of key storage bytes during execution of the program X=3+4, STOP.

INSTRUCTION	ADDRESS	MNEMONIC	BINARY
LDA 6	000	LDA	0011 1010
	001	6	0000 0110
	002	0	0000 0000
JMP 64	003	JMP	1100 0011
	004	64	0100 0000
	005	0	0000 0000
	006	17	0001 0001
	.	.	.
	.	.	.
STA 68	064	STA	0011 0010
	065	68	0100 0100
	066	0	0000 0000
HLT	067	HLT	0111 0110
	068	0	0000 0000

Fig. 16-10. Program using branch instruction.

CODE and then accesses locations 1 and 2 to find out which location's contents to load into the AC. After loading the number 17 into the AC (the contents of location 6), the PC is automatically incremented to 3. Now the "JMP" OP-CODE is read in. The CPU must determine where to jump to; that is, it must determine the address of the next instruction. Therefore, bytes 4 and 5 are read in. At this point the PC reads "5," and normally would be incremented to 6. However, the jump instruction specifies that the next instruction is not located in location 6 but rather is in location 64. What action should the CPU take? We know that at the beginning of each instruction cycle the CPU will look to its PC to determine the address of the next instruction. Thus, all the CPU must do to implement the "JMP 64" instruction is to load the 16-bit address "64" into the PC. No other special action need be taken. Now the CPU finds itself at the beginning of an instruction cycle and places the contents of the PC on the Address Bus. That number is now a 64. The processor has no "memory" that it just performed a jump. It does not know that the last instruction was in location 3 and not in 63. All that matters is that a "64" is on the Address Bus, and so the contents of 64 will be fetched as the first byte of the next instruction. The "STA 68" instruction is read in, and after the number 17 is stored in location 68, the PC gets normally incremented to 67. Normal sequential processing has resumed. The "HLT" instruction in 67 is executed and processing stops.

The effect of an unconditional branch instruction, therefore, is to override the automatic incrementing of the PC that occurs due to the sequential processing convention. The "JMP X" instruction essentially tells the CPU that the next executable instruction is in location X and not necessarily in the location following the jump instruction. The CPU responds by setting the PC to address X. Once again, the PC points to the next instruction. Just by changing the PC, the pointer which tells the machine where it is in the program, unconditional branching has been accomplished.

Conditional Branching

A conditional branch instruction causes a jump to occur only if some specified condition is met. If the condition is not met, the jump is not initiated, and the next instruction to be executed will be the one sequentially following the conditional jump instruction. That is, if the condition is not met, the conditional jump has no effect on the sequential flow of processing. An example of a conditional jump statement in FORTRAN is "IF (J.LT.0) GO TO 20."

The previous section has shown that unconditional branching can be accomplished without any additional hardware over that discussed in Sec. 16.3. This is not the case for conditional branching. The extra hardware devices needed by the CPU to implement conditional branching are a series of single-bit flip-flops called *flags*. Each flag represents some condition that has either occurred or not (thus, each flag need only be one bit). The flags are set to 1 or 0 by certain instructions and tested by other instructions. To understand this, we begin by discussing specific flags and how they get set. We will shortly discuss how they get tested.

Flags are set by the ALU at the completion of certain flag-setting instructions. They are set just before the PC gets incremented to the next instruction location. Flags are implemented to allow us to later test for certain "interesting" results of past instructions. For example, it may be important to know if the result of the last subtract instruction is zero, negative or positive. We may wish to know if two numbers were added together that were so large that the result could not be stored in 8 bits and so an overflow occurred in performing the addition.

The 8080 has flags to report on each of the conditions just mentioned. One of those flags is the ZERO FLAG. If the result of the instruction currently being processed is exactly zero, and if the instruction is a flag-setting instruction, the CPU will set the ZERO FLAG to a 1 and otherwise set the flag to a 0. For example, suppose the ACCUMULATOR contains the number 4 and the current instruction subtracts 4 from the AC. Since the subtract instruction is a flag-setting instruction, after the AC has been decremented by 4, the CPU examines the condition of the AC and sets all the flags accordingly. Since the AC is now exactly zero, the CPU sets the ZERO FLAG to a 1 to indicate the result of the subtraction is 0. If the result had not been 0, the CPU would have set the ZERO FLAG to 0. If the instruction had not been a flag-setting instruction, the CPU would have executed the instruction but would not have changed the value of any of the flags. They would have remained in the same state they were in after the most recent flag-setting instruction was executed.

Another flag implemented on the 8080 is the CARRY FLAG. If a flag-setting instruction has been executed, and if the result caused an overflow out of the 8th bit during an addition or caused a borrow to occur during a subtraction, the CARRY FLAG is set to 1 (otherwise it is set to 0). For example, if register E contains a "250" and the AC contains a "9," and the two registers are added together, an overflow will occur. This is because the sum, 259, cannot be stored in only 8 bits. The number 3 gets stored in the AC, and the CARRY FLAG is set to 1 to indicate that if there were a 9th bit, it would have been set to 1 with the 3 to make the number 259. Similarly, if we subtract 250 from 9, the subtraction can only be performed by borrowing from a "9th" bit. The CARRY FLAG is set to 1 at the completion of the subtraction to indicate that this borrowing took place.

Many more "conditions" can be tested for, with one flag representing each condition. The 8080 itself has three flags in addition to the two just mentioned, but only one other flag will be discussed here. That flag is the SIGN FLAG. The SIGN FLAG will be set to 1 if the 8th bit of the result of a flag-setting instruction is a 1, and will be set to 0 if the 8th bit of the result is a 0. The 8th bit is often thought of as a sign bit, determining whether a number is positive or negative. Negative numbers will be stored with the 8th bit on, in "two's complement" mode. (Those not familiar with two's complement arithmetic should read Appendix A at the end of this chapter.) However, to store a large positive number such as 255, the 8th bit will also be turned on. There is no way the computer can unambiguously distinguish between negative and large positive numbers. It is up to the programmer to interpret the "8th bit on" condition as indicating a negative number in two's complement or a large positive number. The name, "SIGN FLAG," therefore, is perhaps misleading. The flag is actually an "8th-BIT TESTER." For an example of how the SIGN FLAG works, consider subtracting 250 from 9. The result is negative and the 8th bit is turned on. The CPU then sets the SIGN FLAG to 1 as well. For a second example, consider adding 250 to 3. The sum, 253, can only be stored by having the 8th bit on. Thus, at the end of the addition the SIGN FLAG is set to 1, even though the number 253 is positive.

Since it has been mentioned that certain instructions are flag-setting instructions, it is worth pointing out that many instructions are not flag-setting instructions, or at least may not cause the CPU to set every flag. The data-transfer instructions on the 8080 are a good example of instructions which do not affect the flags. For instance, if the instruction "LDA 30" is executed, with location 30 containing all zeros, the AC is effectively set to zero. However, since "LDA 30" is a data-transfer instruction, the ZERO FLAG is *not* set to 1. The ZERO FLAG is unaffected by the "LDA 30" instruction. It will remain in the same state it was in when last set by the execution of the most recent flag-setting instruction. Typical flag-setting instructions are arithmetic instructions like "ADD" and "SUBTRACT" as well as logic instructions like "AND" and "OR." Each computer manufacturer publishes a list of all the instructions on a machine and which flags they affect.

It has been mentioned that setting and testing flags plays a key role in implementing conditional branching. Flags allow the implementation of conditional branching because they can be set by one instruction and tested by another. The 8080 has a pair of conditional jump instructions (flag-testing instructions) for each flag. For example, the instructions "jump to location X if the ZERO FLAG is a 1" and "jump to location X if the ZERO FLAG is a 0" are both available to test the ZERO FLAG (the abbreviations are "JZ X" and "JNZ X" respectively, standing for "jump on zero" and "jump on not zero"). There is also a "JC" and "JNC" pair for testing the CARRY FLAG and "JM" and "JP" for testing the SIGN FLAG (standing for "jump on minus" and "jump on plus" even though the SIGN FLAG is really an 8th-bit tester).

The mechanism for setting and testing flags is best illustrated by an example. The program shown in Fig. 16-11 uses conditional branching.

	FORTRAN	ADDRESS	8080 CODE
	J=4	50	MVI A,4
1000	J=J–2	52	SUI 2
	IF (J.LT.0) GO TO 2000	54	JM 150
	IF (J.EQ.0) GO TO 1000	57	JZ 52
	GO TO 1000	60	JMP 52
		.	.
		.	.
2000	STOP	150	HLT

Fig. 16-11. Program using conditional branching.

In Fig. 16-12 the machine-language code is shown. In this machine-language implementation, the ACCUMULATOR plays the role of the variable "J" in the FORTRAN program. Assume the PC has been set to 50 and let us examine how the branching works.

The first instruction is used to initialize the content of the ACCUMULATOR to 4. The OP-CODE "MVI" stands for "move immediate to the AC." What is moved immediately is the data contained in the next byte of the instruction. Thus, the instruction "MVI A,4" instructs the CPU to move the number "4" into the AC, *not* the content of location 4. Since this is a data-transfer instruction, the instruction does not cause any flags to be set. The PC is automatically incremented to 52. The next instruction read in is "SUI 2," a mnemonic for "subtract immediate the number 2 from the ACCUMULATOR" (again, this does not subtract the content of location 2 but

ADDRESS	CONTENTS (MNEMONIC)	CONTENTS (BINARY)
050	MVI A	0011 1110
051	4	0000 0100
052	SUI	1101 0110
053	2	0000 0010
054	JM	1111 1010
055	150	1001 0110
056	0	0000 0000
057	JZ	1100 1010
058	52	0011 0100
059	0	0000 0000
060	JMP	1100 0011
061	52	0011 0100
062	0	0000 0000
⋮	⋮	⋮ ⋮
150	HLT	0111 0110

Fig. 16-12. 8080 machine-language coding of the program shown in Fig. 16-11.

rather subtracts the actual number 2 from the AC). The ACCUMULATOR is then decremented from 4 to 2. "SUI 2" is a flag-setting instruction. Thus, before the CPU goes on to the next instruction, it sets all the flags based on the outcome of the "SUI 2" instruction. Since no borrowing occurred in performing the subtraction, the CARRY FLAG is set to 0. Since the content of the AC is not 0, the ZERO FLAG is also set to 0, and, because the 8th bit of the AC is a 0, the SIGN FLAG is set to 0. Only after the flags have been set does the CPU increment the PC to 54.

It is at this point that the first flag-testing instruction is encountered. The next instruction reads "JM 150," short for "set the PC to address 150 if and only if the SIGN FLAG is currently set at 1." Since the SIGN FLAG is currently zero, no jump takes place, and the PC is incremented to 57. Note that the conditional jump instruction is not a flag-setting instruction, and so the flags remain in the same state they were in following the completion of the most recent flag-setting instruction encountered, "SUI 2."

The next instruction encountered is also a conditional branching instruction. It reads "JZ 52," short for "set the PC to address 52 if and only if the ZERO FLAG is currently set at 1." Since the ZERO FLAG is set at 0, no jump occurs, and the PC is incremented to 60. Location 60 contains the unconditional jump instruction "JMP 52." To execute this instruction the CPU sets the PC to 52. Thus, the next instruction to be executed will be the one in location 52.

The instruction in location 52 once again subtracts 2 from the AC. The content of the AC is now 0. Since this is a flag-setting instruction, the flags must be updated before processing can proceed. The CARRY and the SIGN FLAGS are set to 0, but since the AC is exactly 0, the ZERO FLAG is set to 1. The PC is then incremented to 54.

The flag testing begins again. No action is taken on the "JM 150" instruction since the SIGN FLAG is still 0. The next instruction, "JZ 52," tests the ZERO FLAG for a setting of 1. Since that flag is in fact currently set at 1, the jump is performed. Thus, the PC is set to 52, and the unconditional jump in location 60 is never even read in.

For the last time, the AC is decremented by 2. The result, −2, is stored in two's complement notation. Thus, the 8th bit of the AC is a 1 and so the SIGN FLAG is set to a 1. The CARRY FLAG is also set to a 1 since it is necessary to "borrow" to perform the subtraction. The ZERO FLAG is set to 0 since the AC does not contain all zeros. The PC is incremented to 54.

Now when the "JM 150" is encountered, the jump will be performed since the SIGN FLAG is currently set at 1. The number 150 is placed in the PC. Location 150 is accessed, and the halt instruction stored there is read in. Processing stops, and the program's execution is complete.

A pictorial representation of the execution of the program is shown in Fig. 16-13. This figure should be studied until the reader understands how the flags are set and then tested to implement conditional branching.

	REGISTERS			FLAGS	
Before any instructions have been executed	AC	RANDOM		SIGN	RANDOM
PC has been set to 50.	PC	0000 0000 0011 0010	(50)	CARRY	RANDOM
				ZERO	RANDOM
After "MVI A,4"	AC	0000 0100	(4)	SIGN	RANDOM
	PC	0000 0000 0011 0100	(52)	CARRY	RANDOM
				ZERO	RANDOM
After "SUI 2"	AC	0000 0010	(2)	SIGN	0
	PC	0000 0000 0011 0110	(54)	CARRY	0
				ZERO	0
After "JM 150"	AC	0000 0010	(2)	SIGN	0
	PC	0000 0000 0011 1001	(57)	CARRY	0
				ZERO	0
After "JZ 52"	AC	0000 0010	(2)	SIGN	0
	PC	0000 0000 0011 1100	(60)	CARRY	0
				ZERO	0
After "JMP 52"	AC	0000 0010	(2)	SIGN	0
	PC	0000 0000 0011 0100	(52)	CARRY	0
				ZERO	0
After "SUI 2"	AC	0000 0000	(0)	SIGN	0
	PC	0000 0000 0011 0110	(54)	CARRY	0
				ZERO	1

Fig. 16-13. Pictorial representation of the execution of the program shown in Figs. 16-11 and 16-12.

Fig. 16-13. (Cont.)

After "JM 150"	REGISTERS			FLAGS	
	AC	0000 0000	(0)	SIGN	0
	PC	0000 0000 0011 1001	(57)	CARRY	0
				ZERO	1

After "JZ 52"	REGISTERS			FLAGS	
	AC	0000 0000	(0)	SIGN	0
	PC	0000 0000 0011 0100	(52)	CARRY	0
				ZERO	1

After "SUI 2"	REGISTERS			FLAGS	
	AC	1111 1110	(−2)	SIGN	1
	PC	0000 0000 0011 0110	(54)	CARRY	1
				ZERO	0

After "JM 150"	REGISTERS			FLAGS	
	AC	1111 1110	(−2)	SIGN	1
	PC	0000 0000 1001 0110	(150)	CARRY	1
				ZERO	0

In Location 150 the Halt Instruction is Stored which Stops the Processing

Subroutine Calls

A subroutine is a block of code separate from the main program. The subroutine often performs a specific task that is frequently needed by the main program such as finding the cosine of an angle or the integral of a function. The subroutine is entered or "called" by a "calling" instruction executed in the main program. Data from the main program are handed ("passed") to the subroutine for processing. When the subroutine has completed execution, control passes back to the place in the main program where the subroutine was called. This is known as "returning" to the main program. In FORTRAN, subroutine X is called simply by executing the instruction "CALL X(L,M,N,O,P)" where L, M, N, O, and P represent data that are to be passed to subroutine X. When this subroutine has finished, the FORTRAN instruction "RETURN" is executed which transfers control back to the main program to the instruction immediately following the "CALL X(L,M,N,O,P)" instruction.

In order to implement subroutine calls, the computer will have to be capable of performing certain operations when a "CALL" instruction is executed. The first of these operations involves saving the address of the instruction immediately following the CALL instruction. This "return address" is saved so that when the call is completed the computer can return and resume executing the instructions which follow the subroutine CALL instruction. Secondly, the computer must be capable of passing data to the subroutine and passing results back to the calling program. Thirdly, the computer must be capable of branching to the first instruction of the subroutine (that is, it must be capable of entering the subroutine). The computer would then begin executing the block of instructions that makes up the body of the subroutine. Finally, when the subroutine has been completely processed (that is, when a "RETURN" instruction is executed), the computer must be capable of retrieving the previously stored return address. The return address is then loaded into the PC. The calling program has now been correctly reentered at the instruction following the CALL instruction. Normal sequential processing resumes once again.

To summarize, in subroutine handling the computer must, **upon executing a "CALL" instruction:**

1. store the address of the instruction immediately following the "CALL" instruction; i.e., save the return address,
2. pass all necessary data to the subroutine,
3. enter the subroutine by branching to the first instruction of the subroutine, and begin executing the subroutine code, and

upon executing a "RETURN" instruction:

4. pass all results back to the calling program,
5. retrieve the previously stored return address, and load this address into the PC. This action will cause the calling program to be reentered at the correct point. Normal sequential processing now resumes.

There are several methods of implementing subroutine calls on a computer. Some methods involve no more than adding a few new instructions to the instruction set of the machine and setting up rules that the subroutine user must follow. However, many machines implement subroutine calls by adding additional hardware to the processor as well as adding some new instructions. The new hardware devices are known as the STACK and STACK POINTER. The 8080 handles subroutine calls through the use of a STACK, and so only this method of subroutine implementation will be discussed here. However, by understanding this widely used method, the reader should be able to see how different methods of subroutine handling could work as well.

The Stack and Stack Pointer

Before discussing what a STACK is and how it works, it is helpful to discuss the problems that arise in subroutine handling. The problems arise primarily because it is desirable to allow one subroutine to call another subroutine. This is called "nesting" subroutine calls. Every time one subroutine calls another, a new return address must be stored. For example, suppose a main program calls subroutine A, and subroutine A then calls subroutine B. When subroutine B has finished its task, it must return to its calling program, which is subroutine A. Thus, one return address has to be stored. But when subroutine A completes its task, control must transfer back to the main program. Thus, a second return address has to be stored somewhere. It is clear that each time a subroutine calls some other subroutine, an additional return address must be saved. Then when a subroutine returns, the correct return address must be retrieved and placed in the PC. This may involve a lot of bookkeeping. The programmer may get bogged down keeping track of where all the return addresses have been stored.

Another problem that arises involves the use of registers by a subroutine. In the course of writing a long program, the programmer will place the intermediate results in various registers. Since the number of registers available is small, there are times when all or almost all of the registers are holding intermediate results for the calling program. When the subroutine is entered it would be very helpful to be able to save all the registers of the calling program and then restore the registers when the subroutine has finished. By saving and restoring the registers, the subroutine could freely use the registers for its own internal computations. If no nesting is allowed, we could just reserve a fixed spot in the memory to store registers in when a subroutine is entered and retrieve the registers from this spot when the subroutine returns. But again, if nesting is permitted much bookkeeping may be needed to keep track of where all the registers have been stored by all the various called subroutines.

The key to solving the bookkeeping problem is to recognize that individual layers of storage are needed by each subroutine to store its return address and save the registers of the calling program. These layers of storage are needed one at a time as the subroutines are called. As the subroutines return, the most recent layer is no longer needed and can be erased. When the last subroutine finally returns, the final layer is no longer needed and the bookkeeping is over. Notice that the first layer of storage required, which stores the return address and registers of the main program itself, will be the last layer of storage to be retrieved. It will be retrieved after all the other layers containing return addresses and registers of other called subroutines have been retrieved.

The concept of layers of storage forming and being erased is more formally called the concept of a STACK STRUCTURE. A STACK is just a block of memory locations used to implement the STACK STRUCTURE concept. The STACK starts out empty and fills up with layers of return addresses and registers of each called program. Each time a program calls another program, a layer of information is added to the STACK. This is known as "pushing" data onto the STACK. As the routines start returning, the calling information is retrieved from the STACK in an operation known as "popping" data off the STACK. Since the first layer of data stored is the last layer to be retrieved, this type of STACK is known as a first-in-last-out (FILO) STACK. Notice that layers are retrieved only from the top of the STACK; that is, since the subroutines always return in the opposite order of which they are called, the next layer to be retrieved when a subroutine returns must be the top layer and not some intermediate layer in the STACK. To keep track of the current top of the STACK, a STACK POINTER is built. The STACK POINTER (SP) is a register which contains the address of the current top of the STACK. If data are to be added to the STACK, the CPU will refer to the SP to find out the address of the top of the STACK. It will then store the N bytes of data to be added to the STACK and change the value in the STACK POINTER so that the SP once again points to the current top of the STACK. If N bytes of data are to be retrieved, the process is reversed. The top N bytes of the STACK, as indicated by the SP, are retrieved and the SP is changed to point to the top of the current, smaller STACK. The function of the SP in STACK maintenance is analogous to the function of the PROGRAM COUNTER in sequential processing; that is, the SP identifies the top of the STACK in the same way the PROGRAM COUNTER identifies the next instruction.

Some machines reserve a special block of memory for the STACK. On these machines the STACK is physically distinct from the core memory where programs and data are stored. On other machines such as the 8080, there is no memory block separate from core memory called the "STACK." Instead, any part of core memory may serve as the STACK. The programmer simply decides he would like the STACK to begin in some location X. An instruction is executed which loads the address "X" into the STACK POINTER. Then as the program and subroutines are executed, the STACK, that is, the amount of core memory being used as a STACK, grows and shrinks as routines are called and exited. When the final "RETURN" is executed to get back to the main program, the SP will once again contain address "X" and the STACK is again of zero length.

Several special instructions are created to facilitate the use of the STACK. To store registers on the STACK, a machine will often have "PUSH" and "POP" instructions. The format for the PUSH instruction on the 8080 is "PUSH REGISTER-PAIR," where "REGISTER-PAIR" means either register B-C, D-E, or H-L. The 8080 PUSH instruction stores both halves of a register pair; that is, one "PUSH" instruction stores two registers. Register pairs are specified by the first register in the pair. Thus, to specify that register B and C should be stored on the STACK, the 8080 mnemonic is "PUSH B," not "PUSH BC." It is understood that "B" in this case stands for both registers B and C. To retrieve the upper two bytes of the STACK and place them back in register pair B-C, the 8080 has the instruction "POP B." Similarly, one can instruct the machine to "POP D" and "POP H." There is also a pair of instructions for storing and retrieving the ACCUMULATOR and the state of all the FLAGS. The instructions are "PUSH PSW" and "POP PSW." "PSW" stands for "processor status word." The name comes from the observation that the information in the ACCUMULATOR and the state of the FLAGS represents, in some sense, the "status" of the machine at the time the "PUSH PSW" instruction is executed.

Subroutine Handling Using A Stack

In the previous section the STACK STRUCTURE and some instructions for manipulating the STACK have been discussed. To use the STACK to implement subroutine calls, two more instructions are needed. The first of these instructions on the 8080 is of the form, "CALL SUB," where "SUB" is the address of the first instruction of subroutine SUB.

To understand what happens when a "CALL SUB" instruction is executed, recall the tasks that must be performed in subroutine handling. First of all, the machine must store the address of the instruction immediately following the "CALL SUB" instruction (the return address). This return address is stored on the STACK. To store this 16-bit return address, the CPU first subtracts 1 from the number stored in the STACK POINTER. The number 1 is subtracted because the STACK grows upward on the 8080 and the CPU wants the SP to point to the next available empty memory location. The contents of the SP are then placed on the ADDRESS BUS and the high-order byte of the PC is placed on the DATA BUS. The CPU then signals the memory with the CONTROL BUS to write the data on the DATA BUS into the specified address. The STACK POINTER is then decremented by 1 again. Once again it points to the next available memory location. The CPU now places the low-order byte of the return address on the DATA BUS, and this is stored in the location above the location where the high-order byte has been stored. Note that at this point the return address has been stored on the STACK and the SP correctly points to the top of the STACK.

After the return address has been stored, data must be passed between the calling program and the subroutine before the latter can be entered. This second task of subroutine handling can be performed in a simple way, without the need for even a special instruction. Argument passing can be handled by convention; that is, the author of the main program simply decides to leave the data in registers B or C or perhaps chooses to leave the data in locations 200-300. The author of the main program then informs the subroutine programmer where the data are, and where he wants the results placed. The subroutine knows where to find the data, and, when the subroutine finally returns to the calling program, the calling program accesses the agreed upon storage area to find the results. Thus, on the assembly-language level, "argument passing" is a matter of setting up data bases in prearranged areas of memory or in certain registers. Nothing special is needed to perform this task. Thus, no special "argument passing" execution phase occurs in processing the "CALL SUB" instruction on the 8080. The arguments are passed by convention.

The third task of subroutine handling, entering the subroutine, is accomplished at the completion of execution of the "CALL SUB" instruction. To enter the subroutine the CPU places the address "SUB" into the PC. Since "SUB" is the address of the first instruction in the subroutine, an unconditional jump to the subroutine has occurred. The body of the subroutine is then processed.

The other special instruction used for subroutine implementation is the RETURN instruction, mnemonically abbreviated as "RET" on the 8080. When the "RET" instruction is executed, the process for retrieving an address off of the STACK is initiated. The location pointed to by the STACK POINTER is accessed and loaded into the lower 8 bits of the PC. The SP is incremented by 1, and the next byte is accessed and is loaded into the upper 8 bits of the PC. The SP is incremented by 1 so that it again points to the top of the STACK. Since the number just loaded into the PC was the return address, the CPU will begin executing instructions back in the calling program. The subroutine has now returned.

The subroutine programmer must be careful when using the STACK to be sure that data are popped off the STACK in the opposite order to which it was stored. For example, if an 8080 programmer wishes to store all six registers by executing PUSH B, PUSH D, PUSH H, he must be sure to restore the registers by executing POP H, POP D, POP B. Executing POP's in any other order would cause a register pair to be restored with data that were originally in another register pair. For another example, recall that return addresses as well as registers are stored on the STACK. If the programmer fails to retrieve all the registers he has saved, it is possible that when a RET instruction is executed the PC will be mistakenly loaded with 16 bits of data from some register pair. The CPU will then branch unpredictably into the memory executing obviously unplanned "instructions." A final word of caution should be said regarding STACK maintenance. The programmer should insure that as each subroutine returns, the STACK is in the same condition it was in when that routine was called. This is known as "cleaning" the STACK. If the STACK is not kept clean, it can keep growing until it finally starts erasing bytes containing the user's program.

As mentioned earlier, the most important advantage of using a STACK STRUCTURE is the ease of implementing nested subroutine calls. This is best shown by an example. Figure 16-14 contains an 8080 program with two levels of nesting. The program takes the data in register D and doubles it N + M times, where M is the number in register C and N is the number in register B. Assume the PC has been set to 0 and let us examine the execution of the program.

The first three instructions should be quite familiar by now. These instructions initialize register D to 32, register C to 2, and register B to 0. The PC reads "6" after processing these instructions. The fourth instruction, LXI SP,1000 is one that has not been mentioned yet. The OP-CODE "LXI" stands for "load immediate the register pair," that is, "LXI" is just like "MVI" except "LXI" loads a 16-bit register pair with data, while "MVI" loads an 8-bit register with data. The purpose of LXI SP,1000 is to initialize the STACK POINTER, a 16-bit register, so that the starting address of the STACK will be defined. The decimal number 1000 is loaded into the SP, and the PC is incremented to 9.

Location 9 contains the first CALL instruction, "CALL 19." To implement the CALL, the CPU first sets the PC to the address of the instruction following the CALL. Since the CALL instruction starts in location 9 and is 3 bytes long, the following instruction must be in location 12. The CPU now accesses the STACK to store the return address. The number in the SP is decremented by 1, to 999. The high-order byte of the PC, 0, is stored in location 999. The SP is decremented again, and the low-order byte, 12, is stored in location 998. The return address has been stored and the first subroutine should now be entered. The CPU accomplished this by loading the PC with 19, the address of the first instruction of "SUB 1."

The subroutine begins by storing registers B and C on the STACK in locations 997 and 996 with a "PUSH B" instruction. The routine now begins its task of computing M + N so that it can double "J" M + N times. The content of register C, 2, is

		IN FORTRAN	IN MACHINE LANGUAGE Address		IN CORE MEMORY Address	
MAIN PROGRAM		J = 32	0	MVI D,32	0	MVI D,
					1	32
		M = 2	2	MVI C,2	2	MVI C,
					3	2
		N = 0	4	MVI B,0	4	MVI B,
					5	0
			6	LXI SP,1000	6	LXI SP,
					7	232
					8	3
		CALL SUB 1 (J,M,N)	9	CALL 19	9	CALL
					10	19
					11	0
		STOP	12	HLT	12	HLT
				⋮		⋮
FIRST SUBROUTINE		SUBROUTINE SUB 1 (J,M,N)	19	PUSH B	19	PUSH B
		INDEX = M + N	20	MOV A,C	20	MOV A,C
			21	ADD B	21	ADD B
	2000	INDEX = INDEX − 1	22	MOV C,A	22	MOV C,A
			23	DCR C	23	DCR C
		CALL DOUBLE (J)	24	CALL 40	24	CALL
					25	40
					26	0
		IF (INDEX.NE.0) GO TO 2000	27	JNZ 23	27	JNZ
					28	23
					29	0
			30	POP B	30	POP B
		RETURN	31	RET	31	RET
				⋮		⋮
SECOND SUBROUTINE		SUBROUTINE DOUBLE (J)	40	PUSH PSW	40	PUSH PSW
			41	PUSH B	41	PUSH B
		J = J + J	42	MOV A,D	42	MOV A,D
			43	ADD A	43	ADD A
			44	MOV D,A	44	MOV D,A
			45	POP B	45	POP B
			46	POP PSW	46	POP PSW
		RETURN	47	RET	47	RET

Fig. 16-14. Program showing nested subroutine implemented by STACK STRUCTURE.

moved to the ACCUMULATOR. The value of "M" has now been placed in the AC. The next instruction adds the content of register B, "N," to the AC. The AC now contains M + N and this sum, 2, is moved back to register C. Register C will now serve as the variable "INDEX" defined in the FORTRAN version of the program. The PC now reads 23.

Location 23 contains another instruction not discussed before, "DCR C." The OP-CODE is the mnemonic for "decrement the specified register by the number 1." Subtracting 1 from the register is such a common operation, that the "DCR" instruction is included in the instruction set. Adding 1 is common as well, so an

"INR -REGISTER-" instruction (increment the register by 1) is also in the instruction set. Both "INR" and "DCR" are flag-setting instructions. Thus, any register, not just the AC, can be loaded with a number, counted down by repeated use of "DCR," and tested for zero. In this way, looping can be implemented. This is precisely the purpose of "DCR C" in this example. Register C is being counted down and tested for 0 by the JNZ instruction in location 27. This will be made clearer later. At this point, register C contains a 1, the FLAGS have been set, and the PC reads 24.

The second CALL instruction, "CALL 40," is now executed. The CPU takes the second return address, 27, and places it on the STACK on locations 995 and 994. The SP reads "994" at this time. The number "40" is then loaded into the PC. The second (nested) subroutine call has been made, and the machine starts executing instructions beginning in location 40.

The second subroutine begins by saving the FLAGS, ACCUMULATOR (PUSH PSW) and registers B and C (PUSH B) of the first routine. The FLAGS are saved by making a byte out of the FLAGS, with one bit for each FLAG. Since there are not 8 FLAGS, certain bits in the byte contain no information and are arbitrarily set. The "FLAG byte" is then stored on the STACK in location 992, and the AC is stored in 993. Registers B and C are stored in 991 and 990, respectively.

Now that the "prologue" of saving the processor FLAGS and some registers has been taken care of, the second subroutine can begin processing data. The number to be doubled, in register D, is moved to the AC. The "ADD A" instruction adds the contents of the AC to itself, thus doubling the number in the AC. The result, 64, is stored back in register D. The PC reads 45. The second routine now begins the return procedure. Registers B and C are restored with the "POP B" instruction. The SP is increased to 992. The "POP PSW" instruction causes the next two bytes from the STACK to be read in. The data in the first byte is broken up into bits representing the state of each FLAG, and the state of the FLAGS is restored. The next byte is loaded into the AC, restoring that value. The SP is again incremented by 2 to 994. The "RET" instruction in location 47 is now executed. Again the STACK is accessed, and the contents of bytes 994 and 995 are loaded into the PC. The SP is again incremented to 996. The PC now reads "27," and the second subroutine has returned.

The instruction in location 27, "JNZ 23," tests the state of the ZERO FLAG. Notice that the intention of this step is to test the state of the ZERO FLAG as it was after the "DCR C" instruction in location 23. If the state of the FLAGS had not been saved and then restored by the second subroutine, the "JNZ" instruction would erroneously be testing the FLAGS as they were after the "ADD A" instruction in location 43 was executed. This is because "ADD A" was the most recent flag-setting instruction, and after "ADD A" was performed the CPU reset all the FLAGS based on the result of the addition. However, since the FLAGS were saved by the "PUSH PSW, POP PSW" pair, the "JNZ" test will be testing the result of the "DCR C" instruction.

Since register C was decreased from 2 to 1, the ZERO FLAG was not set. This means that the jump will be taken since the CPU will jump on "not zero" from this instruction. Thus, the number 23 gets loaded into the PC and the looping begins again.

The "DCR C" instruction is executed, reducing C to 0. The ZERO FLAG is then set to 1 and will be tested later. The next instruction calls the second subroutine again. The return address of 27 gets stored on the STACK in locations 995 and 994. The STACK is growing again. The second subroutine is entered at location 40. The AC, FLAGS, and registers B and C are all placed on the STACK in locations 993, 992, 991 and 990, respectively. The datum in register D is moved to the AC where it is doubled. The result, 128, can only be stored by having the 8th bit on (128 is "1000 0000" in binary). Thus, the SIGN FLAG is set to 1 even though 128 is a positive number. The AC is placed back in register D, and the routine goes through the "epilogue" of restoring the calling program's information. After everything has been restored, the SP contains the number 994. The "RET" instruction pops the return address off of the top of the STACK. The SP is set to 996 and the PC is set to 27. The execution of the second subroutine is now complete.

Because the FLAGS have been restored, the "JNZ" instruction is testing the result of the "DCR C" instruction. Since that instruction previously caused the ZERO FLAG to be set to 1, the test for "not zero" fails, and sequential processing proceeds. The "POP B" instruction in location 30 is executed. The original values of the data in registers B and C are restored. The "RET" instruction now pops the return address of the main program off the STACK. The STACK is now "clean," with the SP pointing again to location 1000. The number 12 is loaded into the PC and the main return has been accomplished. The "HLT" instruction in location 12 is read in and processing stops.

Notice how easy it is to implement nested subroutine calls with a STACK. We see that just by stacking one return address on top of another the CPU can keep track of a virtually unlimited number of nested calls. The only limitation is the amount of memory available for storing calling information. We also see the importance of storing the calling program's registers and FLAGS. In the example given, the "JNZ" test would have been incorrectly processed if the FLAGS had not been saved and restored.

A pictorial diagram of this program's execution is shown in Fig. 16-15. The reader should trace through this rather lengthy example because it is important to understand how the STACK works. The reader should also pay attention to how the FLAGS are set, saved and restored.

In the following example, anything not shown in a particular panel is unchanged from the previous panel.

Before any instructions have been executed; PC has been set to 0.

REGISTERS			STACK	
A	RANDOM		Address	
B	RANDOM		999	RANDOM
C	RANDOM		FLAGS	
D	RANDOM		SIGN	RANDOM
PC	0000 0000 0000 0000	(0)	CARRY	RANDOM
SP	RANDOM		ZERO	RANDOM

After "MVI D,32," "MVI C,2," "MVI B,0" and "LXI SP,1000"

REGISTERS			STACK	
A	RANDOM		Address	
B	0000 0000		999	RANDOM
C	0000 0010			
D	0010 0000			
PC	0000 0000 0000 1001	(9)		
SP	0000 0011 1110 1000	(1000)		

Fig. 16-15. Pictorial diagram of the execution of the program shown in Fig. 16-14.

Fig. 16-15. (Cont.)

After "CALL 19;" PC is loaded with SUB1 address.

REGISTERS		
PC	0000 0000 0001 0011	(19)
SP	0000 0011 1110 0110	(998)

STACK Address	
998	0000 1100
999	0000 0000

Main RETURN address (12) is in locations 998, 999.

After "PUSH B" and "MOV A,C"

REGISTERS		
A	0000 0010	
B	0000 0000	
C	0000 0010	
D	0010 0000	
PC	0000 0000 0001 0101	(21)
SP	0000 0011 1110 0100	(996)

STACK Address	
996	0000 0010
997	0000 0000
998	0000 1100
999	0000 0000

REGISTER C is in location 996; REGISTER B is in location 997.

After "ADD B" and "MOV C,A" [FLAGS are set (for the first time) after "ADD B"]

REGISTERS		
A	0000 0010	
B	0000 0000	
C	0000 0010	
D	0010 0000	
PC	0000 0000 0001 0111	(23)

FLAGS	
SIGN	0
CARRY	0
ZERO	0

After "DCR C" (FLAGS are set)

REGISTERS		
A	0000 0010	
B	0000 0000	
C	0000 0001	
D	0010 0000	
PC	0000 0000 0001 1000	(24)

FLAGS	
SIGN	0
CARRY	0
ZERO	0

After "CALL 40;" PC is loaded with SUB2 address.

REGISTERS		
PC	0000 0000 0010 1000	(40)
SP	0000 0011 1110 0010	(994)

STACK Address	
994	0001 1011
995	0000 0000
996	0000 0010
997	0000 0000
998	0000 1100
999	0000 0000

SUB1 RETURN address (27) is in locations 994, 995.

Fig. 16-15. (Cont.)

After "PUSH PSW" and "PUSH B". On the FLAG word in location 992, bit 8 is the SIGN FLAG, bit 7 is the ZERO FLAG, and bit 1 is the CARRY FLAG (bit 1 = LSB). The other bits are not important.

REGISTERS

A	0000 0010	
B	0000 0000	
C	0000 0001	
D	0010 0000	
PC	0000 0000 0010 1010	(42)
SP	0000 0011 1101 1110	(990)

FLAGS

SIGN	0
CARRY	0
ZERO	0

STACK

Address		
990	0000 0001	C
991	0000 0000	B
992	0000 0010	FLAG word
993	0000 0010	AC
994	0001 1011	
995	0000 0000	
996	0000 0010	
997	0000 0000	
998	0000 1100	
999	0000 0000	

After "MOV A,D," "ADD A" and MOV D,A" (The FLAGS are set after "ADD A.")

REGISTERS

A	0100 0000	
B	0000 0000	
C	0000 0001	
D	0100 0000	
PC	0000 0000 0010 1101	(45)

FLAGS

SIGN	0
CARRY	0
ZERO	0

After "POP B" and "POP PSW" (The old FLAGS are restored after "POP PSW.") REGISTERS A, B and C are restored.

REGISTERS

A	0000 0010	
B	0000 0000	
C	0000 0001	
D	0100 0000	
PC	0000 0000 0010 1111	(47)
SP	0000 0011 1110 0010	(994)

STACK

Address	Contents
994	0001 1011
995	0000 0000
996	0000 0010
997	0000 0000
998	0000 1100
999	0000 0000

FLAGS (Restored)

SIGN	0
CARRY	0
ZERO	0

Fig. 16-15. (Cont.)

After "RET" PC is loaded with the RETURN address.

REGISTERS						STACK	
PC	0000	0000	0001	1011	(27)	996	0000 0010
SP	0000	0011	1110	0100	(996)	997	0000 0000
						998	0000 1100
						999	0000 0000

After "JNZ 23"

REGISTERS						FLAGS	
PC	0000	0000	0001	0111	(23)	SIGN	0
						CARRY	0
						ZERO	0

After "DCR C" (FLAGS are set.)

REGISTERS						FLAGS	
A	0000	0010				SIGN	0
B	0000	0000				CARRY	0
C	0000	0000				ZERO	1
D	0100	0000					
PC	0000	0000	0001	1000	(24)		

After "CALL 40" PC is loaded with SUB2 address.

REGISTERS						STACK	
PC	0000	0000	0010	1000	(40)	994	0001 1011
SP	0000	0011	1110	0010	(994)	995	0000 0000
						996	0000 0010
						997	0000 0000
						998	0000 1100
						999	0000 0000

SUB1 RETURN address (27) is in locations 994, 995.

After "PUSH PSW" and "PUSH B" [FLAG word: Bit 8 = SIGN, Bit 7 = ZERO, Bit 1 = CARRY (Bit 1 is LSB). Other bits are unimportant.

REGISTERS						STACK		
A	0000	0010				990	0000 0000	C
B	0000	0000				991	0000 0000	B
C	0000	0000				992	0100 0110	FLAG word
D	0100	0000				993	0000 0010	AC
PC	0000	000Q	0010	1010	(42)	994	0001 1011	
SP	0000	0011	1101	1110	(990)	995	0000 0000	

Fig. 16-15. (Cont.)

FLAGS	
SIGN	0
CARRY	0
ZERO	1

996	0000 0010
997	0000 0000
998	0000 1100
999	0000 0000

After "MOV A,D," "ADD A" and "MOV D,A" (FLAGS are set after "ADD A.")

REGISTERS		
A	1000 0000	
B	0000 0000	
C	0000 0000	
D	1000 0000	
PC	0000 0000 0010 1101	(45)

FLAGS	
SIGN	1
CARRY	0
ZERO	0

After "POP B" and "POP PSW" (The old FLAGS are restored after "POP PSW.") REGISTERS A, B and C are restored.

REGISTERS		
A	0000 0010	
B	0000 0000	
C	0000 0000	
D	1000 0000	
PC	0000 0000 0010 1111	(47)
SP	0000 0011 1110 0010	(994)

STACK	
994	0001 1011
995	0000 0000
996	0000 0010
997	0000 0000
998	0000 1100
999	0000 0000

FLAGS (restored)	
SIGN	0
CARRY	0
ZERO	1

After "RET" PC is loaded with RETURN address.

REGISTERS		
PC	0000 0000 0001 1011	(27)
SP	0000 0011 1110 0100	(996)

STACK	
996	0000 0010
997	0000 0000
998	0000 1100
999	0000 0000

After "JNZ 23"

REGISTERS		
PC	0000 0000 0001 1110	(30)

FLAGS	
SIGN	0
CARRY	0
ZERO	1

Fig. 16-15. (Cont.)

After "POP B" REGISTERS B and C are restored.

REGISTERS

A	0000 0010	
B	0000 0000	
C	0000 0010	
D	1000 0000	
PC	0000 0000 0001 1111	(31)
SP	0000 0011 1110 0110	(998)

STACK

998	0000 1100
999	0000 0000

After "RET" PC is loaded with RETURN address.

REGISTERS

PC	0000 0000 0000 1100	(12)
SP	0000 0011 1110 1000	(1000)

The "HLT" instruction in location 12 is executed, and processing stops.

Interrupts

All computers have external devices attached to them. These devices may be input/output hardware such as a terminal and a line printer or may be devices the computer is trying to control such as a motor or any automated process. To service all these devices, a computer may periodically check each device in succession to see if it needs attention. This is known as "polling" the devices. There are times, however, when an external device cannot wait for its turn to be polled. For instance, a computer-controlled nuclear reactor which suddenly starts to heat up must have immediate attention. A line printer which has finished printing one line and is now available for printing another must be able to quickly get the computer's attention or the overall system efficiency will be low. Much time may be wasted polling a device over and over again which needs very little attention.

For these and similar reasons it is clear that it would be helpful to have a mechanism for interrupting the computer; that is, a mechanism by which an external device could stop the CPU from doing whatever it is doing and have the CPU immediately service the interrupting device. This mechanism is known as the "interrupt structure." An "interrupt" is a single occurrence of some device using the interrupt structure to get attention. Thus, one says that a device "generates an interrupt" by sending appropriate signals to the interrupt handling part of the CPU. There are so many different ways of setting up an interrupt structure that it is impractical to try to even describe them all here. What follows in this section will be a discussion of what a programmer, and not a hardware designer, must know to successfully handle interrupts.

From the programmer's point of view, an interrupt can be thought of basically as a forced subroutine call which occurs unpredictably during normal program execution. To see why this is so, consider how an interrupt is supposed to work. The CPU is executing some program and suddenly a device determines that it must have immediate attention. The device will send some control signal to the CPU. On the 8080 this would consist of raising the voltage on the "interrupt line" from 0 volts to 5 volts (the interrupt line is one of the 40 pins on the chip). At this point the processor knows it must stop whatever it is doing and service some device. But when it is through servicing the interrupting device it must resume processing at the point where it was interrupted. Thus, the CPU will finish the instruction it was working on if the interrupt occurs in the middle of an instruction, and then store the address of the next instruction. This is a return address. The CPU will then initiate a jump to some part of the memory that hopefully contains the instructions for how to take care of the interrupting device. When the device has been serviced, the CPU will take the stored return address and load it back in the PC. Thus, conceptually the computer has just performed a normal subroutine call. The difference, of course, is that the processor never encountered a "CALL" instruction. The "CALL" is initiated when the external device generates an interrupt. The effect of generating an interrupt is to "insert" a CALL instruction immediately after the instruction the machine is executing at the time of the interrupt. Of course, this "CALL instruction" is not physically inserted into the memory; it is just that the machine responds as if a CALL instruction is suddenly the next instruction.

The reader may be wondering how the machine knows which location to jump to when its interrupt line is signaled; that is, how do we communicate to the CPU the address of the interrupt-handling code? There are many ways of doing this, but only the method used by the 8080 will be discussed as an example.

When the 8080 receives an interrupt signal, the processor first completes the instruction it is executing. At this point the PC contains location "NEXT" where "NEXT" is the address of the next instruction. To begin processing the interrupt, the CPU sets a certain bit to 1 on the CONTROL BUS. This bit is known as the "interrupt acknowledge" bit. The external hardware is looking for this acknowledgement from the CPU. When this acknowledgement is detected, an external device takes control of the DATA BUS away from the memory; that is, the memory can no longer place any information on the DATA BUS at this time. The external device is performing what is known as a "JAM." The "JAM" consists of taking control of the DATA BUS away from memory and "jamming" 8 bits of data on the DATA BUS. Since the CPU is about to enter a fetch cycle, the contents of the DATA BUS will be read in as the next instruction. By jamming a desired OP-CODE on the bus, the external device causes the CPU to read in an instruction that the processor thinks is the next instruction stored in the memory. In actuality the OP-CODE byte has been placed on the DATA BUS by the external hardware during the JAM.

While it is possible to jam any OP-CODE on the bus, the instruction the 8080 designers had in mind when they set up this interrupt structure is a "RESTART" instruction. The "RESTART" instruction is nothing more than a one-byte subroutine call. The reader may be puzzled by this. Since addresses are 16 bits (2 bytes) and the OP-CODE would take another byte, how can one design a one-byte CALL instruction? The only way this is possible is to allow a call to only certain locations in the memory; that is, suppose we allowed a subroutine to start in locations 0, 8, 16, 24, 32, 40, 48, and 56 only, all 8 bytes apart. By allowing only 8 different starting addresses, we could have an instruction in which 5 bits could specify the OP-CODE "RESTART," while the other 3 bits would be a 3-bit coded designation for one of the 8 possible starting locations. For example, if the 3-bit code is "001" that would mean "location 8;" if the code is "010" that would mean "location 16;" if the code is "011" that would mean "location 24." To generalize, the 3-bit code "$N_1N_2N_3$" would mean "location ($8*N_1N_2N_3$)" where N_1, N_2, and N_3 are each either 0 or 1. This is how the 8080 "RESTART" instruction works.

Thus, when the external device detects the "interrupt acknowledge" bit, it will most likely jam a "RESTART $N_1N_2N_3$" instruction on the DATA BUS. Since this instruction is a one-byte CALL, the normal calling procedure is followed. In particular, the address of the next instruction gets stored on the STACK. This address is the address of the instruction following the one in which the interrupt occurred. Thus, the CPU has taken care of the first task of interrupt handling, namely saving the address of where to return to once the interrupt servicing is completed. Since "RESTART $N_1N_2N_3$" is a call to location "$8*N_1N_2N_3$," the CPU will place that number in the PC. The CPU thus jumps to the starting address of the interrupt-handling subroutine, which the programmer should have placed in either location 0, 8, 16, 24, 32, 40, 48, or 56. We see that the 8080 interrupt structure involves both hardware and software; that is, hardware must jam a "RESTART $N_1N_2N_3$" instruction on the bus, and the programmer must know the value of $N_1N_2N_3$ ahead of time so that he can store his interrupt-handling software beginning in location "$8*N_1N_2N_3$."

The last line of most interrupt-handling subroutines is the "RET" instruction. When this instruction is executed the return address is popped off the STACK and placed in the PC. The CPU now resumes processing from the point where it had been interrupted. The entire interrupt has now been processed.

Just as there are times when a device needs immediate attention, there are times when the processor is performing some crucial calculations that it must complete. During these times it may be important to forbid any device from interrupting the CPU. The easiest way of accomplishing this is to have instructions which enable and disable the entire interrupt mechanism. When the system is disabled, the CPU will simply ignore any interrupt request. The two 8080 instructions for doing this are "EI" (enable interrupt) and "DI" (disable interrupt), respectively. As soon as the "DI" instruction is read in, the 8080 cannot be interrupted. However, when "EI" is executed the interrupt structure is not enabled until the completion of the instruction which *follows* the "EI" instruction. One reason for this apparent peculiarity can be seen by considering the mechanism of returning from an interrupt-handling subroutine. Often one does not want to be interrupted away from an interrupt-handling subroutine. For this reason, after each interrupt the CPU automatically turns off the interrupt structure. To allow further interrupts the programmer must execute an "EI" instruction. This is most often done right before the "RET" instruction is executed which causes the return address to be popped off the STACK and placed in the PC. Thus, it is common to find the instructions "EI" and "RET" next to each other. If the interrupt structure is enabled immediately after the "EI" instruction, an interrupt could occur before the "RET" instruction gets executed. The STACK may become cluttered with layers of return addresses. But by not enabling the interrupt structure until after the instruction which follows the "EI" instruction, the 8080 designers guarantee that the "RET" instruction will be executed, thus keeping the STACK clean.

An example of how interrupts work on the 8080 is given in Fig. 16-16. Assume the PC has been set to 0 and let us examine how this program executes.

	MAIN PROGRAM		INTERRUPT-HANDLING ROUTINE	
	0	DI	16	MVI B,8
Interrupt request occurs here →	1	LXI SP,128	18	EI
	4	EI	19	RET
Interrupt allowed here and takes place after "MVI A,3" →	5	MVI A,3		
	7	ADD A		
	8	HLT		

(Instruction to be jammed on bus in the event of an interrupt: "RESTART 2")

Fig. 16-16. Program showing interrupt mechanism on the 8080.

The first instruction the CPU reads in is "DI," which disables the interrupt structure. The structure is disabled because it is crucial that the STACK POINTER be set by the next instruction. If the SP is left uninitialized, it will be random. Then, when an interrupt occurs and the return address is stored on the STACK, the STACK will begin in an unpredictable location. If the STACK POINTER just happens to have the number 9 in it, storing the return address in locations 8 and 7 would erase the last two instructions in the main program.

The second instruction, therefore, sets the SP to 128. We now assume that some time during the execution of "LXI SP,128" an interrupt request is received. Since the interrupt structure is disabled, the CPU ignores the request and continues processing. We now assume that the interrupting device repeatedly tried to interrupt the 8080 and that an interrupt will occur as soon as the 8080 permits it.

The next instruction, "EI," begins the process of enabling the interrupt structure. The CPU enables circuitry that will allow interrupts after the instruction following the "EI" instruction is completed. That next instruction, "MVI A,3," sets the AC to 3. At the conclusion of the move instruction, the interrupt processing begins. The "interrupt acknowledge" bit on the CONTROL BUS is turned on. The external interrupting device then grabs control of the DATA BUS away from the memory and places the bit pattern "11010111" on the BUS, as shown in Fig. 16-17. This is the OP-CODE for "RESTART 2" (the "2" can be read as the 4th, 5th, and 6th bit where bit 1 is the least-significant bit). This instruction is equivalent to "CALL 16" ($8 \times 2 = 16$). The CPU reads the "RESTART 2" ("CALL 16") instruction from the DATA BUS, which appears to the CPU to be just the next stored instruction coming from the memory via the DATA BUS. To process the "CALL," the return address, 7, is stored on the STACK in locations 127 and 126. The number 16 is put in the PC, and the interrupt-handling routine has been entered.

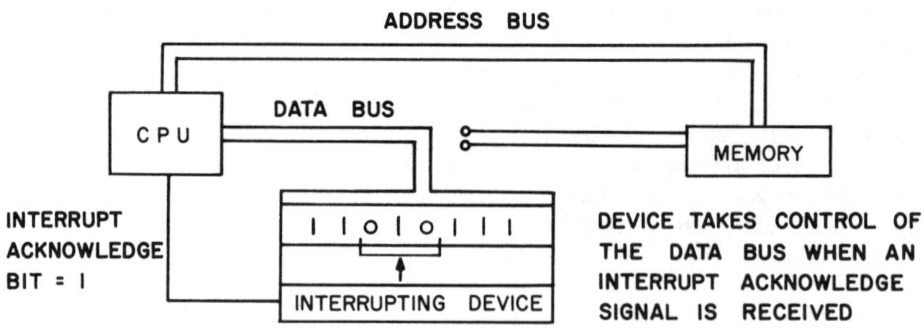

Fig. 16-17. Interrupting device "jamming" a "RESTART 2" instruction onto the DATA BUS.

For this example, the interrupt-handling routine consists of nothing more than loading the number 8 into register B. This is accomplished with the "MVI B,8" instruction in location 16. Recall that after an interrupt occurs the CPU automatically disables its interrupt structure just as surely as if a "DI" instruction had been executed. Thus, to allow for future interrupts, the interrupt mechanism, disabled since the PC read "16," must be enabled. The EI instruction in location 18 begins this process. The actual enabling does not take place until after the "RET" instruction in location 19 is completed. The "RET" instruction causes the return address, 7, to be loaded into the PC. The SP is incremented back to 128, and the STACK is clean. If another interrupt request had occurred at any time during the execution of the instructions in locations 16-19, it would have been ignored. Now that the "RET" instruction has been executed, an interrupt request would be acknowledged. At this point processing resumes from location 7, the address of the instruction the machine was about to execute when the interrupt occurred. The "ADD A" instruction is executed, followed by the execution of the "HLT" instruction and processing stops.

A pictorial diagram of the program's execution is given in Fig. 16-18.

	Interrupts Allowed?	SP	A	B	126	127	PC
After "DI"	NO	RANDOM	RANDOM	RANDOM	RANDOM	RANDOM	1
After "LXI SP,128" (Interrupt request occurs during this instruction)	NO	128	RANDOM	RANDOM	RANDOM	RANDOM	4
After "EI"	NO	128	RANDOM	RANDOM	RANDOM	RANDOM	5
After "MVI A,3"	YES	128	3	RANDOM	RANDOM	RANDOM	7

Since an interrupt request has been made, the interrupt occurs now. The external hardware jams a "RESTART 2" on the DATA BUS.

	Interrupts Allowed?	SP	A	B	126	127	PC
After "RESTART 2" ("CALL 16")	NO	126	3	RANDOM	7	0	16
After "MVI B,8"	NO	126	3	8	7	0	18
After "EI"	NO	126	3	8	7	0	19
After "RET"	YES	128	3	8	7	0	7
After "ADD A"	YES	128	6	8	7	0	8

The halt instruction in location 8 is read in and processing stops.

Fig. 16-18. Pictorial execution of program of Fig. 16-16.

Addressing Modes

In the course of discussing examples of machine-language instructions, we have seen that the operands can come from several different places. For example, in the "MOV A,B" instruction the datum is moved from one register to another. On the other hand, in the "LDA 3" instruction, datum is moved from a memory location (3) to a register (AC). Specifying where the operand is coming from is known as specifying the "addressing mode." Since we have seen that data could come from different places, it is important to understand the addressing mode of each instruction. Knowing the type of addressing mode answers the question, "Where is the datum?".

There are several addressing modes which designers have found useful to implement. The 8080 has four types, all of which will be discussed in this section. Other modes found on different machines are usually related to the four general types discussed here.

Register Mode

One type of addressing is called the "register" mode. In this mode, the operand is in some register and will end up in some register. Examples of this are "MOV A,B" and "ADD B." With instructions in this group, the answer to the question, "Where is the datum?" is, "It is in a register.".

Immediate Mode

Instructions using immediate addressing have the operand specified as part of the instruction. For example, in the instruction "MVI C,4" it is desired to move the *number* 4 to register C. The operand, 4, is specified right in the instruction. The datum is stored immediately after the OP-CODE byte. We see that the address of the datum is given relative to the OP-CODE byte; that is, if the OP-CODE byte is stored in location X, the datum is stored in location X + 1. Thus, the answer to the question, "Where is the datum?" is, "It is stored in the location immediately following the OP-CODE byte.".

Direct Mode

Instructions using direct addressing specify the address of the datum as a part of the instruction. For example, the instruction "LDA 24" informs the CPU that the address of the operand is location 24. The CPU must then access location 24 to find the actual datum to be moved to the AC. The answer to the question, "Where is the datum?" is, "It is in some memory location whose address is given directly as part of the instruction.".

Indirect Mode

Indirect addressing is harder to explain than the other three modes, and may at first seem very peculiar to the reader. Instructions using indirect addressing specify neither the datum nor the address of the datum directly. Instead, these instructions specify some register or location which contains the *address* of the datum. To see how this may work, assume register pair H-L contains the 16-bit address "40." Further, assume that location 40 contains the number 92. To move the datum in location 40 to some register, say register D, one could use the 8080 instruction "MOV D,M." In this case, "M" is a special mnemonic, representing "memory," which really just converts the normal register addressing of the move instruction into indirect addressing. When the "MOV D,M" is encountered, the CPU is informed that an indirect move is to be performed. The CPU will look at, by convention, the contents of register pair H-L. It will interpret the contents of H-L as the high- and low-order address bytes of the datum. The CPU will then access this location, "40," in our example, and will interpret the contents of this address as the datum. Since location 40 contains a 92, the number 92 will be loaded into register D. Notice that the address, 40, is not specified directly as part of the instruction. The "MOV D,M" instruction merely tells the CPU where to find the address of the datum (it will look at the contents of H-L). The names of registers "H" and "L" come from this special use of these registers in indirect addressing; that is, in indirectly-addressed instructions, H and L contain the high- and low-order address bytes of the datum.

Other forms of indirect addressing are found on various computers. One variation would be to allow the programmer to specify which register pair contains the address of the datum instead of fixing it to one particular register pair as the 8080 does. Another variation is to allow the programmer to specify the address of a location whose contents are the *address* of the datum. In this variation, some specified memory location would serve the same purpose as the H-L pair does on the 8080 during an indirectly-addressed instruction. We see that for indirect addressing the answer to the question, "Where is the datum?," is somewhat complicated. The answer is, "A register (H-L on the 8080) or a memory location will be specified which contains an address. The content of that address is the datum.".

The basic reason why indirect addressing is useful is that it is often desirable to let the *address* of the datum become a variable. To see an example of this problem, consider the problem of adding an array of 100 numbers stored in 100 successive memory locations. Two 8080 programs to perform this task are shown in Fig. 16-19.

ADDING LOCATIONS 0-99
WITHOUT USING INDIRECT ADDRESSING

a)

	ADDRESS	CONTENTS
	200	LDA 0
	203	MOV B,A
	204	LDA 1
	207	ADD B
	208	MOV B,A
	209	LDA 2
	212	ADD B
	213	MOV B,A
99 of	214	LDA 3
these	217	ADD B
blocks	.	.
of codes	.	.
	.	.
	692	MOV B,A
	693	LDA 99
	696	ADD B
	697	HLT

ADDING LOCATIONS 0-99
USING INDIRECT ADDRESSING

b)

	ADDRESS	CONTENTS
	200	LXI H,0
Initialization	203	MVI B,99
	205	LDA 0
	.	.
	.	.
	.	.
	208	INX H
	209	ADD M
Main Program	210	DCR B
	211	JNZ 208
	214	HLT

	ADDRESS	CONTENTS
	0	2
Data	1	5
	2	8
	3	4
	4	0
	.	.
	.	.
	.	.
	99	9

Fig. 16-19. Two ways of adding locations 0-99, with and without indirect addressing.

The first of these programs does not use indirect addressing. Thus, to specify 100 different locations, 100 location-specifying instructions must be used. In this case, "LDA X" is the location-specifying instruction. No looping is possible so 99 repetitions of essentially the same code must be written. The program occupies 499 bytes of memory. The second program, on the other hand, uses indirect addressing. Let us carefully examine how this 15-byte program works.

The first three instructions initialize the registers. Register pair H-L is set to 0, the starting address of the array of numbers. Register B is set to 99, one less than the number of elements in the array. The AC is loaded with the first piece of data, 2, located in location 0.

The main looping program begins in location 208. The instruction "INX H" is a register-pair increment. Just as "LXI H,0" operates on both registers of a register pair, "INX H" adds the number 1 to the 16-bit number stored in H and L. This is in contrast to "INR H" which adds 1 to the 8-bit number stored in register H. At this point, H-L contains the number 1.

The next instruction, "ADD M," uses indirect addressing. Once again, the "M" stands for memory. To find out where the datum is, the CPU looks at the address stored in H-L. Since H-L now contains a "1," location 1 is accessed. The datum in location 1, 5, is then read in and added to the AC.

The next two instructions are used for finding out when to stop looping through the array. "DCR B" decrements register B by 1. Since B contains one less than the number of elements in the array, 99, B will contain 0 when the loop has been executed 99 times. Testing for zero is done with the "JNZ 208" instruction. As long as B does not contain 0, the CPU will jump back to location 208 to execute the loop again. The "INX H" instruction in location 208 increases the contents of H-L to 2. Now when the "ADD M" instruction is encountered, the CPU is instructed to add the contents of location *2* (an "8") to the AC, since H-L now contains a "2." If the reader hand-simulates the rest of the looping, he will see how all 100 elements in the array are added.

The program using indirect addressing is obviously much shorter than the program in Fig. 16-19(a). The reason is that the address of the next data byte to be added to the AC varies from address 0 to address 99. Indirect addressing allows us to vary the address of the operand in a very simple way. By simply incrementing the contents of H-L, the computer can be made to loop through the array. A pictorial diagram of a part of this program's execution is shown in Fig. 16-20.

Finally, it should be noted that we have seen another use of indirect addressing, namely the "RET" instruction. The instruction does not specify where to return to but rather informs the CPU to look at the STACK POINTER which contains an address. To execute the "RET" instruction the CPU accesses this address, takes the contents of this address, and places it in the PC, causing a jump to that location. Thus, "RET" may be considered to be a "jump indirect" instruction utilizing the SP.

Input/Output

Many kinds of input/output (I/O) devices have been built for digital computers. The reader is probably familiar with disks, tapes, teletypes, and line printers, to name just a few. Conceptually, however, all these devices are the same to the computer in the sense that they all either transmit to or receive information from the processor. From the point of view of the programmer, it does not matter what I/O device is linked to the CPU; there is simply an I/O structure and some new instructions associated with using that structure that he must learn about. The structure and instructions will be the same regardless of what particular I/O device is attached to the computer.

From the programmer's standpoint, each I/O device possesses one or more I/O

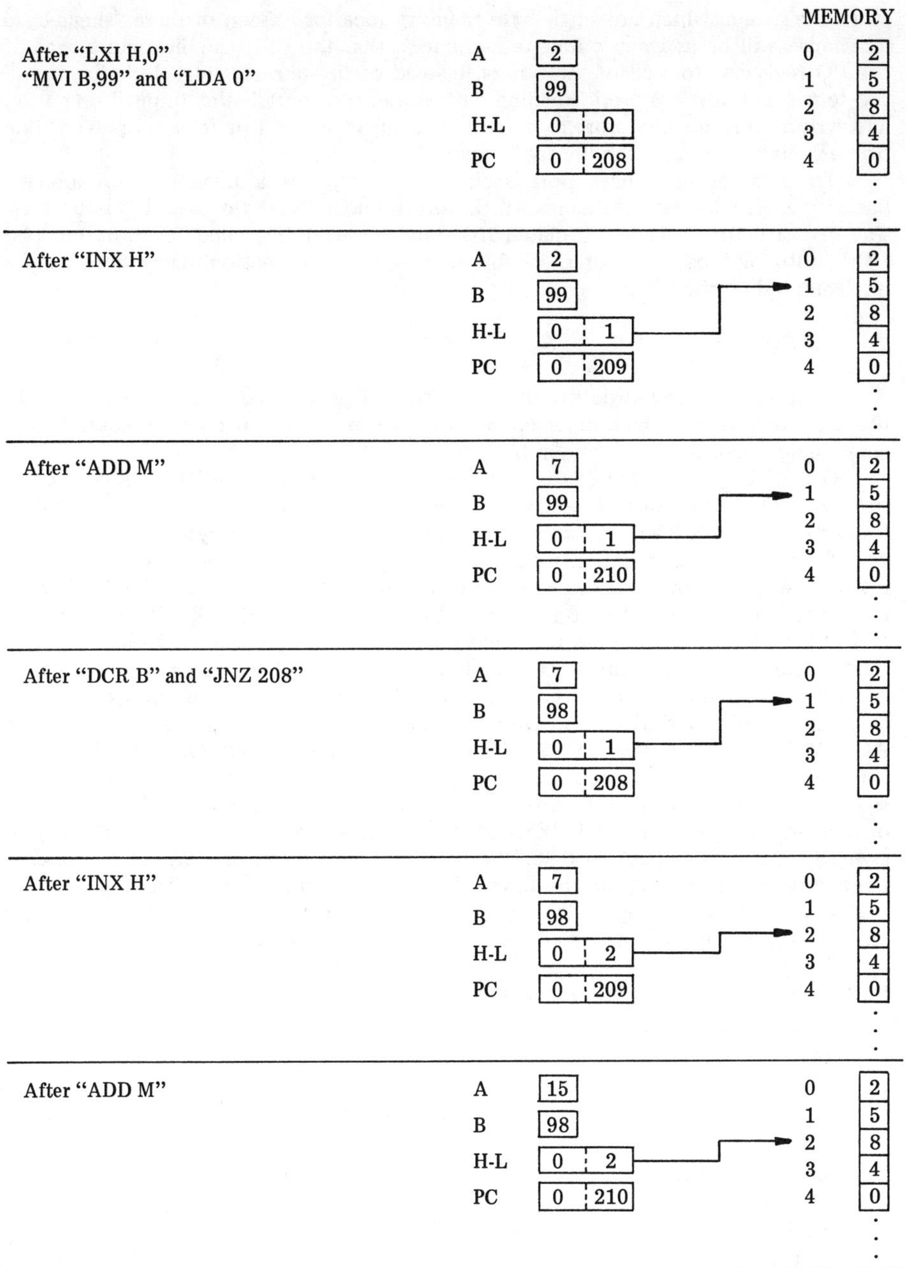

Looping continues until B contains 0. Then the halt instruction in location 214 is read in and processing stops.

Fig. 16-20. Diagram of partial execution of program given in Fig. 16-9(b), using indirect addressing.

storage registers, which are single-byte memory locations. Each of these "single-byte locations" will be assigned a unique number so that the CPU can distinguish each of the I/O registers from all of the others. Instead of the phrase "single-byte location," the term "I/O port" is used. Reading data stored in a port is the "input" operation and writing a result to a port to be stored (disk, tape, etc.) or to be displayed (line printer, teletype, etc.) is the "output" operation.

The numbering of each port is obviously a type of addressing. There are two basic types of addressing schemes which correspond to the two basic I/O structures. The first structure is called "isolated I/O" and the second is called "memory-mapped I/O." Both of these structures will be discussed in this section. They both can be implemented on the 8080.

Isolated I/O

In an *isolated I/O* structure the addressing of ports is completely separate from the addressing of core memory. Separate instructions will be needed to access data in core memory and to access data in I/O devices.

On the 8080, isolated I/O is implemented through the ACCUMULATOR using the two I/O instructions "IN--PORT#--" and "OUT--PORT#--." For example, when the instruction "IN 8" is executed, the 8080 reads in the data byte that is stored in port 8 and places that datum into the AC. To be more specific, when the "IN 8" instruction is encountered, the 8080 turns on an "IN" bit on the CONTROL BUS. It then places the number "8" on the first 8 bits of the ADDRESS BUS. Because the "IN" bit is on, the memory is informed that the number on the ADDRESS BUS is a port number and not a memory location. The memory, therefore, will relinquish control of the DATA BUS to the ports. Additional decoding logic will inform port 8 to place its information on the DATA BUS. The information is then read into the ACCUMULATOR and the "IN 8" instruction has been executed. The "OUT" instruction works similarly. For example, to execute the "OUT 59" instruction, the CPU first turns on the "OUT" bit on the CONTROL BUS before it places the number 59 on the first 8 bits of the ADDRESS BUS. This bit is turned on to inform memory to give up control of the DATA BUS. Port 59 then takes control of the BUS and latches on to the datum coming on the DATA BUS. This datum is the content of the ACCUMULATOR. Thus, we see that on the 8080, input and output using isolated I/O consists of single-byte transfers between the AC and I/O ports.

A brief example of this is given by the program shown in Fig. 16-21. Assume the PC has been set to 0, and let us examine the execution of this program.

IN FORTRAN	IN 8080 CODE		IN CORE MEMORY	
	Address	Contents	Address	Contents
J = 3	0	MVI B,3	0	MVI B
			1	3
READ K	2	IN 14	2	IN
			3	14
L = J + K	4	ADD B	4	ADD B
PRINT L	5	OUT 192	5	OUT
			6	192
STOP	7	HLT	7	HLT

PORT #14	PORT #192
35	RANDOM

Fig. 16-21. Program using isolated I/O structures.

The first instruction, "MVI B,3" initializes register B to 3. The next instruction, "IN 14," is the first I/O instruction of this program. When "IN 14" is encountered the CPU turns on the "IN" bit on the CONTROL BUS and places the number 14 on the first 8 bits of the ADDRESS BUS. Hopefully, port 14 is the specific device the FORTRAN programmer had in mind when he wrote the "READ K" instruction. Port 14 will now place its contents, 35, on the DATA BUS. This datum will be loaded into the AC. The next instruction, "ADD B," adds the AC to register B so that the AC now contains 38. The "OUT 192" instruction causes the CPU to turn on the "OUT" bit on the CONTROL BUS and places "192" on the first 8 bits of the ADDRESS BUS. Port 192 responds by taking control of the DATA BUS and latching on to the contents of the BUS. Since the contents of the AC will be placed on the DATA BUS during an "OUT" instruction, the contents of the AC, 38, will be written into port 192. After this, the "HLT" instruction in location 7 is read in and processing stops.

The coded version of this 8080 program drawn in "core memory" shows that the "IN" and "OUT" instructions are two-byte instructions. The first byte is the OP-CODE byte, and the second byte is used to specify a particular port number. Since an entire byte (8 bits) is available to specify a port number, we see that up to 256 different input devices and 256 output devices may be addressed using this isolated I/O structure on the 8080. This is more than adequate for most microprocessor applications.

A pictorial diagram of the program's execution is given in Fig. 16-22.

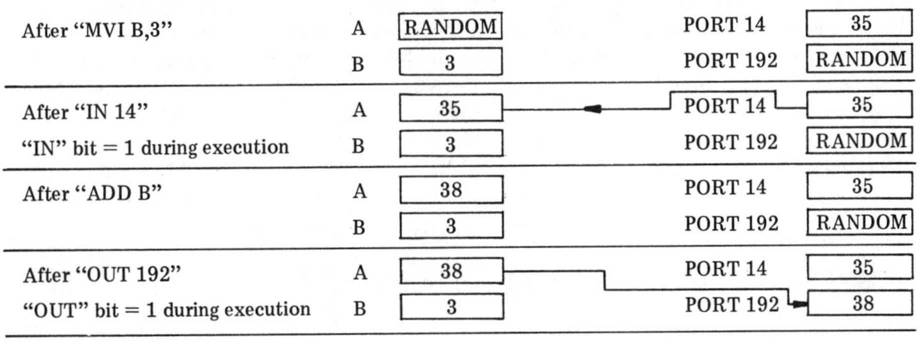

Halt instruction is read in and processing stops.

Fig. 16-22. Diagram of execution of program shown in Fig. 16-21, using isolated I/O.

Memory-Mapped I/O

The section on isolated I/O has shown that input/output operations are essentially read and write operations to special memory locations (ports). Since the ports are single-byte addressable locations, it is possible to consider them as part of the core memory. We can reserve locations in the core memory and put ports in them instead of the ordinary memory. For example, consider a machine such as the 8080 which can individually address 65,536 memory locations. On such a machine, we may choose to purchase and install only 65,000 bytes of core memory. The other 536 locations then become available for use as I/O ports. Using them as I/O ports requires only additional decoding logic. To see this, consider how the instruction "LDA 65001" would be executed. The CPU would place the number 65001 on the ADDRESS BUS, but the 65,001st byte of the core memory was never installed. Instead, decoding logic would recognize that an I/O operation is being requested. The logic would signal port 1 (now addressed as "location 65001") that it must place its contents on the DATA BUS. The datum on port 1 would then be loaded into the AC. As far as the CPU is concerned, it just reads in the datum stored in location 65001. In actuality, there is no core location 65001. Through decoding logic, the address "65001" becomes the numbering of an input device (port 1). This technique of addressing I/O devices by using core locations for which the core memory has not been installed is known as the technique of *memory-mapped I/O.*

In memory-mapped I/O, the addressing of ports and the addressing of core memory are exactly the same. The disadvantage of using memory-mapped I/O is that it limits the number of locations that may be used for storing programs and data. The advantage is that all the instructions that operate on core memory can now be used to facilitate I/O operations. For example, consider the 8080 instruction "MOV B,M." Suppose the number in H-L is 65001. Then "MOV B,M" will cause the contents of "location 65001" to be moved into register B (see the section on indirect addressing for details). If "location 65001" is really just the code for port 1, then the effect of "MOV B,M" is to have a byte transferred between an input device and register B. Notice that previously, using isolated I/O, data could only come from an input device to the ACCUMULATOR. Other "I/O instructions" are also created. For example, the instruction "ADD M" now can be used to add the contents of the ACCUMULATOR to the data in an input device. The instruction "MOV M,C" can be used to output the contents of register C to an output device. In effect, a new "I/O instruction set" has been created just by treating an I/O port as another memory location.

Let us reprogram the example given in the last section using memory-mapped I/O. The program may look like the one shown in Fig. 16-23. Assume that the PC has been set to 0 and see how this program works.

IN FORTRAN	IN 8080 CODE Address	Contents	IN CORE MEMORY Address	Contents
J = 3	0	MVI B,3	0	MVI B
			1	3
READ K	2	LDA 65001	2	LDA
			3	233
			4	253
L = J + K	5	ADD B	5	ADD B
PRINT L	6	STA 65002	6	STA
			7	234
			8	253
STOP	9	HLT	9	HLT

PORT 14 [35] PORT 192 [RANDOM]

Fig. 16-23. Program using memory-mapped I/O structure.

The first instruction "MVI B,3" initializes register B to 3. The next instruction "LDA 65001" normally would cause the contents of location 65001 to be loaded into the AC. On this machine, however, there is no "location 65001." When the address "65001" is placed on the ADDRESS BUS, it is decoded into an I/O operation involving port 14. Port 14 then places its contents, 35, on the DATA BUS, and this number is read into the AC. The next instruction adds B to the AC so that the AC now contains 38. The instruction "STA 65002" is another instruction using the memory-mapped I/O concept. When the address "65002" is placed on the ADDRESS BUS, it is decoded to mean "port 192." The content of the AC is then written into output device 192. The last instruction, "HLT," then stops further processing. A pictorial diagram of the execution of this program is shown in Fig. 16-24.

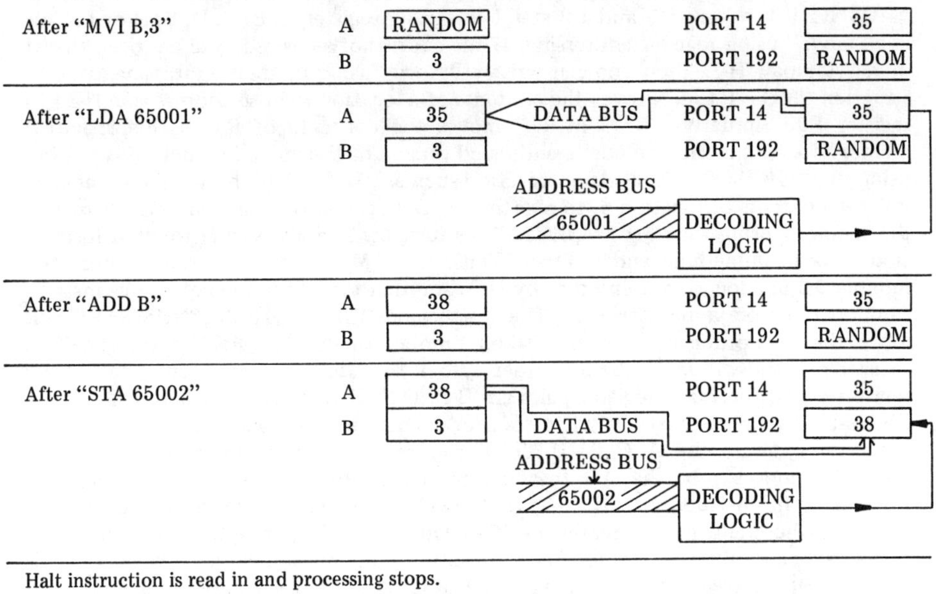

Fig. 16-24. Diagram of execution of program shown in Fig. 16-23, using memory-mapped I/O.

Classification Of Machine-Language Instructions

Throughout the examples of the 8080 machine code given so far, we have seen a wide variety of instructions. It would perhaps be useful at this time to try to classify the apparently endless list of instructions into groups. We will do so by presenting part of the 8080 instruction set since it is fairly typical of computer instruction sets. We will not list every instruction the 8080 is capable of executing. Instead, we will discuss enough instructions so that the reader will have a good idea of what to expect when he learns some complete instruction sets.

As the reader will see, many of the instructions from each group can be written using any of the four types of addressing modes discussed earlier (register, immediate, direct and indirect). To help classify the long list of instructions within each group, the reader should keep in mind that most major operations (e.g., add, move) will be listed as 3 or 4 separate instructions, one instruction corresponding to each type of addressing. Thus, there may not be as prohibitively many instructions as there first appear to be.

Data Transfer Group

Instructions in the *data transfer* group are concerned with moving data from place to place. On the 8080, none of the instructions in this group are flag-setting instructions.

The three basic operations of instructions in this group are moving data from the memory to the registers, moving data among the registers, and moving data from the registers back to the memory. Depending on the addressing mode used, these operations generate a variety of instructions.

The basic 8080 instruction for moving data between one register and another is "MOV R1,R2" where R1 and R2 stand for either register A, B, C, D, E, H or L. This is a "move" using register addressing. If the address of a desired byte has been stored in register pair H-L, then one can write a "move" using indirect addressing. The instruction "MOV R1,M" moves the contents of a location (whose address is in H-L) to register R1. Similarly, "MOV M,R1" moves the contents of R1 to that specified memory location. The immediate-addressed version of the move instruction is written using the "MVI" OP-CODE. An instruction such as "MVI R1,6" moves the number 6, not the contents of location 6, to register R1. Data can also be moved immediately to the memory using the "MVI M,DATA" instruction. The addressing for that instruction is both immediate and indirect. Thus, "MVI M,12" can be used to store the number 12 in a location pointed to by H-L. Another type of immediate move instruction operates on a register pair. The instruction "LXI RP,16-BIT NUMBER" will move the high-order byte of the 16-bit number to the first half of the specified register pair and will move the low-order byte of the number to the second half of the register pair. (The 8080 register pairs are B-C, D-E, H-L, and SP.) When using direct addressing in a move instruction to transfer a single byte between the processor and the memory, only the ACCUMULATOR can be used. The instruction "LDA --ADDRESS--" moves into the AC the contents of the byte whose address is directly specified. The instruction "STA --ADDRESS--" moves the contents of the AC into the byte whose address is directly specified. On the 8080, there is no direct-addressed move between a byte of memory and any other 8-bit register besides the ACCUMULATOR. Such a transfer involving a register other than the AC can only be done using indirect addressing.

Arithmetic Group

Instructions in this group are concerned with the basic arithmetic operations such as addition and subtraction. On some machines there are also multiplication and division instructions included in this group. The 8080, however, has no multiply or divide instructions.

The operations in this group on the 8080 basically have three addressing modes. These are the register, immediate, and indirect modes. Most of the instructions in this group set all the FLAGS.

One of the basic arithmetic operations is addition. The "ADD R1" instruction is the register-addressed version of the add. The content of R1 is added to the AC, and the result is placed in the AC. Similarly, "ADI --DATA--" (add immediate) causes the CPU to add the number given in the instruction to the AC. This is addition using immediate addressing. The instruction "ADD M" is addition using indirect addressing. The content of the byte pointed to by H-L is added to the AC.

Another group of addition instructions involve the CARRY FLAG. In these instructions, the contents of the AC, the data specified, and the value of the 1-bit CARRY FLAG (1 or 0) all get added together. This operation is useful in simulating multiple-byte additions on a one-byte operand machine. The three instructions which add with the CARRY are, "ADC R1" (add register R1 and the CARRY to the AC), "ADC M" (add the contents of the byte pointed to by H-L to the AC and add the CARRY), and "ACI DATA" (add the number immediately specified to the AC, and add the CARRY). It is obvious that these three instructions correspond to three of the addressing modes.

The analogous subtract instructions are implemented. There are three instructions to subtract some byte from the AC. These are "SUB R1" (subtract register R1 from the AC and put the result in the AC), "SUB M" (subtract memory from the AC), and "SUI --DATA--" (subtract immediate from the AC).

Another group of subtract instructions subtract the operand from the AC and then subtract the value of the CARRY FLAG. Since this is subtraction and not addition, the CARRY FLAG is called the "borrow" (but it is physically the same FLAG). There are three instructions that subtract with a borrow (that is, the content of the AC, minus the operand, minus the value of the CARRY FLAG, is placed on the AC). These three instructions are "SBB R1" (subtract register R1 with borrow), "SBB M" (subtract memory with borrow), and "SBI --DATA--" (subtract immediate with borrow).

All six subtraction operations are done using the two's complement arithmetic; that is, the two's complement of the number to be subtracted is taken and that number is then added to the contents of the AC (see Appendix A).

The other operations in this group are the increment (by 1) and the decrement (by 1) operations. These instructions are included because they often simplify counting procedures. On the 8080, two addressing modes for each operation are implemented, namely the register and indirect modes. The instruction "INR R1" adds 1 to the contents of register R1 while "INR M" adds 1 to the contents of the location pointed to by H-L. Similarly, "DCR R1" subtracts 1 from a register and "DCR M" subtracts 1 from memory. It is also often useful on the 8080 to increment or decrement a register pair. That is, it is useful to be able to add and subtract 1 from the two numbers stored in a register pair, now considered to be a single 16-bit number. For example, when the numbers in H and L are the two halves of a 16-bit address, this register-pair increment and decrement can be used to loop through an array of data [see Fig. 16-19(b)]. The instructions which do this are "INX RP" (increment the register pair) and "DCX RP" (decrement the register pair).

Logical Group

Instructions in this group perform logical (Boolean) operations. Most of the instructions in this group set all the FLAGS.

The Boolean logic is performed bit for bit between two operands, one of which is in the AC; that is, bit 1 of the AC is logically operated on with bit 1 of the other operand, bit 2 of the AC is operated on with bit 2 of the operand, and so forth.

Three basic logic operations, "AND," "OR" and "EXCLUSIVE OR" are each implemented on the 8080 in three addressing modes: register, indirect, and immediate. The register-addressed instructions are "ANA R1" ("AND" the AC with R1), "ORA R1" ("OR" the AC with R1) and "XRA R1" ("EXCLUSIVE OR" the AC with R1). The indirect-addressed instructions are "ANA M" ("AND" memory to the AC), "ORA M" ("OR" memory to the AC) and "XRA M" ("EXCLUSIVE OR" memory to the AC). The immediate-addressed instructions are "ANI --DATA--" ("AND" the data specified to the AC), "ORI --DATA--" ("OR" the data specified to the AC) and "XRI --DATA--" ("EXCLUSIVE OR" the data specified to the AC). There is only one "NOT" instruction, "CMA," which takes the bit-for-bit complement of the ACCUMULATOR.

Another type of logical operation is the rotate operation. To understand the rotate operation, imagine that bit 1 of the AC is connected to bit 8. Now suppose that each bit position gets moved by one position to the left or to the right. This is the rotate operation. For example, after executing the "ROTATE RIGHT" operation, the new bit 1 used to be bit 2, bit 2 used to be bit 3, and so on. Bit 1 gets rotated around and becomes bit 8. In the "ROTATE LEFT" the direction of bit rotation is reversed. For example, if the AC contained "11001100" and a "ROTATE LEFT" instruction is executed, the AC would contain "10011001." These instructions may seem strange, but they are useful in multiplication and division operations and have other uses as well. Two rotate instructions on the 8080 are "RLC" (rotate the AC left by 1 position) and "RRC" (rotate the AC right by 1 position). Sometimes it is useful to include the CARRY FLAG in the rotation; that is, the bit that gets shifted off the right or left end is moved into the CARRY FLAG. Meanwhile, the old value of the CARRY FLAG is rotated into the left or right end, respectively. Conceptually, this kind of rotate is the same that would occur if the CARRY FLAG were the 9th bit. To give an example, if the CARRY FLAG is set to 1 and the "RIGHT ROTATE THROUGH THE CARRY" instruction is executed, the AC will change from "11001100" to "11100110" and the CARRY FLAG will now contain a 0. The two 8080 rotate instructions which use the CARRY are "RAL" (rotate the AC left through the CARRY) and "RAR" (rotate the AC right through the CARRY).

A final type of operation in this group is the "COMPARE" operation. The compare operation is a pure flag-setting instruction. That is, no data in the memory or in the registers are changed by the compare instructions; only the FLAGS are affected. The actual comparison is done by subtracting the operand from the content of the AC, and not placing the result *anywhere.* The FLAGS, however, are all set according to the usual rules of subtraction.

To see the use of the compare instruction, suppose it is desired to determine whether or not the content of the AC is 39. One way to do this would be to execute the instruction "compare the content of the AC with the number 39." The number 39 is subtracted internally from the AC, but the result is not stored back in the AC; that is, the old value in the AC is not erased or changed in any way. If the AC does contain 39, the result of the subtraction will be zero, and the ZERO FLAG will be set to 1. Testing the ZERO FLAG after executing such a compare will establish whether or not the AC contains a 39 without altering the value of the AC or any other memory or register byte. The three 8080 compare instructions are "CMP R1" (subtract R1 from the AC and set the FLAGS), "CMP M" (subtract memory from the AC and set the FLAGS) and "CPI --DATA--" (subtract the number specified from the AC and set the FLAGS).

Branch Group

Instructions in the branch group alter the normal sequential processing. None of the instructions in this group affect the FLAGS, in the 8080.

Two basic "JUMP" instructions belong to this group, conditional and unconditional jumps. The unconditional jump instruction, "JMP --ADDRESS--" causes the 8080 to begin executing code starting in the location specified. The jump is accomplished by placing the desired address in the PC. The conditional jump is a flag-testing instruction. The jump to the specified address will only occur if the state of a FLAG is the same as the test condition in the instruction. For example, "JZ 100" will cause a transfer of control to location 100 only if the ZERO FLAG is currently set to 1. Each FLAG can be tested for a 1 or a 0 condition. If the condition is not met, sequential processing proceeds as usual.

There are two basic "CALL" operations, conditional and unconditional. The 8080 unconditional call instruction "CALL --ADDRESS--" works just like an unconditional jump except that a return address (the address of the instruction immediately following the CALL) is stored on the STACK. For the conditional call instruction, the CALL will occur only if the condition is met. Otherwise, normal sequential processing continues.

The 8080 also has a conditional and unconditional return instruction. The unconditional return instruction ,"RET," causes the top of the STACK to be placed in the PC. The conditional return causes a return only if the condition is met, and, if not, normal sequential processing continues.

Two other special branch instructions are available on the 8080. The "RESTART $N_1 N_2 N_3$" instruction is a one-byte call instruction to location "$8{*}N_1 N_2 N_3$." This instruction is used primarily in interrupt handling. The other branch instruction is "PCHL." This instruction causes the contents of register pair H-L to be loaded into the PC. Since the PC points to the next instruction, "PCHL" is really a "JUMP INDIRECT" instruction using H and L to store the address of the next instruction.

Stack, I/O and Machine Control Group

The STACK, I/O and Machine Control Group are somewhat of a miscellaneous collection of instructions that do not fit into the other four groups. Most instructions in this group do not affect the FLAGS.

Typical STACK-manipulating instructions are the "PUSH" and "POP" instructions. Any of the registers and the FLAGS can be stored on the STACK with a push instruction. "PUSH B" stores register B and C, "PUSH D" stores D and E, "PUSH H" stores H and L, and "PUSH PSW" stores the AC and the FLAGS. "POP B," "POP D," "POP H," and "POP PSW" take the top two bytes off the STACK and place the bytes in the specified register pair.

There are two I/O instructions on the 8080 when isolated I/O is used. The instructions are "IN --PORT#--" and "OUT --PORT#--." The input instruction is performed by having the input byte read in from the specified port and placed in the AC. The output instruction is performed by taking the content of the AC and sending it to the specified port.

There are several 8080 machine control instructions. For example, the programmer can control whether or not he wishes to allow interrupts by executing the "DI" (disable interrupts) or "EI" (enable interrupts) instruction. Another machine control instruction is "NOP" (no operations). When this instruction is encountered, no operation whatever is performed. "NOP" is sometimes useful in debugging a program, but seldom appears in a final version of a program. A final machine control instruction on the 8080 is the "HLT" (halt) instruction. Execution of this halt instruction causes the 8080 to stop processing.

Miscellaneous

The Reset Control Line

Every computer has some set of wires going into the machine called control lines which can affect the operation of the computer. For example, we have discussed the interrupt control line, the line on which an external device signals the CPU that it needs immediate attention. Other control lines have not been discussed in this chapter because the emphasis here is on programming (software) and not hardware. There is one more control line, however, that is worth discussing, that is, the reset control line.

In all the 8080 examples given, it has been repeatedly stated that the reader should "assume the PC has been initially set to 0." The careful reader may have been puzzled as to how this was done. It is clear that no software instruction could initialize the PC. This is true since the PC always points to the next instruction. If the PC is random when we first turn on the processor, how could we get our PC-initializing instruction executed? The PC would have to contain the address of the PC-initializing instruction for it to ever get executed!

Thus, when power is first applied to the processor, there must be an electronic method of initializing the PC. The physical mechanism of initialization uses the reset control line. Some machines may be constructed so that when the reset signal is detected the CPU will read in some switch settings and set the PC to the binary configuration on the switches. This would allow a reset-jump to any location. On the 8080 the reset-jump location is fixed. Whenever the voltage on the reset line is changed from 0 volts to 5 volts, the PC is electronically forced to 0. Thus, the PC can only be reset to 0, and so the first instruction in a power-up situation must be located in location 0. Of course, if it is desired to actually begin the program in location 32531, the instruction in location 0 can be "JMP 32531." After the PC is initially set to 0, the value of the PC will depend on the sequential flow of processing and on any branching or halt instructions encountered.

Real-Time Programming

To computer-simulation programmers, the amount of time the computer takes to process a program only matters financially. That is, the more "computer time" which is used to process a program, the more money it will cost. A much more serious "computer time" problem is faced by the real-time programmer.

The real-time programmer is concerned with using a computer to control a process that is now taking place. The process is not being simulated; it is real and will need updated control every so often so that it will perform properly. The problem of real-time control is that the computer takes a finite amount of time to execute each instruction. A certain control algorithm may be a stroke of genius, and a particular program may be brilliantly written, but, if the program takes so long to execute that the process goes out of control, the program is useless.

In order to figure out how long a program will take to execute, the programmer must first understand a little about the computer's *clock*. Throughout this section we have been stating that the events involved in computer processing take place in an orderly, step-by-step manner. We have mentioned that to implement this orderly processing, the CPU has a finite-state machine (FSM). As various control signals are detected by the FSM, the latter will order fetch cycles, data transfers, input operations, etc. The "ordering" will be done with control signals consisting of logic levels and pulses. However, it should be understood that the FSM of a digital computer is not continuously ordering new operations. A digital computer is a synchronous machine, which means that its finite-state controlling circuitry can change its state (and thus issue new commands) only when it receives a *clock pulse*. A "clock pulse" is a very brief occurrence of one of the control lines (the "clock" or "timing" line) being forced from logical 0 to logical 1 and back again to logical 0 (or vice versa on some machines). These clock pulses are continuously being generated by a device known as the *system clock*. Only when the FSM receives another clock pulse can a new computer action take place. Obviously, it is desirable to have as high a clock rate as possible, since instructions clearly *cannot be executed faster than the clock rate.* The maximum clock rate of a computer is determined by inherent limitations (such as capacitance) in the electronics.

On the 8080, the clock rate can be set by the designer over a range of values. Thus, it is impossible for the manufacturer to specify in units of seconds how long each instruction takes to perform. What they can specify, however, is how many clock pulses it will take to execute each instruction. Then, by knowing the pulse rate, the user can figure out the execution time of each instruction on his system. The basic unit of "time" of each instruction given by the manufacturer on the 8080 is the "state." The number of "states" that each instruction takes to execute is listed in the operations manual. The actual time an instruction takes to execute is found by multiplying the number of states by the clock period. Thus, using a clock producing pulses at a rate of 2 megahertz, having a period of 0.5 microsecond per pulse, each "state" would correspond to 0.5 microsecond of real time. By adding up the total number of states that a program requires to execute and multiplying by the clock period, the user can determine how long his program will take to execute. If the program takes too long, the programmer must shorten his program, use a faster processor, or try some fancy multiple-processor scheme to cut the computation time.

Summary

In Sec. 16.4 we have seen how the computer architecture and the underlying basic principles of programming combine to form the modern digital computer.

We began Sec. 16.4 with a simple example of a program which used no branching. Many assumptions were made about the CPU for the sake of simplicity. As the section progressed these assumptions were relaxed. For example, it was initially assumed that data could only be loaded from the memory into the AC. After indirect addressing was discussed, it was pointed out that instructions such as "MOV C,M" cause data from the memory to be transferred to a register other than the AC. For another example, it was initially assumed that in all arithmetic operations, one of the operands had to be in the AC. Later, in discussing instructions such as "INR B" and "DCR B" we saw examples of instructions in which none of the operands were in the AC.

Branching was discussed early. While unconditional branching involved no new hardware, conditional branching required the introduction of FLAGS, set by the CPU after certain instructions are executed. Related to branching was the topic of subroutine calls. To show how to efficiently implement nested subroutine calls, the STACK and STACK POINTER were discussed. It was later seen that the STACK had other uses as well, such as its use in interrupt processing.

Input/output structures were discussed. It was shown that I/O devices can be thought of as containing single-byte memory locations (ports) which must be addressed. The addressing of ports led to the development of the two basic I/O structures, isolated and memory-mapped I/O.

We turned our attention away from hardware for a while to classify and categorize machine-language instructions. Four different addressing modes were discussed. It was shown that for each basic group of instructions, up to four of the instructions within each group could actually be performing the same major operation. The difference between the four instructions would be just the type of addressing mode used to specify where the operand is.

Finally, we returned to discuss hardware briefly by describing the reset control line and the system clock. It was seen that using the reset line is the only way to initialize the PC from a cold start. Knowing the clock rate is important to the real-time programmer.

We have now built up a rather complete picture of a computer. All the major building blocks have at least been mentioned. The reader should study the block diagram of a computer given in Fig. 16-25, and be certain that the role each part plays in computer processing is understood.

At long last we are ready to begin our discussion of assembly-language programming.

16.5 ASSEMBLY-LANGUAGE PROGRAMMING

The Need For An Assembler

Up to now every instruction presented has been written in machine code. In writing the examples certain mnemonics for OP-CODES have been used, and data have not always been written as a string of 1's and 0's, but the binary translation of

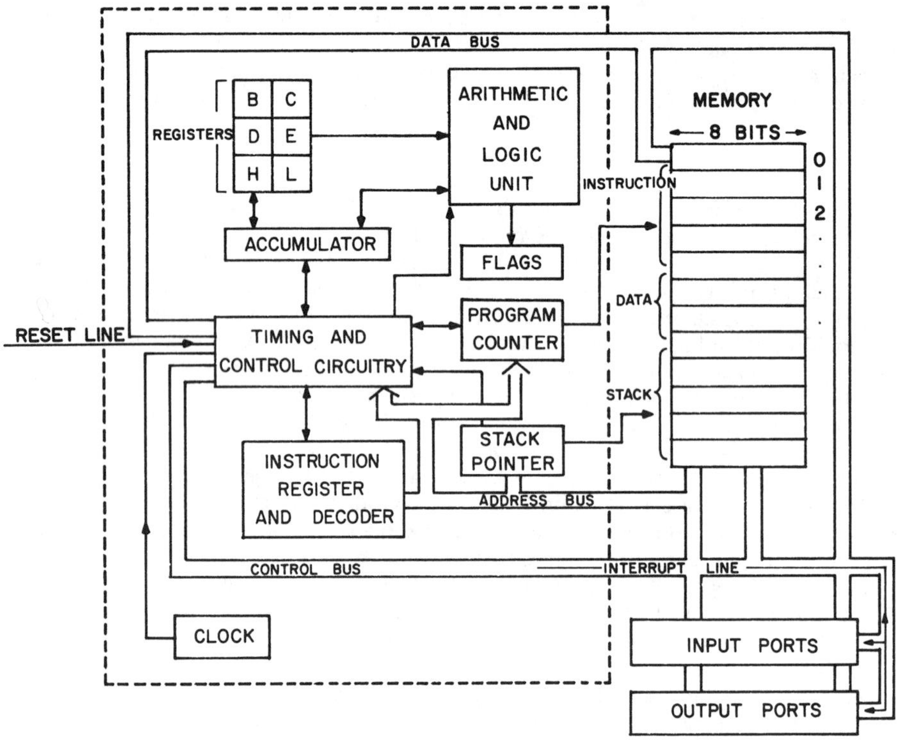

Fig. 16-25. The digital computer, exemplified by the 8080 microprocessor.

the instructions and data have been often given to emphasize that the computer can only understand binary codes. The mnemonics have been used for our convenience in writing the examples, but it was always understood that once the code was written we would have to look up the binary equivalents of the OP-CODES and write the data as binary numbers as well.

Computer scientists soon realized that this was very tedious to do. Programs were written whose inputs were letters of the alphabet and numbers like "MVI A,7" and whose outputs were the binary codes that the machine could understand. Such programs came to be known as "assemblers" because they assembled a machine-language program out of the more complicated input. If an assembler is not available to a computer user, he himself must translate the mnemonic code into binary. For those who have ever attempted to do this tedious translation, the need for an assembler is obvious.

A Primative Assembler

The simplest assembler is one which can translate OP-CODE mnemonics into the correct binary OP-CODES and translate data given in decimal into binary. This assembler is too primative to be very useful, but it will be discussed for the purpose of illustration as we add to our knowledge on assemblers.

In order to be able to read in letters of the alphabet and digits, these symbols must be coded into binary. Two commonly used codes are "EBCDIC" (which uses an 8-bit code) and "ASCII" (which uses a 7-bit code). Each letter, digit, and punctuation symbol is assigned a unique 7- or 8-bit code. For example, the letter "A" may look like "11000001" in some binary code. With all the symbols coded, it is possible to design input devices like teletypes in which the user can type letters and other symbols. The output of the device is the correct binary-coded version of the symbol, which is sent to the computer.

Using such an input device we can write a primative assembler. Suppose such a device is linked to the CPU, and the CPU started reading in a string of bytes in succession. Suppose the bytes are the binary code for "A," "D," "I," "-BLANK-," "9," "8," "-BLANK-." The assembler program must keep reading data until it can determine the instruction. The assembler program will look at the first letter read in, "A." The program will now jump to an internal part of code that contains a list of all the instructions that mnemonically begin with the letter "A." Now the next letter, "D," is used to branch to the part of code that contains a list of all instructions that begin with an "A" followed by a "D." Next, the "I" is used to further narrow down the instruction. When "-BLANK-" is read in, the assembler is informed that the OP-CODE has been completely specified. At this point, there must be no ambiguities about what instruction is being requested. If the indicated branching has been set up properly, the assembler should now know that the instruction requested is "ADI" (add immediate). A table of OP-CODES stored within the assembler is now accessed. The assembler finds out that the OP-CODE for "ADI" is "11000110." It outputs this byte to some storage device. The assembler is also programmed to recognize that after the "ADI" OP-CODE some data will be specified. Therefore, the assembler looks at the next few bytes until it finds another "-BLANK-." It now knows the data are "9" and "8." Some subroutine will be called to convert the symbolic code for the 2 bytes "9" and "8" into the single-byte binary number "98" ("01100010"). This byte will also be sent to the storage device. The entire instruction "ADI 98" has now been successfully translated into binary by the assembler. Continuing in this manner, the assembler can translate all the OP-CODES and data, now mnemonically written using letters and decimal digits, into the binary string of 1's and 0's that the CPU can understand. We see that our primative assembler is a fancy look-up table and an ASCII- or EBDCIC-to-binary data converter.

The One-Pass Assembler

Now that we have found a way to avoid writing all binary numbers for our programs and data, it would be nice to find a way of symbolically writing addresses. For example, up to now if we want to call a subroutine in location 200 we have been writing "CALL 200." It would be easier to program if we could just write "CALL SUB," and then tell the machine that "SUB" is the symbol for "location 200." For another example, if we want to store the contents of the AC in some variable, we first have to pick an available memory location, say location 50. We then remember that location 50 would be the address of the changing data. To actually store the AC, we use a "STA 50" instruction. It would be much nicer if we could just write "STA X," and then let the machine know that address "X" is location 50. It would be even better if we did not have to explicitly pick any particular location, since we really do not care where address "X" is as long as data can be stored and retrieved from "X" properly.

One program which will allow us to do some of the things just mentioned is called a *one-pass assembler*. The assembler's input is a block of instructions written using mnemonics and symbolic addresses. Its output is a binary translation of the instructions. For example, the assembler might encounter a statement "JNZ LOOP1" where "LOOP1" is a symbol representing some address and "JNZ" is the mnemonic OP-CODE for "11000010." Translating the OP-CODE involves nothing more than searching a look-up table. The assembler is also supposed to take out the letters "LOOP1" and replace them with the numerical value the symbol actually represents. For example, if the programmer has named location 504 "LOOP1," then the instruction "JNZ LOOP1" is actually "JNZ 504." The assembler must find this out. One way the assembler could find out the value of "LOOP1" is by the programmer explicitly informing the assembler. A statement could be written saying that location 504 has been given the name "LOOP1." Therefore, every time the symbol "LOOP1" is used, the assembler would know it should simply replace these letters with the number 504. This information would be stored in a *symbol table*. The symbol table is an assembler-generated list of the names of all symbols used in a program, and the numerical value of each symbol. In the above example, when the assembler is informed that the symbol "LOOP1" has the value "504," the assembler would record an entry in the symbol table of the form "LOOP1 — 504." Then whenever it encounters the letters "LOOP1" it will go to its symbol table, find the name "LOOP1" and look up its value (504). It will then substitute the number 504 for the symbol.

Besides being explicitly informed by the programmer that a symbol has a certain value, there is another important way the assembler can determine the value of a symbol. This second way involves computing the address of any instruction in a program just by knowing the byte length of each instruction and knowing where some byte of the program is. For example, suppose we are told that the instruction "MVI B,3" is in location 450. Now suppose a program contains the statements:

```
       MVI B,3
LOOP:  DCR B
       JNZ LOOP
```

We can look up the length of the "MVI B,3" instruction. It is two bytes long. Therefore, the next byte must be in location 452. The programmer has indicated he wishes to give the location which contains the "DCR B" instruction the name "LOOP." He has indicated this by writing the name "LOOP," followed by a colon, followed by the instruction to be stored in address "LOOP" ("DCR B"). By writing "LOOP: DCR B" the programmer has defined the value of the symbol "LOOP" to be the address of the instruction "DCR B." We have figured out that the address of this instruction is location 452. Thus, an entry should be made in the symbol table "LOOP — 452." We are able to determine the value of the symbol "LOOP" *without explicitly being told* that "LOOP" represents the number "452." The programmer himself need not have calculated where "LOOP" is. By simply knowing the address of some byte of the program, an "origin" is established. In this case, we are told that the "MVI B,3" instruction is in location 450, and this establishes the origin. After the origin is known, the address of any instruction can be calculated by counting how many bytes of program have been used since the start. This requires knowing the length of each instruction, information which is always available. Now that we know the value of "LOOP," the instruction "JNZ LOOP" is easy to translate. We first translate the OP-CODE "JNZ" into the byte "11000010." We next look up the name "LOOP" in the symbol table. The numerical value of "LOOP" is listed as being "452." Thus, we replace the symbol "LOOP" with the number "452." The instruction now reads "JNZ 452." The number 452 is then translated into binary.

The one-pass assembler works in this manner. The assembler is given the address of the first byte of the program. This establishes the origin. Only the origin need be specified numerically. All other addresses can be specified symbolically. This is done by attaching a symbolic name to a byte to define the symbol as being the address of that byte. The named byte may contain data or an instruction. The assembler will automatically compute the value of the symbol by counting the length of all the instructions consecutively specified from the origin. As the assembler determines that a symbol is being defined (that is, when it finds a name followed by a colon), the symbol name will be entered in a symbol table along with the symbol's numerical value. Thus, the assembly-language programmer need not specify (or need not even know) the value of each symbolic address. It can be computed for him by the assembler.

To see in more detail how the one-pass assembler works, consider the program shown in Fig. 16-26. Let us call the location of the first byte of the program the "origin," abbreviated "ORG." Suppose we know the numerical value of address "ORG" is 200. Let us see how we could "assemble" the code, that is, generate the binary code that represents the instructions.

```
Y:      0          ←The "origin" is the address of this first byte.
X:      6
        MVI B,2
        LDA X
LOOP:   ADD A
        DCR B
        JNZ LOOP
        STA Y
        HLT
```

Fig. 16-26. Program demonstrating one-pass assembler code.

The first byte of code contains a piece of datum, the number 0. The programmer wants to give this address of the datum the name "Y." This is indicated by writing the name "Y" followed by a colon. Since the origin has been specified as being at location 200, and since address "Y" is at the origin, address "Y" must be at location 200. We must keep track of this information. To do so, we make an entry in the symbol table of the form "Y − 200." The programmer wants location Y (200) to contain a 0. Therefore, we write out to some device the information that location 200 should contain all zeros. We have now translated the first line of the assembly code.

The next byte is also named by the programmer. The name given is "X." Since the first line of the code consists of a single byte of datum, the address of this next byte must be "ORG + 1," or 201. Therefore, the entry "X − 201" will be made in the symbol table. The programmer wishes to store the number 6 in this byte, so we write down the information that location 201 should contain the number 6. Since the second line of code also consists of a single byte of datum, it is known that the next byte is in location "ORG + 2," or 202.

The next byte is the first actual instruction of the program. Since the data in addresses X and Y take up two bytes, the "MVI B,2" instruction is in location ORG + 2 (202). To translate this instruction this instruction, we look up the binary code for "MVI B" and then write the 8-bit binary representation of "2." Since "MVI B,2" is a two-byte instruction, the next instruction must be in location ORG + 4 (204).

The next instruction, "LDA X," makes use of the symbol table. When we encounter the "LDA X" instruction, we look up "X" in the symbol table. We find that "X" has the value of 201. We, therefore, can take the instruction "LDA X" and rewrite the instruction as "LDA 201." We know how to translate "LDA 201" into binary. The reason we were able to successfully translate "LDA X" is that, although "X" is a symbol, we have stored the numerical value represented by X in the symbol table. Thus, we can look up "X" to find what binary bit pattern it represents in exactly the same way we can look up the "LDA" OP-CODE to find what bit pattern it represents. The "LDA X" instruction takes three bytes, so the next byte is in location ORG + 7 (207).

The next byte is given the name "LOOP." We enter this in the symbol table as "LOOP — 207." Notice that every time a colon is encountered, it means another byte is being given a name. Thus, symbol table entries are made whenever a colon is encountered, and the symbol table is used whenever a symbolic reference is made in an instruction. The instruction in "LOOP: ADD A," uses no symbolic addressing and can be immediately translated to binary. Similarly, the next instruction, "DCR B," can be translated immediately into binary. These instructions take two bytes, so that the next byte is in location ORG + 9 (209).

To translate the instruction "JNZ LOOP" again requires the use of the symbol table. We look up "LOOP" in the table and find the value of the symbol "LOOP" is 207. Thus, "JNZ LOOP" is really "JNZ 207." Having made the substitution, we can now readily translate the three-byte instruction "JNZ 207" into binary.

Similarly, we can translate the next instruction, "STA Y," with the help of the symbol table. The symbol "Y" is found to have the value 200. The instruction "STA 200" must, therefore, be translated into binary. We have no problem immediately translating this or the last instruction, "HLT." The program has now been translated into binary and is ready for execution.

We see that by storing symbols in a table as the symbols are defined and looking up the value of the symbols as they are used, we can implement symbolic addressing. The one-pass assembler works exactly in this fashion. Using this assembler, we no longer have to write instructions in the form, "JNZ 207," specifying the exact address to jump to (this is called the *absolute address*). Instead, we can write "JNZ LOOP," with "LOOP" known to be seven locations past the origin, and the origin known to be the absolute location 200. Specifying addresses relative to an origin is known as *relative addressing.*

The method the assembler uses to compute the value of a defined symbol can be conceptually thought of in terms of relative addressing; that is, we can think of the assembler counting the number of bytes away from the origin of each line of assembly code. When a symbol is defined as being the address of a particular line, the assembler knows the value of the address relative to the origin. By also knowing the numerical value of the origin, the assembler can compute the actual numerical value of the symbol.

Figure 16-27 shows how the one-pass assembler builds up and then uses the symbol table. Note that after the assembler has read in the last instruction, the binary translation is immediately available for execution. The assembler needed to read the assembly code only one time to translate the code into binary. It is because the assembly code is only read once that this type of assembler is called a "one-pass" assembler.

Pseudo-Ops

Pseudo-ops (pseudo-operations) are instructions to the assembler only. They are not instructions that ever get executed by the computer. These nonexecutable instructions inform the assembler of facts it must know to correctly translate the user's assembly code. There are many pseudo-ops on a variety of assemblers. Only a few of them will be discussed here.

One common pseudo-op informs the assembler what the absolute origin of the program is. An instruction like "ORG 200" could be used to tell the assembler that the next byte is intended to be stored in location 200. If parts of the program are in nonconsecutive locations (such as a subroutine being in a different part of memory), multiple "ORG" statements may be used. Each time the assembler sees "ORG --ADDRESS--" it knows it is being told that the next byte is meant to be stored in location --ADDRESS--. Unless another "ORG" statement is encountered, the assembler will

The Assembler is informed that the Origin is Location 200.

After Processing "Y: 0"

SYMBOL TABLE

Name	Value
Y	200

ASSEMBLER OUTPUT

Address	Contents	
200	0000 0000	(0)

BYTE CONTAINED IN RELATIVE ADDRESS [0]

After Processing "X: 6"

SYMBOL TABLE

Name	Value
Y	200
X	201

OUTPUT

201	0000 0110	(6)

RELATIVE ADDRESS [1]

After Processing "MVI B,2"

SYMBOL TABLE

Name	Value
Y	200
X	201

OUTPUT

202	0000 0110	(MVI B)
203	0000 0010	(2)

RELATIVE ADDRESS [2]

After Processing "LDA X"

SYMBOL TABLE

Name	Value
Y	200
X	201

→ "LDA X" = "LDA 201"

OUTPUT

204	0011 1010	(LDA)
205	1100 1001	(201)
206	0000 0000	

RELATIVE ADDRESS [4]

After Processing "LOOP: ADD A"

SYMBOL TABLE

Name	Value
Y	200
X	201
LOOP	207

OUTPUT

207	1000 0111	(ADD A)

RELATIVE ADDRESS [7]

Fig. 16-27. Steps involved in the assembly of the program shown in Fig. 16-26.

Fig. 16-27. Cont.

After Processing "DCR B" and "JNZ LOOP"

SYMBOL TABLE

Name	Value
Y	200
X	201
LOOP	207

"JNZ LOOP" = "JNZ 207"

RELATIVE ADDRESS OF "JNZ LOOP" 9

OUTPUT

Address	Contents	
208	0000 0101	(DCR B)
209	1100 0010	(JNZ)
210	1100 1111	(207)
211	0000 0000	

After Processing "STA Y"

SYMBOL TABLE

Name	Value
Y	200
X	201
LOOP	207

"STA Y" = "STA 200"

RELATIVE ADDRESS 12

OUTPUT

Address	Contents	
212	0011 0010	(STA)
213	1100 1000	(200)
214	0000 0000	

After Processing "HLT"

SYMBOL TABLE

Name	Value
Y	200
X	201
LOOP	207

RELATIVE ADDRESS 15

OUTPUT

Address	Contents	
215	0111 0110	(HLT)

---

assume that all instruction and data bytes are stored in consecutive locations after the origin location.

Another common pseudo-op initializes a data byte to a number. In the example given in Fig. 16-27, no special pseudo-op was listed; to initialize the contents of address Y to 0, we just wrote "Y: 0." More commonly some code like "DC" (define constant) or "DB" (data byte) precedes the number. Thus, it would be common to see "Y: 0" written as "Y: DC 0" or "Y: DB 0" in various assembly languages.

Reserving storage is often done with a pseudo-op. In FORTRAN, the "DIMENSION" statement is a storage-reserving pseudo-op to the FORTRAN compiler. A code like "DS" (define storage) is sometimes used. Thus, to reserve 10 bytes of storage for an array, with the address of the first byte given the name "ARRAY," one might see the pseudo-op "ARRAY: DS 10."

Many assembly languages have the pseudo-op "EQU" (equate). This pseudo-op instructs the assembler to make an entry in the symbol table, where the name of the symbol and its value are given by the "EQU" instruction. For example, the instruction "X EQU 15" tells the assembler that the user wishes the name "X" to be placed in the symbol table, and the value of X should be recorded in the table as being 15.

This pseudo-op has a variety of uses. To give an example of one, suppose we need to write the instruction "MVI B,6" many times in a program. Suddenly we find that the number to be moved to B should not be 6, but instead should be 7. We would normally have to go back and change all the "MVI B,6" instructions. But suppose we had written the "MVI" instruction in the form "MVI B,Z" and also had written the pseudo-op "Z EQU 6." This would have created an entry in the symbol table informing the assembler that every time it saw the symbol "Z," it should replace it by the number 6. Now to change all the "MVI B,6" instructions to "MVI B,7," we only have to change the "EQU" instruction to "Z EQU 7," and then reassemble the program. Another way the "EQU" statement is used is in descriptive labeling; that is, through use of the "EQU" statement numbers can be given descriptive names that help document a program. For example, suppose a teletype is assigned output port number 135. To transmit a character to the teletype, the programmer would normally write the instruction "OUT 135." However, using the "EQU" statement the programmer can assign the number 135 to the symbol "TELETYPE" (by writing "TELETYPE EQU 135"). Then the programmer can write an output instruction that looks like "OUT TELETYPE." It is much easier to understand the programmer's intentions when reading a program that has been written using descriptive labels such as this one.

Finally, the "END" pseudo-op is often found. This instruction tells the assembler that there are no more instructions to translate. When a particular assembly language requires an "END" statement, it is always the last instruction of the assembly-language program.

The Two-Pass Assembler

Consider an attempt by a one-pass assembler to assemble the simple program shown in Fig. 16-28. The first instruction is a pseudo-op telling the assembler that the address of the next byte is location 200. The next instruction, however, is "LDA X." The one-pass assembler will look at the symbol table to try to find what "X" is. But no entry has been made yet for X. The one-pass assembler cannot continue the translation. Yet, the problem is not that "X" is never defined. "X" certainly is defined; it is the address of the byte after the "HLT" instruction. The problem is that the symbol "X" is being used *before* it has been defined. This is called a *forward reference*. From this example, we see that the one-pass assembler can only handle "backward references." That is, it can only translate symbolic addresses that have already been defined. Since it puts too much of a burden on the programmer to expect him to use only backward references, the most common assembler used is another type of assembler called the two-pass assembler. As the name implies, the two-pass assembler reads the assembly code twice. By making two passes through the code, forward references can be handled.

FORTRAN		8080 ASSEMBLY CODE
		ORG 200
		LDA X
		ADI Y
		STA Z
		HLT
X = 1	X:	DB 1
		Y EQU 3
		ORG 500
Z = X + 3	Z:	DS 1
STOP		END

Fig. 16-28. Program showing the need for a two-pass assembler.

The first pass is devoted to defining all the symbols. Every time a statement with a symbol definition on it or an "EQU" statement is encountered an entry is made in the symbol table. No attempt is made to generate binary code. At the end of the first pass, the name of every symbol defined in the entire program will have been stored in the symbol table, along with the value of the symbol. The value is determined by knowing the origin and knowing the length of each instruction.

The second pass is devoted to generating code. All OP-CODES and data bytes are translated into binary. All symbolic references are replaced with their numerical equivalent as recorded in the symbol table on PASS 1.

To see how this works, let us use a two-pass assembler to translate the program shown in Fig. 16-28.

On the first pass, symbols are defined. The assembler starts counting the number of bytes from the top, where the first byte is known to be in location 200. The first instruction takes three bytes, the second takes two, the third takes three, and so on. By the time the first symbol is encountered, nine bytes have been used. The address, "X," therefore, is location ORG + 9 (209). An entry is made in the symbol table that looks like "X — 209." Next the pseudo-op "Y EQU 3" is processed. The assembler generates the symbol table entry "Y — 3." The pseudo-op takes *no* bytes of storage. The 8080 will never execute this nonexecutable pseudo-op. The next instruction, "ORG 500," is another pseudo-op. The assembler is being told that the address of the next byte is 500. The origin has been changed. The entry "Z — 500" is made. If another byte of data had followed "Z," it would have been assigned location 501. When the assembler sees the "END" statement it knows that PASS 1 is over. All symbols have now been defined.

PASS 2 is the code-generating phase. The assembly instructions are reread in the second pass. Now when the instruction "LDA X" is encountered, X is perfectly well-defined. The assembler looks up "X" in the symbol table and finds that X has the value 209. At this point, the assembler knows that "LDA X" is really "LDA 209." The code for "LDA 209" is generated. Similarly, the "ADI Y" instruction is changed to read "ADI 3" since "Y" has been assigned the value "3" in the symbol table. The "STA Z" instruction is then translated into "STA 500" by the same procedure. The "HLT" instruction is directly translated into binary. Finally, the "END" pseudo-op tells the assembler that PASS 2 is over. The program is now completely translated into binary. The two-pass assembler has correctly handled forward references.

A diagram of the two-pass assembler's actions are given in Fig. 16-29. The reader should study this figure and try to understand how the two-pass assembler generates binary code. The code itself should be hand-simulated by the reader to verify its correctness.

PASS 1

STATEMENT READ IN	ACTION TAKEN	SYMBOL TABLE
ORG 200	Assembler is informed the next byte is in address 200. Internal byte counter is initialized to 0.	EMPTY
LDA X	Internal byte counter increased to 3.	EMPTY
ADI Y	Internal counter increased to 5.	EMPTY

Fig. 16-29. List of actions taken by two-pass assembler.

Fig. 16-29. Cont.

STA Z	Counter increased to 8.	EMPTY
HLT	Counter increased to 9.	EMPTY
X: DB 1	Entry made in symbol table. Counter increased to 10.	X − 209
Y EQU 3	Entry made in symbol table.	X − 209 Y − 3
ORG 500	Assembler is informed that next byte is in address 500. Byte counter is reset to 0, since a new origin has been defined.	X − 209 Y − 3
Z: DS 1	Counter is increased to 1. Entry is made in symbol table.	X − 209 Y − 3 Z − 500
END	PASS 2 is started.	X − 209 Y − 3 Z − 500

During all of PASS 2, symbol table is: X − 209
Y − 3
Z − 500

PASS 2

STATEMENT READ IN	ACTION TAKEN	CODE GENERATED Address	Contents	
ORG 200	None	None		
LDA X	"X" looked up, 209 substituted. Code is generated.	200	0011 1010	(LDA)
		201	1101 0001	(209)
		202	0000 0000	
ADI Y	"Y" looked up, 3 substituted. Code is generated.	203	1100 1110	(ADI)
		204	0000 0011	(3)
STA Z	"Z" looked up, 500 substituted. Code is generated.	205	0011 0010	(STA)
		206	1111 0100	(500)
		207	0000 0001	
HLT	Code is generated.	208	0111 0110	(HLT)
X: DB 1	Code is generated.	209	0000 0001	(1)

Fig. 16-29. Cont.

Y EQU 3	None	None
ORG 500	None	None
Z: DS 1	None	None
END	Assembler signals that binary version of program is available for execution.	None

Summary

Section 16.5 marks the culmination of our presentation of assembly-language programming. In this section the previous work of machine-language programming was used to explain assembly-language instructions. It was seen that assembly language is just a convenient way of programming machine-level instructions.

The section began by showing how mnemonics and decimal data may be translated by an assembler program into the correct binary string of 1's and 0's. We next saw a way of getting away from the drudgery of keeping track of where each address and variable is located, namely, using symbolic addressing. The "symbol table" and "relative addressing" concepts were introduced as a way of implementing symbolic addressing. Relative addressing involved keeping track of how far down, in bytes, a symbolic address was located from some origin defined numerically in the program.

In the first scheme presented, only backward references could be handled. That is, if a symbol was referenced in an instruction, it must have been previously defined. A better scheme, the two-pass assembler, was then presented. This scheme allowed forward references because there was an entire pass devoted to searching through the program to define all the symbols. It was only on the second pass, after the symbol table had been completed by the first pass, that code generation was attempted. Since there can be no undefined symbols when the second pass begins, the two-pass assembler will be able to successfully translate the assembly code into binary by the end of the second pass. The binary code is then ready to be executed on a digital computer.

16.6 MICROPROCESSOR CONTROL OF DC MOTORS

The preceding sections have shown what a microprocessor is and how it works. The purpose of this section is to suggest several ways in which a microprocessor can be used in the control of dc motors. Advantages and limitations of microprocessor control will be discussed. Only a few control strategies will be covered since it is impossible to discuss general control theory exhaustively in limited space. However, a sufficient amount of material will be presented so that the reader will be made aware of the problems that may arise in microprocessor control of dc motors. Hopefully, the reader will learn about some of the common solutions to these problems as well.

As before, all examples concerning microprocessors will deal with the Intel 8080 microprocessor. The reader should be certain that he is familiar with the 8080 STACK structure, INTERRUPT structure, and INPUT/OUTPUT structure before reading this section.

Advantages And Limitations Of Microprocessor Control

There are many features of microprocessor control that make it attractive to the design engineer. For example, because a microprocessor-based control system can be constructed out of only a few integrated circuits, the microprocessor controller is often smaller and more lightweight than an analog controller. In addition, the microprocessor system is often cheaper than other controllers. Furthermore, because there tend to be fewer components in a microprocessor-based system, there are fewer things to go wrong, leading to increased system reliability. Also, the reduced number of system components can make it easier for beginning engineers to learn about work already completed on a project which has been designed using microprocessor control.

Other important benefits are realized when using microprocessor control. One of the most important benefits is the ease at which a system can be modified to perform new functions. Often this will involve no more than writing a new assembly-language program for the already-existing controller. Having a programmable controller provides an element of flexibility that is hard to match with other control systems. Another important benefit is that, by continually using microprocessor controllers, a company can unify and standardize its design procedures. For example, in a company using microprocessor control for much of its work, a library of programs can be built up. These standard programs represent and summarize the efforts of many engineers, and may allow results of projects to be more easily applied to other projects within a company. Still another benefit is that a microprocessor helps to make a system more modular, which can make the system easier to troubleshoot and debug.

The microprocessor, of course, is not without its limitations. The engineer must understand these limitations if he is to design a successful microprocessor-based controller.

The problems that arise when using a microprocessor to control a dc motor encompass all the problems that arise when using any digital computer to control an analog process. Some of the problems are due to sampling; that is, in order to gather data on a process, the computer will periodically sample the process. This sampled data will be used by the processor to compute a new value of control. The control will be updated periodically as well. But between the sampling instants, a control is being applied to the process which is out-of-date. This can lead to stability problems if the sampling rate is too slow. In addition to not continuously updating the control, there are problems that arise due to the finite word length in the computer. To see this, note that when a signal is sampled, it must be converted from an analog signal to a digital number. The number it is converted to will only have as much resolution as the analog-to-digital converter (ADC) allows. This number may be truncated still further, depending on the bit length of the computer word (e.g., if the ADC has 12 bits of resolution but the computer has only an 8-bit word length, then the computer must be programmed to either store the number in two words or simply ignore the least significant four bits of the number. The latter course of action truncates the data further, in addition to the truncation caused by the ADC). Thus, the data used in the computer's control algorithm are not perfect digital versions of the actual sampled data. Furthermore, when the control is being computed by the processor, round-off and truncation errors will occur, again due to the finite word length. The end result is a control that is not necessarily the same as an analog controller would produce. This can again lead to stability problems, cause an undesirable transient response, or an unacceptable steady-state error.

Because a microprocessor is a digital computer, a microprocessor-based controller will suffer from all of the problems mentioned above. In fact, these problems are likely to be more serious with a microprocessor than with a minicomputer or full-sized computer. This is to be expected for several reasons. First of all, most microprocessors are substantially slower than large computers (each instruction takes longer to execute). This will increase the minimum possible sampling period. Secondly, the microprocessor usually has a smaller word length than a large computer. This typically means that numerical quantization effects will be more pronounced with a microprocessor. Of course, a larger word length can be simulated on a microprocessor by storing data in multiple bytes.* However, when this is done it will take more instructions to process the data than it would take if only single-byte operands are allowed. The extra instructions will require more time to execute, slowing down the processor further.

Another problem with using extra instructions to overcome word length limitations is that these extra instructions will require additional storage locations, and the memory is often of very limited size in a microprocessor. The memory size is limited for two reasons: address bit length and cost. The 8080, for example, is capable of addressing only 65,536 memory locations due to its 16-bit address length. Certain sophisticated control algorithms cannot fit in 64K bytes. But even if they could, processor memory is still limited because it is frequently the most expensive part of the microprocessor system (for example, at the present time, 1K bytes of 8080 memory can cost more than the 8080 CPU chip). Thus, the microprocessor programmer has the burden of writing a control program that must not only work, but that also is as short as possible.

Keeping the program short is further complicated by the fact that most microprocessors have an instruction set which is substantially inferior to that found on larger computers. For example, there are no multiplication or division instructions on the 8080. While multiplication and division can be simulated, say, by using repeated addition and subtraction, the simulation programs use up execution time and memory space. One way out of this problem is to limit the multiplications and divisions to those involving only powers of 2. This is because on a binary computer it is very easy

*To see how double precision additions may be performed on the 8080, suppose it is desired to add the 16-bit number stored in registers B and C with the 16-bit number stored in D and E and place the result back in B and C. Let the first number be 1018 and the second number be 2321. Since these are stored as 16-bit numbers, register B will contain "3" and C will contain "250" (since $3 \times 256 + 250 = 1018$) and register D will contain "9" and E will contain "17" (since $9 \times 256 + 17 = 2321$). To add these 16-bit numbers, we first add the lower-order bytes. The instructions "MOV A,C; ADD E; MOV C,A" accomplish this on the 8080. The sum of 250 and 17 is 267. This number will not fit in 8 bits. Thus, the CARRY FLAG will be set to 1 and the remainder, 11 ($267 - 256$), is stored in the AC and moved to register C. Now we add the high-order bytes and add in the one-bit CARRY FLAG. The instructions "MOV A,B; ADC D (add with carry); MOV B,A" accomplish this. The sum of 3 and 9 is 12, and since the CARRY FLAG is 1, the final sum is 13. Thus, "13" is stored in register B. Note that the "add with a carry" instruction enables us to keep track of how adding the lower-order bytes affects the sum in the higher-order bytes. The number in B and C is $13 \times 256 + 11$, or 3339, the correct sum of 1018 and 2321. Thus, through proper use of the CARRY FLAG to "link" the high- and low-order byte operations, double precision operations can be simulated on the computer.

Still another way of representing multiple-byte data is to use floating-point arithmetic. Floating-point arithmetic can be simulated on an integer machine such as the 8080 in a number of ways. One common way involves the use of a convention for storing floating-point numbers in two bytes. The first byte contains the significant digits of a number and the second byte contains the exponent of the number, here considered to be in scientific notation. The bytes can be stored consecutively in the memory. To multiply two numbers stored in this fashion, we add the exponents and multiply the significant digits. For example, bytes 200 and 201 may contain the numbers 50 and 10, bytes 300 and 301 may contain the numbers 3 and 6, respectively. If these are considered to be floating-point numbers stored in our special two byte convention, the numbers represent 50×10^{10} and 3×10^{6}, respectively. To multiply these numbers we simply multiply the first bytes (locations 200 and 300) to get 150, and then add the second bytes (locations 201 and 301) to get 16. Interpreting these two answer bytes as a floating-point number, we see that the answer we get is 150×10^{16}, the correct result. Other floating-point operations can be similarly simulated.

(and takes little time) to multiply or divide a number by a power of 2.* However, rounding off all of the control algorithm's multiplicative gains to the neareast power of 2 may lead to a set of gains that do not produce the most desirable response. The trade-off between reduced execution time along with saved memory space (due to simple "multiplication and division" routines), and possible undesirable system performance (due to artificially rounding off the control gains) is another of the problems that the microprocessor programmer must face.

Direct Microprocessor Control Of DC Motors

The last section discussed some of the advantages and limitations of microprocessor control. The rest of this chapter will be spent discussing two different approaches to microprocessor control. It will be seen that each approach takes advantage of the benefits and tries to overcome the limitations of the microprocessor in different ways.

One approach to using a microprocessor may be called the "direct" approach. The direct approach involves designing a controller that basically contains no analog control circuitry. In the direct approach, the system data are fed into a microprocessor, which then computes a new value of control. This control is then applied to the system through the use of digital-to-analog (DAC) converters. This scheme is depicted in Fig. 16-30. The second approach to microprocessor control, which is called the "indirect" method, involves designing a microprocessor-based controller that, none the less, contains some analog controller as part of the control implementation. The indirect approach will be discussed in the next section.

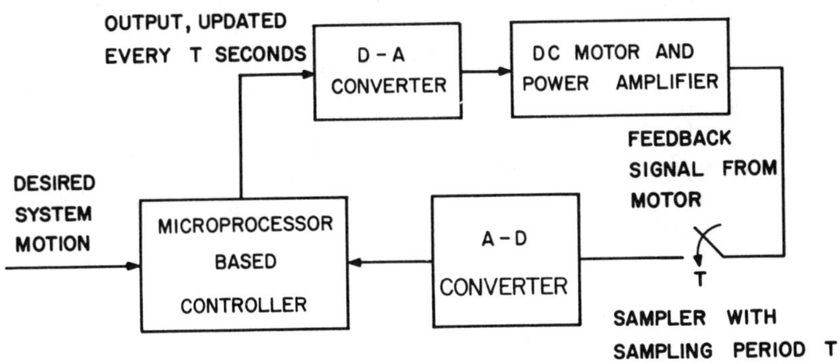

Fig. 16-30. Block diagram of direct method of microprocessor control.

*Consider a number stored in binary. If the bits are shifted over one position to the left and a zero is shifted in to the least significant bit (LSB) position, then the new number is the old number multiplied by 2. For example, in binary, the number 6 can be written as "00000110." Shifting each bit one position to the left, and setting the LSB to 0, we get "00001100," which is "12" or 6 × 2. Shifting the bits of 12 gives "00011000" which is 24, or 12 × 2. Thus, a simple way of multiplying a binary number by some power of 2, is to shift the number to the left the correct number of times. Division by powers of 2 is accomplished by shifting to the right. For example, shifting the number "00000110" (6) to the right gives "00000011" (3). Shifting the bit pattern for 3 to the right gives "1," the correct answer for 3/2 using integer division. If the number is a two's complement negative number, the CPU should shift in 1's, not 0's, into the most significant bit (MSB) position. Thus, to divide −6 by 2, one takes the bit pattern for −6, "11111010," and shifts it to the right. With a "1" shifted into the MSB position, the bit pattern is "11111101," the two's complement bit pattern for −3.

In the direct approach, the microprocessor is responsible for not only determining what the motor velocity and position should be, but also for insuring that the motor actually attains the required velocity and position; that is, the microprocessor must insure that the motor-amplifier system is stable, and that any steady-state error is within tolerable limits.

One way to accomplish this is to design a familiar analog controller, and then "digitize" or "discretize" the analog controller as the final step in the design procedure. For example, suppose it is desired to control the velocity of a dc motor, whose modeling equations are given by

$$T(t) = K_t i(t) \tag{16-1}$$

$$v(t) = Ri(t) + L\frac{di(t)}{dt} + K_b\omega(t) \tag{16-2}$$

$$T(t) = J\frac{d\omega(t)}{dt} + B\omega(t) \tag{16-3}$$

where
T = torque
K_t = torque constant
$i(t)$ = armature current
$v(t)$ = applied voltage
R = resistance
L = inductance
K_b = back emf constant
J = system inertia
B = viscous damping constant
$\omega(t)$ = angular velocity

Consider that the steady-state velocity error is required to be zero and that a tachometer is attached to the motor shaft to measure the velocity. An analog controller that could accomplish this goal is the PI (proportional-plus-integral) controller [3]. Using a PI controller, the voltage v(t) takes the form,

$$v(t) = g_1[\omega_d - \omega(t)] + g_2\int[\omega_d - \omega(t)]dt \tag{16-4}$$

where g_1 and g_2 are constant feedback gains, and ω_d, a constant, is the desired motor velocity.

The block diagram representing the dc motor system with the PI controller is shown in Fig. 16-31.

The PI controller consists of a term which is proportional to the error $[\omega_d - \omega(t)]$, and a term proportional to the integral of the error. To see how the integral term affects the system, let $x_3(t)$ be defined as

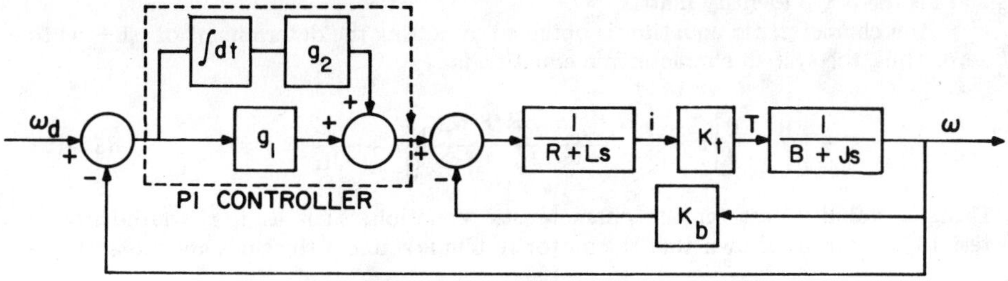

Fig. 16-31. Block diagram of dc motor system with PI controller.

$$x_3(t) = \int_{t_0}^{t} [\omega_d - \omega(t)]\,dt + x_3(t_0) \tag{16-5}$$

where t_0 is the initial time, and $x_3(t_0) = 0$.

Then it is clear that

$$\dot{x}_3(t) = \omega_d - \omega(t) \tag{16-6}$$

Thus, the dynamic equations of the system become

$$\dot{\omega}(t) = -\frac{B}{J}\omega(t) + \frac{K_t}{J}\,i(t) \tag{16-7}$$

$$\dot{i}(t) = -\frac{R}{L}\,i(t) - \left[\frac{K_b + g_1}{L}\right]\omega(t) + \frac{g_2}{L}x_3(t) + \frac{g_1}{L}\omega_d \tag{16-8}$$

$$\dot{x}_3(t) = \omega_d - \omega(t) \tag{16-9}$$

Now if the gains g_1 and g_2 are chosen so as to make the system described above asymptotically stable, then eventually the system will reach its equilibrium point. The equilibrium point is characterized by the fact that the derivatives of the state variables are zero. In particular, this means that $\dot{x}_3$ is zero in the steady state. But this means that

$$\dot{x}_3\big|_{\text{steady-state}} = 0 = \omega_d - \omega_{\text{steady-state}} \tag{16-10}$$

or $\omega_{\text{steady-state}} = \omega_d$, as required.

In order to choose the proper feedback gains, the system characteristic equation must be derived. This can be done in a variety of ways, such as by computing the determinant of (sI − A), where the A matrix of the dynamic equations is given by

$$A = \begin{bmatrix} -\frac{B}{J} & \frac{K_t}{J} & 0 \\ -\frac{K_b + g_1}{L} & -\frac{R}{L} & \frac{g_2}{L} \\ -1 & 0 & 0 \end{bmatrix} \tag{16-11}$$

and I is the 3×3 identity matrix.

The characteristic equation is obtained by setting the determinant of $(sI - A)$ to zero. Thus, the system characteristic equation is

$$s^3 + \left[\frac{R}{L} + \frac{B}{J}\right] s^2 + \left[\frac{K_t K_b + BR + K_t g_1}{JL}\right] s + \frac{K_t g_2}{JL} = 0 \tag{16-12}$$

Using a stability test on the characteristic equation, such as the Routh-Hurwitz test [3], it can be shown that the motor system is stable if the gains are chosen to be

$$g_1 > \frac{JLg_2}{RJ + BL} - \frac{K_t K_b + BR}{K_t}$$
$$g_2 > 0 \tag{16-13}$$

To discretize the PI controller, we need to separately discretize the two terms of Eq. (16-4). The first term, $g_1 [\omega_d - \omega(t)]$ is easy to discretize. Let T be the chosen sampling period. Then, $\omega(kT)$ is the discretized value of $\omega(t)$ at $t = kT$, $k = 0, 1, 2, \ldots$. Therefore, the discretized form of the first term is simply $g_1 [\omega_d - \omega(kT)]$.

The second term of Eq. (16-4) contains an integral which may be approximated digitally by some integral algorithm such as the trapezoidal rule. Letting $t = kT$ and $t_0 = (k - 1)T$, the integral in Eq. (16-4) can be approximated as

$$\int_{(k-1)T}^{kT} [\omega_d - \omega(t)]\, dt \cong \omega_d T - \frac{T}{2}\Big[\omega(kT) + \omega[(k-1)T]\Big] \tag{16-14}$$

for $k = 1, 2, \ldots$. Therefore, the integral in the last equation at $t = kT$ can be computed based on the given constant value of ω_d and the input data for $\omega(kT)$, $k = 0, 1, 2, \ldots$. However, it takes the microprocessor a finite amount of time to compute the integral in Eq. (16-14), the result is not readily available at $t = kT$. In general, one has to add up all the times required to execute the OP-CODES of the integration subroutine on the microprocessor to find out this time delay. For convenience, we shall assume that this time delay is equal to the sampling period T. Then, the second term of Eq. (16-5) is discretized as

$$x_3[(k+1)T] \cong x_3(kT) + \omega_d T - \frac{T}{2}\Big[\omega(kT) + \omega[(k-1)T]\Big] \tag{16-15}$$

Thus, to discretize the second term of Eq. (16-4) the microprocessor must store two pieces of the past data, $x_3(kT)$, and $\omega[(k-1)T]$. Then, at $t = kT$, the processor reads in the latest value of the motor speed $\omega(kT)$, and then computes the next value of $x_3(t)$ at $t = (k + 1)T$.

The discretized version of Eq. (16-4) is now written as

$$v[(k+1)T] = g_1 [\omega_d - \omega(kT)] + g_2 x_3 [(k+1)T] \tag{16-16}$$

This control is then applied to the dc motor system at $t = (k + 1)T$. The control is updated every T seconds, and is held constant between the sampling instants.

A typical system used to implement the discretized PI controller is shown in Fig. 16-32. The system shown in this figure uses an analog timer to determine the start of the next sampling period (alternatively, the processor can use a software

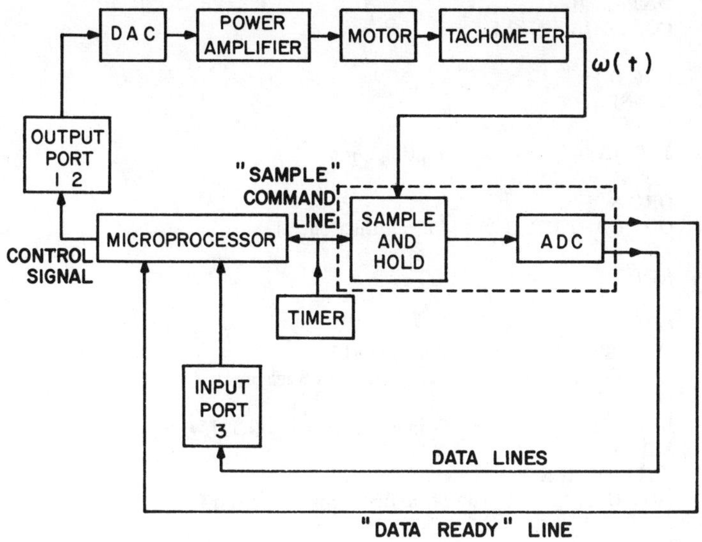

Fig. 16-32. Block diagram of implementation of discretized PI controller.

timing loop to keep track of when T seconds have elapsed). The timer outputs a pulse every T seconds. This pulse is used in two ways. Firstly, the pulse is applied to the processor's interrupt line. This will cause the processor to stop what it is doing and execute the interrupt routine, which in this case would be a routine to output the next value of control. This updated value of control is sent to a DAC, and then sent to the power amplifiers. Meanwhile, the timing pulse has also been sent to the "sample" command line of the ADC. A pulse on the "sample" line triggers the sample-and-hold circuitry within the ADC. At this time, $t = kT$, the motor velocity, $\omega(t)$, is sampled and held at $\omega(kT)$. The analog voltage value of $\omega(kT)$ is then converted to an N-bit binary number by the rest of the ADC circuitry. There is a finite conversion time associated with the conversion process. During this conversion time, the output of the ADC is unpredictable. Therefore, the ADC must signal the processor, via a "data ready" line, that the sampled data has been converted. The "data ready" line may also be attached to the processor's interrupt line. This second interrupt will cause the processor to read in the value of $\omega(kT)$ and compute the next value of control, $v[(k+1)T]$. After the control has been calculated, the processor waits for another interrupt from the timer before it outputs the control at time $t = (k+1)T$.

A program to perform the PI controller is shown in Fig. 16-33. The program assumes that using $g_1 = 1/2$ and $g_2 = 4$, both powers of 2, will result in a stable system. The sampling period has been chosen to be 1 second. The program further assumes that the data from the ADC is available on input port 3, and that the control is to be sent out on output port 12. The system has been set up so that when an interrupt occurs from the timer, a "RESTART 1" ("CALL 8") instruction will be jammed on the DATA BUS. Also when an interrupt occurs from the ADC, a "RESTART 7" ("CALL 56") will be jammed on the BUS. Let us briefly examine how this program works.

The first instruction, in location 0, initializes the STACK POINTER. This is necessary because the STACK will be needed in subroutine calls and interrupt handling. and the "top of the STACK" must be defined. The next instruction, "EI," enables the interrupt structure. The next instruction, "WAIT: JMP WAIT," is an infinite waiting loop. The only way out of the waiting loop is by an interrupt.

```
            DATA EQU 3
            CONTROL EQU 12

            ORG 0
            LXI SP,1000
            EI
     WAIT:  JMP WAIT                 Waiting Loop

            ORG 8
            OUT CONTROL              Output the Control
            EI
            RET

            ORG 56
            IN DATA                  Read in ω(kT)
            MOV D,A                  Store ω(kT) in Register D

            LDA WD                   Compute g1[ωd − ω(kT)] with g1 = 1/2
            SUB D
            CALL DIVIDE 2
            MOV C,A                  Store first term in Register C

            LDA WOLD                 Compute g2[x3(k + 1)T] with g2 = 4
            MOV B,A
            MOV A,D
            STA WOLD                 Replace stored value of ω(kT) by latest value
            ADD B
            CALL DIVIDE 2
            MOV B,A
            LDA WD
            SUB B
            MOV B,A
            LDA X3OLD
            ADD B
            STA X3OLD                Replace stored value of x3[(k + 1)T] by latest value
            CALL MULTIPLY 4

            ADD C                    Form v[(k + 1)T] = g1[ωd − ω(kT)] + g2x3[(k + 1)T]
            EI
            RET

 DIVIDE 2:  JM NEGATIVE
            ORA A
            RAR
            RET
NEGATIVE:   STC
            RAR
            RET

MULTIPLY 4: ORA A
            RAL
            ORA A
            RAL
            RET

    WOLD:   DB 0
      WD:   DB 25
   X3OLD:   DB 0
```

Fig. 16-33. 8080 assembly-language program implementation of a PI controller. Gains are fixed at $g_1 = 1/2$, $g_2 = 4$, and $T = 1$.

Assume an interrupt occurs from the timer. Then the processor will see a "RESTART 1" instruction and so will go to location 8. In location 8, the simple interrupt handling code shown is stored. The code instructs the processor to send the contents of the ACCUMULATOR, which presumably contains the new value of control, to OUTPUT PORT 12. The interrupt structure is then reenabled and the program returns to the waiting loop.

Now assume that an interrupt occurs from the ADC. In this case the processor will begin executing the code starting in location 56. The code begins by reading in the new value of $\omega(kT)$, which should be stored in INPUT PORT 3, and by saving this number in register D. The next block of codes computes $1/2[\omega_d - \omega(kT)]$ with the aid of a subroutine called "DIVIDE 2." The subroutine performs the division by 2 by shifting the number to the right one bit. If the number is negative a "1" is shifted into the high-order bit position; otherwise a zero is shifted in.*

After the first term in the control is computed, it is stored in register C. The processor next computes $x_3[(k+1)T]$ using Eq. (16-16), with $T = 1$. Once $x_3[(k+1)T]$ has been computed, it is saved in the location used to save the "old" value of $x_3[(k+1)T]$. Similarly, the current value of $\omega(kT)$ has been saved for the next cycle by the instruction "STA WOLD" (executed previously). $x_3[(k+1)T]$ is multiplied by 4 by the "MULTIPLY 4" subroutine. The multiplication is accomplished by shifting to the left twice.** At this point both terms of the control have been computed. They are then added together and left in the ACCUMULATOR. The program then reenables its interrupt structure and returns to the waiting loop.

The preceding program is meant to be a very simple example of a PI controller and should not be copied and used by the reader, due to the program's flaws. For example, no provision is made in the program if an "overflow" occurs in any of the computations. Also, there is no guarantee that the value of gains chosen here or the sampling period specified will lead to a stable system. Still another problem will arise if the execution time of the program is too long, say longer than the desired sampling period. Hopefully, however, in spite of the flaws, the program has given the reader an idea of what is involved in setting up a PI controller on a microprocessor.

This section began by discussing ways of "discretizing" an analog controller. We have so far discussed ways of discretizing linear gain terms and integrals. It should be clear that derivatives can be approximated in a similar manner as integrals. For example, one common approximation to the derivative of a signal is the "backward difference" equation

$$\dot{x}(t) \cong \frac{x(kT) - x[(k-1)T]}{T} \tag{16-18}$$

Thus, if some analog controller feeds back a term of the form

$$v(t) = -g_3\dot{\omega}(t) \tag{16-19}$$

this can be discretized by the equation

*The instruction "ORA A" is used as a trick to set the CARRY FLAG to zero. The instruction itself appears to do nothing, "OR-ING" the AC with itself, but on the 8080 all logic operations just happen to clear the CARRY FLAG. Thus, "ORA A" can be viewed as a simple code to clear the CARRY FLAG.

**The code "ORA A; RAL" is repeated twice because two shifts to the left are required. It turns out that it takes fewer instructions to repeat the code than to try to set up a loop to execute twice.

$$v[(k+1)T] \cong -\frac{g_3\left[\omega(kT) - \omega[(k-1)T]\right]}{T} \tag{16-20}$$

Again, we have assumed that a one-sampling-period delay is incurred in the computation of the right-hand side of Eq. (16-20).

The preceding controllers are relatively easy to discretize because they are "time-domain" controllers, involving gains, integration and differentiation. Sometimes, however, a controller is designed in the analog frequency domain. For example, a controller might be designed to be a low-pass filter or perhaps designed as a lead-lag network to give a system a certain gain and phase margin [3]. These types of controllers are characterized by a transfer function in the s-domain.

The discretization of an s-domain transfer function can be conducted in a number of ways. One method depends on the fact that there is an s-domain transfer function corresponding to Eq. (16-4). Taking the Laplace transform on both sides of Eq. (16-4), we get

$$\frac{V(s)}{E(s)} = g_1 + \frac{g_2}{s} \tag{16-21}$$

where E(s) is the Laplace transform of $\omega_d - \omega(t)$, and V(s) is the Laplace transform of v(t). Thus, the right-hand side of Eq. (16-21) is recognized as the transfer function of the PI controller.

Now it is apparent that given the transfer function of an analog controller, we can obtain the time-domain integro-differential (dynamic) equations relating the input and output of the controller, and then perform the discretization as illustrated above. In general, the procedure of going from the transfer function to the dynamic equations is called "decomposition" [3].

Another method of discretizing the transfer function in Eq. (16-21) is to use the z-transformation (see Chapter 4). The output of an analog system can be approximated by a staircase function with jump discontinuities at the sampling instants. Therefore, the transfer function of the PI controller in Eq. (16-21) can be discretized by the following z-transform:

$$\frac{V(z)}{E(z)} = \mathcal{Z}\left[\frac{1 - e^{-Ts}}{s}\left(g_1 + \frac{g_2}{s}\right)\right] \tag{16-22}$$

where $(1 - e^{-Ts})/s$ is known as the transfer function of the sample-and-hold device. The z-transform in the last equation is carried out to give

$$\frac{V(z)}{E(z)} = g_1 + \frac{g_2 T}{z - 1} \tag{16-23}$$

Since the z-transform can be viewed as $z = e^{Ts}$, substituting this relation into Eq. (16-23) we can show that

$$\lim_{T \to 0} \frac{V(z)}{E(z)} = \lim_{T \to 0}\left[g_1 + \frac{g_2 T}{e^{Ts} - 1}\right] = g_1 + \frac{g_2}{s} \tag{16-24}$$

which verifies that V(z)/E(z) is a discretized approximation to V(s)/E(s).

Once we have the z-transform expression of the controller transfer function, we can decompose it to obtain the dynamic equations in the discrete-time domain.

Equation (16-23) is written

$$\frac{V(z)}{E(z)} = \frac{g_1 + (g_2T - g_1)z^{-1}}{1 - z^{-1}} \frac{X(z)}{X(z)} \tag{16-25}$$

where X(z) is a variable which is introduced just for convenience.

It follows from Eq. (16-25) that

$$V(z) = [g_1 + (g_2T - g_1)z^{-1}]X(z) \tag{16-26}$$

and

$$X(z) = E(z) + z^{-1}X(z) \tag{16-27}$$

The state diagram portraying Eqs. (16-26) and (16-27) is shown in Fig. 16-34. Since z^{-1} corresponds to a time shift of one sampling period, if X(z) corresponds to the time-domain signal x(kT), then $z^{-1}X(z)$ pairs with x[(k − 1)T]. The dynamic equations of the discretized PI controller are written directly from Fig. 16-34.

$$v(kT) = g_1[\omega_d - \omega(kT)] + g_2Tx[(k-1)T] \tag{16-28}$$

$$x(kT) = x[(k-1)T] + \omega_d - \omega(kT) \tag{16-29}$$

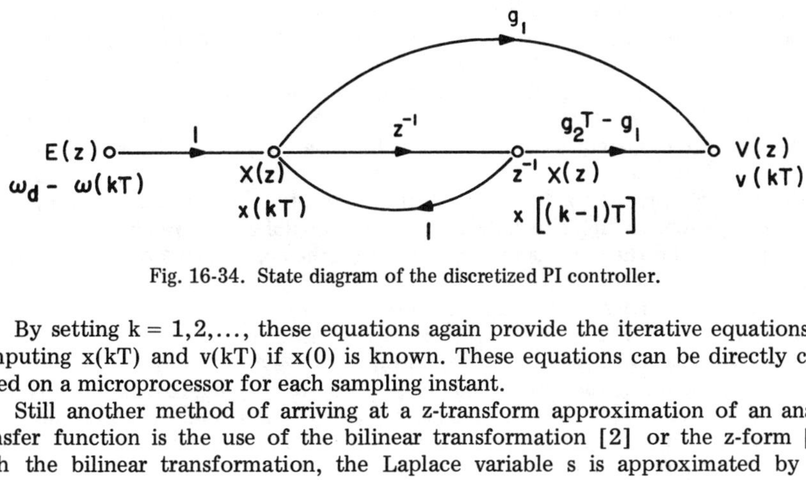

Fig. 16-34. State diagram of the discretized PI controller.

By setting k = 1,2,..., these equations again provide the iterative equations for computing x(kT) and v(kT) if x(0) is known. These equations can be directly computed on a microprocessor for each sampling instant.

Still another method of arriving at a z-transform approximation of an analog transfer function is the use of the bilinear transformation [2] or the z-form [1]. With the bilinear transformation, the Laplace variable s is approximated by the following relation:

$$s = \frac{2}{T}\left[\frac{1 - z^{-1}}{1 + z^{-1}}\right] \tag{16-30}$$

Applying this transformation to Eq. (16-21), we get

$$\frac{V(z)}{E(z)} = g_1 + \frac{g_2T}{2}\left[\frac{1 + z^{-1}}{1 - z^{-1}}\right] \tag{16-31}$$

By decomposition we can show that the dynamic equations which correspond to the last equation are

$$v(kT) = (g_1 + \frac{g_2T}{2})[\omega_d - \omega(kT)] + g_2Tx[(k-1)T] \tag{16-32}$$

and

$$x(kT) = x[(k-1)T] + \omega_d - \omega(kT) \tag{16-33}$$

In addition to the three methods described above, other methods such as the impulse-invariant technique [2] and the digital redesign method [1] can also be used for the discretization of analog systems.

It should be pointed out that when a continuous-time controller is discretized by any method, the resulting discrete-time controller may not retain all of the properties of the analog controller. Two things in particular that will affect the performance of the discretized controller are the sampling period and the choice of the feedback gains. It can be shown using sampled-data theory that the stability of a discretized system generally depends explicitly on the sampling period T. That is, for some values of T the system will be stable while for other values of T, with all other system parameters held fixed, the system will be unstable. Since there is no parameter T in the continuous-time controller, it is clear that the discretized controller behaves somewhat differently from the analog controller. Similarly, it can be shown that if the controller feedback gains are chosen so as to make the analog system stable, these gains may or may not make the corresponding discretized system stable. This would mean, for example, that choosing the discretized PI controller gains according to the restrictions given in Eq. (16-13) may not lead to a stable discretized system. This is because Eq. (16-13) is derived using a continuous-time stability test (Routh-Hurwitz) which is not the correct stability test to apply to the discretized system. The stability of the discretized system should be tested by an appropriate sampled-data stability test such as Jury's test [1]. Using the criteria given by Jury's test, equations can be derived that will specify a range of controller feedback gains which will make the discretized system stable. The restrictions on the digital gains given by Jury's test will not, in general, be the same as the restrictions on the analog gains given by the Routh test.

Certain properties of the analog controller, of course, are preserved in the discretized controller. For example, consider the steady-state velocity error that exists when the discretized PI controller given in Eqs. (16-32) and (16-33) is used to control a dc motor. Assuming the motor and controller system is made asymptotically stable (by an appropriate choice of feedback gains), the system will reach equilibrium. For a discrete-time system, equilibrium is characterized by the fact that $x(k+1) = x(k)$. Referring to Eq. (16-33), it is clear that in equilibrium

$$\begin{aligned} x(kT) &= x[(k-1)T] + \omega_d - \omega(kT) \\ &= x(kT) + \omega_d - \omega(kT) \end{aligned} \tag{16-34}$$

Subtracting $x(kT)$ from both sides of Eq. (16-34), it is seen that in the steady state

$$\omega_d - \omega(kT)\big|_{\text{steady-state}} = 0 \tag{16-35}$$

or

$$\omega(kT)\big|_{\text{steady-state}} = \omega_d$$

Thus, the analog PI controller property of zero steady-state error is preserved in the discretized controller. To properly investigate which properties of an analog controller are preserved in the discretized controller and which are not, one should analyze the microprocessor-motor system by using sampled-data control theory [1].

The methods of designing the digital controller presented in the previous sections are all based on the discretization of an analog controller which has been arrived at by continuous-data control system design theory. A more appropriate way, however, is to treat the overall system as a digital control system and design the controller using digital control theory. For the dc motor control system described by Fig. 16-32 and Eqs. (16-1) through (16-3), the digital control system block diagram representation of the system is shown in Fig. 16-35. In this case the ADC and the DAC are modeled by samplers and sample-and-hold devices. The microprocessor is represented by the digital controller with the transfer function D(z). The closed-loop transfer function of the system in Fig. 16-35 is

$$\frac{\Omega(z)}{\Omega_d(z)} = \frac{D(z)G_p(z)}{1 + D(z)G_p(z)} \tag{16-36}$$

where $\Omega(z)$ and $\Omega_d(z)$ are the z-transforms of $\omega(t)$ and ω_d, respectively.

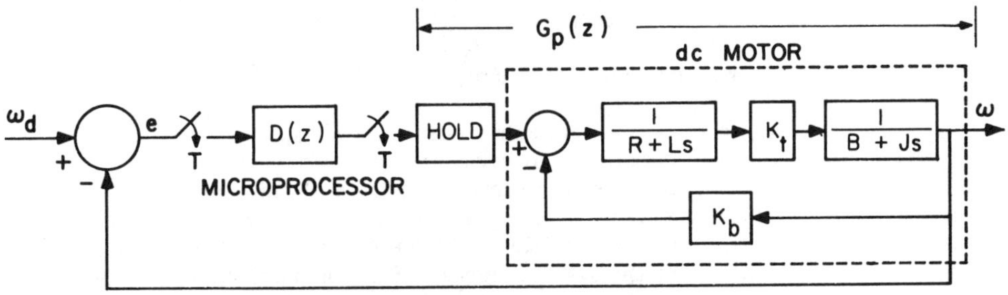

Fig. 16-35. Block diagram of digital control of dc motor.

There are an abundance of design techniques available for the design of digital control systems [1]. The crux of the problem is that the transfer function of the digital controller, D(z), is designed to satisfy the performance requirement of the discrete-time system, so that the sampling period T is taken into consideration at the outset of the design.

Summary of the Direct Control Method

The preceding sections are designed to give the reader an overview of the techniques involved in designing a direct microprocessor control of a dc motor. The section begins by discussing ways of discretizing an already designed analog controller. Various methods of discretizing time-domain controllers (such as amplifiers, integrators, differentiators) as well as frequency-domain controllers (that is, controllers specified by transfer functions) are given. It is pointed out that the discretized controller may behave differently than the corresponding analog controller. For example, the stability of the sampled-data system generally depends on the sampling period, T, which does not even exist for a continuous-time system. Also the gains chosen so as to make the continuous-time system stable, may not be appropriate for the sampled-data system. The preferred way of designing the digital controller is to treat the overall system as a digital control system.

The direct method of microprocessor control offers some clear-cut benefits. The benefits exist partly due to the fact that the microprocessor replaces so many of the costly components of an analog controller. For example, an analog integrator within a controller can be replaced by a difference equation stored in the processor memory which mimics the behavior of the integrator. Similarly, linear amplifiers in the analog controller are replaced by multiplication and division subroutines in the microprocessor. The flexibility of computer control is also realized using the direct control method. Changing a feedback gain or even changing the design of the entire controller involves no more than writing a new assembly-language program.

The direct control method suffers from some limitations, however. The limitations arise due to the fact that the microprocessor is often too slow and has too small a word length for certain control applications. For example, it was mentioned several times that the sampling period T played an important role in system stability. Specifically, large values of T usually lead to unstable systems. It is clear that one basic limitation on the minimum choice of T is the time it takes the microprocessor to compute a new control, given the data at time $t = kT$. Another limitation arises due to the fact that the system stability is sometimes very sensitive to the values of the feedback gains. Some controllers require the gains to be carried out to five or more decimal places in order to attain the desired controller performance. Clearly in cases like these, rounding off the calculated feedback gains to the nearest power of 2 (to save execution time) is totally unacceptable.

Indirect Microprocessor Control Of DC Motors

In the preceding sections it is pointed out that in certain applications it may not be possible to use a microprocessor to directly control a dc motor, due to the computation time and numerical accuracy problems associated with microprocessors. One way to get around these problems is to use the processor indirectly in the motion control solution. In the "indirect" method of microprocessor control, the analog servo controller is still part of the system. However, the command input to the servo now comes from a microprocessor. In a sense, the processor is being used to turn the servo on and off, while the servo is involved in the detailed motion control of the dc motor.

The indirect method may seem to be an inefficient use of a microprocessor, but the scheme has a wide range of practical applications. For example, in some applications there may be many parts of a system (other than just one dc motor) that need to be controlled. It may make good sense to use the microprocessor as a central master controller, with the processor issuing control commands to many subsystems. It would then be up to each subsystem to receive the processor commands and carry them out. In such a system setup, it would almost be impossible for the processor to be fast enough to directly control the motors and some large subsystems simultaneously. Of course, more than one microprocessor can be used, but it might be just as feasible to use an analog controller for the motion control part of the system.

One such system that uses indirect microprocessor control is the computer terminal partially shown schematically in Fig. 16-36. The terminal uses a daisywheel in its printing operations. A daisywheel consists of spokes emanating from the center, with one character attached to each spoke. A hammer is attached to some fixed part of the terminal. When the hammer is energized, it slams into the spoke at the edge of the daisywheel, causing that character to be printed on a page. By rotating the daisywheel so that the desired character lines up with the hammer, all the characters can be printed.

It is clear from Fig. 16-36 that the processor has many tasks to perform. The

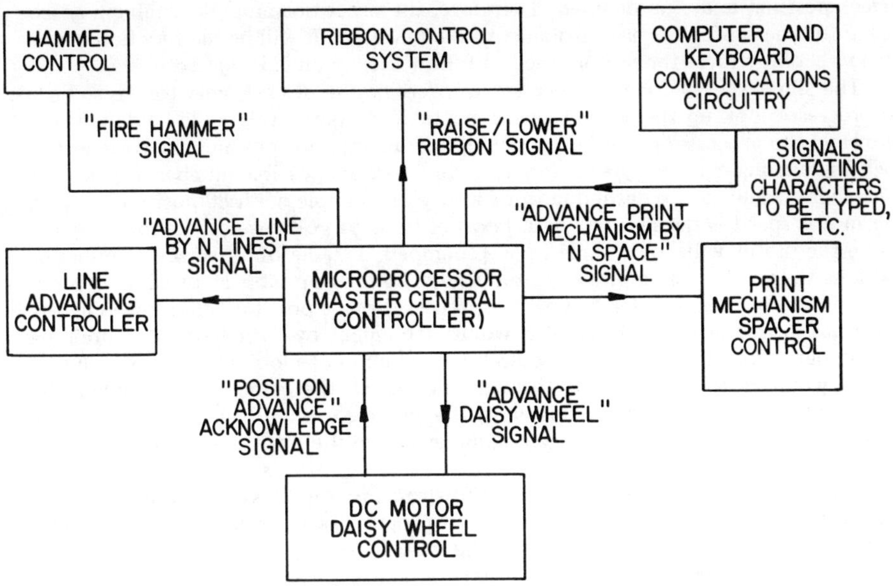

Fig. 16-36. Schematic diagram of part of a microprocessor-controlled computer terminal.

command to print a character comes from external communications circuitry. Once this command is received, the processor must move the daisywheel to the correct position, raise the typewriter ribbon, fire the hammer, and move the entire printing mechanism to the next space. Additionally, the processor will have to periodically advance the platen when such action is requested. In order to insure that the processor has sufficient time to perform these and other tasks, indirect control of the dc motor that runs the daisywheel is used. Let us examine in some detail how the daisywheel controller works.

Suppose the daisywheel has 96 spokes. We may consider that these spokes are individually numbered. Of course, to refer to the Nth spoke implies that there is a way of knowing where the first spoke is. This can be done simply by attaching a small magnet to any one of the spokes and positioning a detector near the hammer. This will identify "spoke #1." When the daisywheel rotates, the processor monitors the detector, and when a signal is detected the processor knows that spoke #1 is in line with the hammer. This is known as "homing" the daisywheel. By now stopping the daisywheel in the "home" position, the processor will be able to figure out how far to move the wheel and in which direction to type any given character. Once the daisywheel has been homed to establish a wheel orientation there is no need to ever again return to the home position, provided that the processor keeps track of the last position it moved the daisywheel to.

Assume that the daisywheel is at the home position, and the processor gets a command to type a letter located on spoke 21. The wheel must be advanced 20 positions, say, clockwise. We now assume that a position encoder is mounted on the motor shaft that produces a pulse each time another spoke moves directly in line with the hammer. This position pulse will be sent to the microprocessor. We also assume that a servo controller exists which accepts a command speed, ω_d, and which then tries to make the motor speed equal to ω_d. The function of the microprocessor is to send a set of ω_d's to the analog servo such that the motor will rapidly move to the correct position and stop damped. Therefore, the processor must determine a speed profile for the motor for each distance to be traversed. It will be the function of the servo to actually force the motor speed to follow the command trajectory.

The simplest way for the processor to produce the desired trajectory is to have the processor look up the desired speeds in a table. One way this could be done is the following: the processor could keep track of how many positions must be traversed to reach the character to be typed. Corresponding to each possible number of positions to go there would exist a desired motor speed. For example, with 80 positions to go a high motor speed is desired, but with two positions to go the velocity should be low or else the motor will not be able to stop damped. A table of ω_d versus "number of positions to go" can be empirically produced. The look-up table is stored as an array in the processor memory. Each time the processor gets a position pulse (meaning the motor has moved another position) it would decrement by 1 the location containing the number of positions to go. It would use this number to look up the next value of ω_d and would output this value, through a DAC, to the velocity servo. When the number of positions to go reaches zero, the processor would initiate the printing sequence (i.e., lift the ribbon, fire the hammer, move the print mechanism over one space).

Let us examine the 8080 program shown in Fig. 16-39 which implements this indirect control scheme. A flowchart of the program is given in Fig. 16-37, and a diagram of the print controller is given in Fig. 16-38.

The first 96 locations in the memory (0-95) are reserved for the program's look-up table. These locations contain the values of the desired ω_d, given the number of positions yet to be travelled. The table is arranged in the following way: location 1 contains the value of the desired motor speed when there is one position to go (thus, location 1 contains a small number). Location 94 contains the value of ω_d when there are 94 positions left to go (thus, location 94 contains a large number). In general, location X, with $0 \leqslant X \leqslant 95$, contains the value of ω_d desired with X positions yet to be travelled. The look-up table arranged in this special way will simplify the search for ω_d later on in the program.

The program to perform the printing begins in location 200. The processor first reads in the ACTION-REQUEST (AR) FLAG byte (the number in PORT 1). The instruction "CPI 0" is used simply to set the 8080 FLAGS so the ACTION-REQUEST byte can be examined. If no action is requested ($AR = 0$) the FLAG is read in again. When there is a character to be printed ($AR \neq 0$), a data byte is read in (from PORT 2) which contains the spoke position (1-96) of the character. The next few instructions determine how far to move the motor by subtracting the old spoke position from the newly requested one. If the difference is zero, the motor is already in the correct position so the character is immediately printed. If the difference is not zero, the program stores the number of positions to go (which we will call "POSGO") in register L. Register H is then set to zero. Note that the effect of these instructions is that the H-L register pair now contains an address which corresponds to the correct place to go in the look-up table to find ω_d(POSGO). Thus, the instruction using indirect addressing, "MOV A,M" is all that is needed to get the correct value of ω_d out of the look-up table. ω_d(POSGO) is then sent out, via PORT 20, to the DAC. The program now waits for the motor to advance one position.

The program knows that the motor has advanced one position by testing the position latch (on PORT 10). The position latch works in the following way. When the motor has advanced one position, the motor's position encoder sends out a pulse. This pulse sets a flip-flop latch to 1. The latch is then tied to PORT 10, which the processor repeatedly reads. As soon as the port reads "1," the motor has advanced. The flip-flop latch must first be reset to 0 so that it can be used again. This is accomplished by sending a "1" out on PORT 15, which is connected to the reset line on the

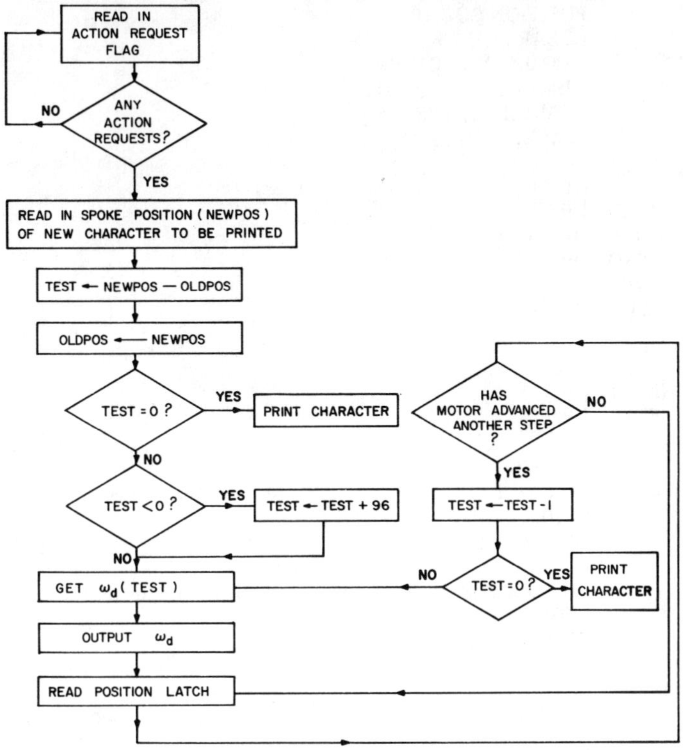

Fig. 16-37. Flow chart of program shown in Fig. 16-39.

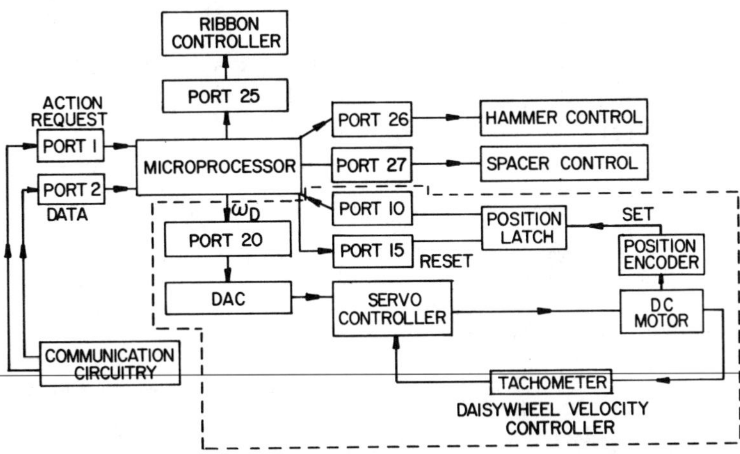

Fig. 16-38. Block diagram of printing controller.

```
             REQUEST EQU 1
             CHAR EQU 2
             POSITION EQU 10
             CLEAR EQU 15
             WCOMMAND EQU 20
             RIBCONTROL EQU 25
             HAMMER EQU 26
             SPACER EQU 27

             ORG 0
       ZERO: DB 0
        ONE: DB 2
        TWO: DB 5
      THREE: DB 7
       FOUR: DB 10
          .
          .
          .
    NINETY4: DB 250
    NINETY5: DB 255

             ORG 200
  NEWACTION: IN REQUEST          Check for character to be printed
             CPI 0
             JZ NEWACTION

             LDA OLDPOS          Character printing routine
             MOV B,A
             IN CHAR
             STA OLDPOS          Update newest "last position" of the daisy-
             SUB B                 wheel
             JZ PRINT            If NEWPOS = OLDPOS, immediately print
                                   character
             JM ADD96            If NEWPOS < OLDPOS, add 96 to the
             JMP SETHL             difference

      ADD96: ADI 96
      SETHL: MOV L,A             Load H-L with number of positions to go. This
             MVI H,0               will be used in the look-up table.

     LOOKUP: MOV A,M             Get ωd from the look-up table
             OUT WCOMMAND        OUTPUT ωd to DAC
TESTPOSITION: IN POSITION        Has motor moved?
             CPI 0
             JZ CHORES           No? Then do chores
             MVI A,1             Yes? Then first reset position latch, then get
             OUT CLEAR             new value of ωd
             DCR L
             JZ PRINT            ωd = 0 when printing
             JMP LOOKUP

      PRINT: OUT WCOMMAND
             MVI A,1
             OUT RIBCONTROL      Raise ribbon, fire hammer, advance spacer
             OUT HAMMER
             CALL STALL
```

Fig. 16-39. 8080 Assembly-language program demonstrating indirect microprocessor control of a dc motor.

Fig. 16-39. Cont.

```
          OUT SPACER
          CALL STALL
          MVI A,0
          OUT CLEAR            Lower ribbon; reset all output ports
          OUT RIBCONTROL
          OUT HAMMER
          OUT SPACER
          JMP NEWACTION

  STALL:  MVI B,250            Program does nothing, but takes 2.5 milli-
   LOOP:  MOV C,C                seconds to execute, assuming a 0.5 micro-
          DCR B                  second/state clock (used for software timing)
          JNZ LOOP
          RET

 OLDPOS:  DB 1

 CHORES:    .                  Any tasks can be performed while waiting for
            .                    the motor to move to the next position
            .
          JMP TESTPOSITION     When chores are done, go back to check
                                 position latch
```

latch. Now register L, which contains POSGO, is decremented by 1. If (POSGO − 1) = 0, then it is time to print the character. Otherwise, it is time to go back to the look-up table and get a new value for ω_d.

Printing the character is accomplished by a series of output instructions which turn on other controllers. For example, a "0" to "1" transition on PORTS 25 and 26 causes the ribbon to be lifted and the hammer to be fired. It is assumed that it will take 2.5 milliseconds for the hammer to hit. Thus, a subroutine named "STALL" is called which does nothing, but takes 2.5 milliseconds to return.* Now that the character is printed, the processor causes the printing mechanism to be moved one space. Another 2.5 milliseconds is stalled for while the spacing takes place. The final action of the processor is to reset all output ports to zero so that they will be ready for the next printing sequence.

This case study of a computer terminal has provided a detailed look at a system setup under microprocessor control, but in which the microprocessor does not control every detail of the dc motor's speed trajectory. It is clear from the example that the designer of such a system no longer has to be concerned with some of the numerical quantization effects and discrete-time stability problems encountered earlier. Thus, the indirect control method is one way to "get around" some of the inherent microprocessor limitations. The price paid, of course, is that now many of the complex and expensive analog components are once again part of the final controller. However, the microprocessor still replaces some of the control circuitry, and still provides the flexibility of a programmable controller.

*The "STALL" subroutine is just a loop that executes 250 times, taking 5000 "states" to complete. Assuming that the processor clock is adjusted so that 1 state = 0.5 microseconds, it is seen that "STALL" will not return for 2.5 milliseconds. A subroutine such as "STALL" used for timing purposes is called a "software timing loop."

APPENDIX A

TWO'S COMPLEMENT ARITHMETIC

Most modern computers store negative numbers in a format called "two's complement." This format is used because it simplifies the hardware needed to add negative numbers and to perform subtraction.

To understand two's complement it is perhaps easier to first understand what might be called ten's complement. Ten's complement will work on decimal numbers while two's complement works for binary numbers. Suppose we wanted to subtract 4 from 27. One way to do this of course is to simply write: $27 - 4 = 23$. But there is an alternative way of doing the subtraction. Suppose we subtract 4 from 100 and look at the remainder, 96. Let us define 96 as somehow representing the number -4 in ten's complement, and see what happens when we *add* 27 and 96. In adding 27 and 96, we observe that it will take 4 from the 27 to make the sum go over 100. The part over 100 is then 23. That is, $27 + 96 = (23 + 4) + 96 = 23 + (4 + 96) = 23 + 100 = 123$. If we now ignore the "100" part, we have the number 23, the correct difference between 27 and 4. The above example shows how the correct answer was obtained. We presubtracted 4 from the "base number" 100 to obtain the "ten's complement" version of -4, namely 96. We then added 27 to 96 and noted that 4 out of the 27 would add to the 96 to make exactly 100, leaving 23 left over in the ten's and one's digits. Now by simply ignoring the hundred's digit we have effectively *subtracted* 4 from 27 by *adding* to 27 the number that was 4 less than 100, and ignoring the hundred's digit.

We could make this a convention for storing negative numbers. That is, to store -4, we subtract 4 from 100 and store 96. The advantage is we then can use the same hardware to perform addition and subtraction because *subtracting* B from C will become *adding* the ten's complement of B to C. To subtract three-digit numbers, we presubtract from the base number 1000. For example, to perform $352 - 216$, we take the ten's complement of 216 and add it to 352. The ten's complement of 216 is found from the presubtraction $1000 - 216$, which is 784. Adding 784 to 352 we get $784 + 352 = 784 + (216 + 136) = (784 + 216) + 136 = 1000 + 136 = 1136$. Now by ignoring the thousand's digit, we correctly find the difference between 352 and 216, namely 136.

The binary version of the above subtraction and negation algorithm is called two's complement arithmetic. To see how this works in binary, consider subtracting the four-bit number 2 from 5. The operation we want to perform is $5 - 2$, or $0101 - 0010$. Instead of directly subtracting, let us find the two's complement of 2 and add it to 5. The base number in this case is "10000" binary, or 16 decimal. The two's complement of 2 is, therefore, $10000 - 0010 = 1110$ (binary). To see that "1110" is the number desired, let us add 1110 to 0101. We get: $1110 + 0101 = 1110 + (0010 + 0011) = (1110 + 0010) + 0011 = 10000 + 0011 = 10011$. Ignoring the 16's bit we get the correct answer to the original problem, 3.

A major reason why two's complement arithmetic is used on computers is that there is a very simple algorithm for finding the two's complement of a number. Before we state the algorithm, let us motivate it's discovery. If we study the examples given we will find that the two's complement (or ten's complement) of a number x is the number that when added to x gives the base number. Thus, the ten's complement of 4 is 96, since $4 + 96 = 100$ (the base number), and the two's complement of 0010 is 1110 since $0010 + 1110 = 10000$ (the base number). We see that the base number contains as many zeros as the operands contain digits or bits, with the zeros being preceded by a 1. Now, given an N-bit binary number, x, how can we find the N-bit binary number, y, that when added to x produces a sum that consists of a 1 followed by N zeros?

There is an easy way to find this number y. To see how, we begin with any N-bit number x, say $x_1x_2x_3...x_N$, where $x_i = 1$ or 0; $i = 1,2,...,N$. We form the number z by taking the bit-for-bit complement of x. That is, $z = z_1z_2z_3...z_N$, where $z_i = 1$ if $x_i = 0$ and $z_i = 0$ if $x_i = 1$. Now let us add x and z. Since x_i is the complement of z_i, the sum of x and z is guaranteed to be an N-bit number consisting of *all* 1's. This is almost the result we wanted, but not quite. If we now add x + z + 1, we see that the sum must be an N + 1 bit number consisting of a leading 1 followed by N zeros. It must be true, therefore, that z + 1 is the two's complement representation of −x. That is, y = z + 1. To summarize our procedure in a simple algorithm, it can be stated that to find the two's complement representation of some N-bit number, −x, we write down x, take the bit-for-bit complement, and add 1 to the result. The new result is the two's complement representation of −x. If we now add the two's complement representation of −x to some number w and ignore the N + 1 bit of the sum, we will have effectively subtracted x from w.

One final point should be noted. Suppose we have an N-bit byte and store a number x in the byte which is small enough so that the Nth bit is a 0. Then it will always be true that to store the two's complement representation of −x, the Nth bit will be a 1 (the reader should convince himself of this by trying a few examples). In this sense, the Nth bit is a "sign" bit. That is, if +x is small enough that it can be stored with the Nth bit equal to 0, then −x will be some number whose Nth bit is a 1. Therefore, looking at the Nth bit of a stored number will determine whether such a number is positive or negative (if it's negative, the stored number is a two's complement representation). However, to store a large positive number, it will be necessary to turn on the Nth bit. The two's complement number will have the Nth bit equal to 0 even though a two's complement number is supposed to be negative. There is no way a computer can unambiguously distinguish between negative and large positive numbers, or conversely between positive and large negative numbers. For example, with an 8-bit byte, the bit pattern "1111 1110" could be the positive number 254 or could be the two's complement representation of −2. Thus, in testing the 8th bit to see if it is 1 or 0 and finding it is a 1, the programmer must be certain he will be able to tell if the 8th bit is a 1 because the *positive number* stored is larger than 127, or because the number stored is *negative*, in two's complement.

APPENDIX B

HEXADECIMAL AND OCTAL NUMBERS

The hexadecimal (base 16) and octal (base 8) number systems are often used in representing the contents of a byte of information. The digits in octal are 0, 1, 2, 3, 4, 5, 6, and 7 while the hexadecimal digits are 0, 1, 2, 3, 4, 5, 6, 7, 8, 9, A, B, C, D, E, and F. In hexadecimal, the letters A through F stand for the decimal numbers 10 through 15, respectively.

It is usually much more convenient to represent a long binary number by a few hexadecimal or octal digits. To see why, consider the eight-bit binary byte "0110 1010." It takes eight binary digits to represent the information in this byte. But now suppose we take each group of four binary bits and write down the one hexadecimal digit that it represents. That is, we take the first four bits ("0110") and write down the hexadecimal number "6," and then take the second four bits ("1010") and write down the hexadecimal number "A." The eight bits of information can now be conveniently written as "6A." To find out what binary pattern is represented by some hexadecimal digits, say "51," we write out the four-bit binary expansion for each hexadecimal digit. Thus, "51" in hexadecimal is "0101 0001" (0101 = 5; 0001 = 1) in binary.

The octal system works similarly, except that we take each group of three binary bits together to make one octal digit. Thus, the six-bit binary byte "111 010" would become "72" in octal, since "111" (binary) is "7" (octal) and "010" (binary) is "2" (octal). To expand an octal digit into binary we write out the three-bit binary expansion. Thus, "51" in octal becomes "101 001" (101 = 5; 001 = 1) in binary.

Because most machines have a byte bit length that is evenly divisible by four, the hexadecimal system is more often used. It is very common to see the output of an assembler (the binary translation of a program) printed in hexadecimal instead of binary to save space in printing. It is also often easier to think of addresses in hexadecimal rather than in binary or decimal. For example, consider the 8080 instruction "LDA 367." In binary, the instruction looks like:

"LDA" OP-CODE	0011 1010
ADDRESS 367	0000 0001 0110 1111

The bit pattern "0000 0001 0110 1111" looks nothing like its decimal equivalent, 367. On the other hand, consider the hexadecimal representation of the same instruction:

"LDA" OP-CODE	3A
ADDRESS 367	016F

The address "016F" is easily reconverted into binary by simply writing down the four-bit binary expansion of each digit. The hexadecimal representation "looks like" the binary representation. For this reason, many programmers prefer to actually think of addresses in hexadecimal. That is, they will think of "LDA 367" (where 367 is in decimal) as "LDA 016F" (where 016F is in hexadecimal). This can have advantages when debugging a program on a microprocessor. For example, suppose the user wishes to examine the contents of location 367 to see if his program is executing properly. He will probably have to hand-translate decimal address 367 into binary and then tell whatever debugging aids he has built that they should display the contents of location "0000 0001 0110 1111" (367 decimal). If the address is written from the very beginning as "016F," the hand-translation to binary becomes trivial.

Because of the conceptual and space-saving advantages of using hexadecimal (and octal), many assemblers allow a user to specify data in hexadecimal (and octal). Thus, it is often permissible to write instructions such as "ANI 00001111" (AND-IMMEDIATE 00001111 with the AC) in the more compact form "ANI 0F" or "ANI 0FH," with H specifying that the preceding number is in hexadecimal.

APPENDIX C

8080 INSTRUCTION SET [4]

Summary of Processor Instructions

Mnemonic	Description	Instruction Code (1) D_7	D_6	D_5	D_4	D_3	D_2	D_1	D_0	Clock (2) Cycles	FLAGS Set (3)
$MOV_{r1,r2}$	Move register to register	0	1	D	D	D	S	S	S	5	None
MOV M,r	Move register to memory	0	1	1	1	0	S	S	S	7	None
MOV r,M	Move memory to register	0	1	D	D	D	1	1	0	7	None
HLT	Halt	0	1	1	1	0	1	1	0	7	None
MVI r	Move immediate register	0	0	D	D	D	1	1	0	7	None
MVI M	Move immediate memory	0	0	1	1	0	1	1	0	10	None
INR r	Increment register	0	0	D	D	D	1	0	0	5	PSZ
DCR r	Decrement register	0	0	D	D	D	1	0	1	5	PSZ
INR M	Increment memory	0	0	1	1	0	1	0	0	10	PSZ
DCR M	Decrement memory	0	0	1	1	0	1	0	1	10	PSZ
ADD r	Add register to A	1	0	0	0	0	S	S	S	4	CPSZ
ADC r	Add register to A with carry	1	0	0	0	1	S	S	S	4	CPSZ
SUB r	Subtract register from A	1	0	0	1	0	S	S	S	4	CPSZ
SBB r	Subtract register from A with borrow	1	0	0	1	1	S	S	S	4	CPSZ
ANA r	And register with A	1	0	1	0	0	S	S	S	4	LPSZ
XRA r	Exclusive Or register with A	1	0	1	0	1	S	S	S	4	LPSZ
ORA r	Or register with A	1	0	1	1	0	S	S	S	4	LPSZ
CMP r	Compare register with A	1	0	1	1	1	S	S	S	4	CPSZ
ADD M	Add memory to A	1	0	0	0	0	1	1	0	7	CPSZ
ADC M	Add memory to A with carry	1	0	0	0	1	1	1	0	7	CPSZ
SUB M	Subtract memory from A	1	0	0	1	0	1	1	0	7	CPSZ
SBB M	Subtract memory from A with borrow	1	0	0	1	1	1	1	0	7	CPSZ
ANA M	And memory with A	1	0	1	0	0	1	1	0	7	LPSZ
XRA M	Exclusive Or memory with A	1	0	1	0	1	1	1	0	7	LPSZ
ORA M	Or memory with A	1	0	1	1	0	1	1	0	7	LPSZ
CMP M	Compare memory with A	1	0	1	1	1	1	1	0	7	CPSZ
ADI	Add immediate to A	1	1	0	0	0	1	1	0	7	CPSZ
ACI	Add immediate to A with carry	1	1	0	0	1	1	1	0	7	CPSZ
SUI	Subtract immediate from A	1	1	0	1	0	1	1	0	7	CPSZ
SBI	Subtract immediate from A with borrow	1	1	0	1	1	1	1	0	7	CPSZ
ANI	And immediate with A	1	1	1	0	0	1	1	0	7	LPSZ
XRI	Exclusive Or immediate with A	1	1	1	0	1	1	1	0	7	LPSZ
ORI	Or immediate with A	1	1	1	1	0	1	1	0	7	LPSZ
CPI	Compare immediate with A	1	1	1	1	1	1	1	0	7	CPSZ
RLC	Rotate A left	0	0	0	0	0	1	1	1	4	C
RRC	Rotate A right	0	0	0	0	1	1	1	1	4	C
RAL	Rotate A left through carry	0	0	0	1	0	1	1	1	4	C
RAR	Rotate A right through carry	0	0	0	1	1	1	1	1	4	C

APPENDIX C (Cont.)

Mnemonic	Description	Instruction Code (1) D_7	D_6	D_5	D_4	D_3	D_2	D_1	D_0	Clock (2) Cycles	FLAGS Set (3)
JMP	Jump unconditional	1	1	0	0	0	0	1	1	10	None
JC	Jump on carry	1	1	0	1	1	0	1	0	10	None
JNC	Jump on no carry	1	1	0	1	0	0	1	0	10	None
JZ	Jump on zero	1	1	0	0	1	0	1	0	10	None
JNZ	Jump on no zero	1	1	0	0	0	0	1	0	10	None
JP	Jump on positive	1	1	1	1	0	0	1	0	10	None
JM	Jump on minus	1	1	1	1	1	0	1	0	10	None
JPE	Jump on parity even	1	1	1	0	1	0	1	0	10	None
JPO	Jump on parity odd	1	1	1	0	0	0	1	0	10	None
CALL	Call unconditional	1	1	0	0	1	1	0	1	17	None
CC	Call on carry	1	1	0	1	1	1	0	0	11/17	None
CNC	Call on no carry	1	1	0	1	0	1	0	0	11/17	None
CZ	Call on zero	1	1	0	0	1	1	0	0	11/17	None
CNZ	Call on no zero	1	1	0	0	0	1	0	0	11/17	None
CP	Call on positive	1	1	1	1	0	1	0	0	11/17	None
CM	Call on minus	1	1	1	1	1	1	0	0	11/17	None
CPE	Call on parity even	1	1	1	0	1	1	0	0	11/17	None
CPO	Call on parity odd	1	1	1	0	0	1	0	0	11/17	None
RET	Return	1	1	0	0	1	0	0	1	10	None
RC	Return on carry	1	1	0	1	1	0	0	0	5/11	None
RNC	Return on no carry	1	1	0	1	0	0	0	0	5/11	None
RZ	Return on zero	1	1	0	0	1	0	0	0	5/11	None
RNZ	Return on no zero	1	1	0	0	0	0	0	0	5/11	None
RP	Return on positive	1	1	1	1	0	0	0	0	5/11	None
RM	Return on minus	1	1	1	1	1	0	0	0	5/11	None
RPE	Return on parity even	1	1	1	0	1	0	0	0	5/11	None
RPO	Return on parity odd	1	1	1	0	0	0	0	0	5/11	None
RST	Restart	1	1	A	A	A	1	1	1	11	None
IN	Input	1	1	0	1	1	0	1	1	10	None
OUT	Output	1	1	0	1	0	0	1	1	10	None
LXI B	Load immediate register Pair B & C	0	0	0	0	0	0	0	1	10	None
LXI D	Load immediate register Pair D & E	0	0	0	1	0	0	0	1	10	None
LXI H	Load immediate register Pair H & L	0	0	1	0	0	0	0	1	10	None
LXI SP	Load immediate stack pointer	0	0	1	1	0	0	0	1	10	None
PUSH B	Push register Pair B & C on stack	1	1	0	0	0	1	0	1	11	None
PUSH D	Push register Pair D & E on stack	1	1	0	1	0	1	0	1	11	None
PUSH H	Push register Pair H & L on stack	1	1	1	0	0	1	0	1	11	None
PUSH PSW	Push A and Flags on stack	1	1	1	1	0	1	0	1	11	None
POP B	Pop register pair B & C off stack	1	1	0	0	0	0	0	1	10	None
POP D	Pop register pair D & E off stack	1	1	0	1	0	0	0	1	10	None

APPENDIX C (Cont.)

Mnemonic	Description	Instruction Code (1) D_7	D_6	D_5	D_4	D_3	D_2	D_1	D_0	Clock (2) Cycles	FLAGS Set (3)
POP H	Pop register pair H & L off stack	1	1	1	0	0	0	0	1	10	None
POP PSW	Pop A and Flags off stack	1	1	1	1	0	0	0	1	10	CPSZ
STA	Store A direct	0	0	1	1	0	0	1	0	13	None
LDA	Load A direct	0	0	1	1	1	0	1	0	13	None
XCHG	Exchange D & E, H & L Registers	1	1	1	0	1	0	1	1	4	None
XTHL	Exchange top of stack, H & L	1	1	1	0	0	0	1	1	18	None
SPHL	H & L to stack pointer	1	1	1	1	1	0	0	1	5	None
PCHL	H & L to program counter	1	1	1	0	1	0	0	1	5	None
DAD B	Add B & C to H & L	0	0	0	0	1	0	0	1	10	D
DAD D	Add D & E to H & L	0	0	0	1	1	0	0	1	10	D
DAD H	Add H & L to H & L	0	0	1	0	1	0	0	1	10	D
DAD SP	Add stack pointer to H & L	0	0	1	1	1	0	0	1	10	D
STAX B	Store A indirect	0	0	0	0	0	0	1	0	7	None
STAX D	Store A indirect	0	0	0	1	0	0	1	0	7	None
LDAX B	Load A indirect	0	0	0	0	1	0	1	0	7	None
LDAX D	Load A indirect	0	0	0	1	1	0	1	0	7	None
INX B	Increment B & C registers	0	0	0	0	0	0	1	1	5	None
INX D	Increment D & E registers	0	0	0	1	0	0	1	1	5	None
INX H	Increment H & L registers	0	0	1	0	0	0	1	1	5	None
INX SP	Increment stack pointer	0	0	1	1	0	0	1	1	5	None
DCX B	Decrement B & C	0	0	0	0	1	0	1	1	5	None
DCX D	Decrement D & E	0	0	0	1	1	0	1	1	5	None
DCX H	Decrement H & L	0	0	1	0	1	0	1	1	5	None
DCX SP	Decrement stack pointer	0	0	1	1	1	0	1	1	5	None
CMA	Complement A	0	0	1	0	1	1	1	1	4	None
STC	Set carry	0	0	1	1	0	1	1	1	4	C
CMC	Complement carry	0	0	1	1	1	1	1	1	4	C
DAA	Decimal adjust A	0	0	1	0	0	1	1	1	4	CPSZ
SHLD	Store H & L direct	0	0	1	0	0	0	1	0	16	None
LHLD	Load H & L direct	0	0	1	0	1	0	1	0	16	None
EI	Enable Interrupts	1	1	1	1	1	0	1	1	4	None
DI	Disable Interrupts	1	1	1	1	0	0	1	1	4	None
NOP	No operation	0	0	0	0	0	0	0	0	4	None

NOTES:
1. DDD or SSS — 000 B — 001 C — 010 D — 011 E — 100 H — 101 L — 110 Memory — 111 A.
2. Two possible cycle times, (5/11) indicate instruction cycles dependent on condition flags.
3. FLAG Codes: C = CARRY, P = PARITY (this FLAG is set to 1 on even parity; i.e., the number of 1's in the result of an operation is an even number), S = SIGN, Z = ZERO, L = CARRY FLAG is cleared to zero at the end of the operation, D = CARRY FLAG is set to 1 if a carry occurs out of the double precision addition.

REFERENCES

1. Kuo, B. C., *Digital Control Systems,* SRL Publishing Company, Champaign, Ill., 1977.

2. Oppenheim, A. V. and Schafer, R. W., *Digital Signal Processing,* Prentice-Hall, Inc., Englewood Cliffs, N.J., 1975.

3. Kuo, B. C., *Automatic Control Systems,* 3rd. ed., Prentice-Hall, Inc., Englewood Cliffs, N.J., 1975.

4. *Intel 8080 Microcomputer Systems User's Manual,* Intel Corporation, Santa Clara, Calif., Sept. 1975.

17 DC Motor Stepping Systems In Computer Peripherals

F. G. MORITZ

17.1 INTRODUCTION

The computer industry has been, and continues to be, one of the most dynamic industries in existence today. This is especially true with respect to the use by this industry of the latest and most sophisticated technical developments in many areas. These involve the use of step motors, clutches and brakes, and high-performance dc motors to achieve various types of incremental motion.

As the industry grows, we see an ever expanding variety of electromechanical devices being used to enter, store, retrieve, print, and display data in one form or another. Almost every one of these devices has at least one subsystem that must generate incremental motion with a fast response, good accuracy and repeatability. Also, these systems are required to be reliable and their mean-time-to-failure is expected to be in the range from 2000 to 5000 hours.

In this chapter we review the applications of dc-motor stepping systems in computer peripherals. We start with a general discussion of various types of equipment, and then consider two design examples.

17.2 EXAMPLES OF INCREMENTAL DRIVES IN PERIPHERALS

The following sections briefly describe incremental systems as they are used in some of the more commonly known peripherals.

Perforated Tape Equipment

A perforated tape reader (Fig. 17-1) is required to move a tape which contains data in the form of punched holes past a read head, usually in an incremental fashion.

The data bytes are punched 1/10 inch apart, along with a sprocket or feed hold, and the data rates of such a device (10 to 300 characters per second (CPS)) are slow when compared to magnetic-tape equipment (5×10^3 to 1.25×10^6 CPS), for example.

For many years these devices achieved their incremental motion by the use of pinch roller/capstans, clutches and brakes, or cams, star wheels, and Geneva mechanisms. Such devices usually had maximum speeds of 10 CPS.

At the present time, most perforated tape readers use step motors, which provide as high as 300 CPS rates at about the same price as a 30 CPS star-wheel reader of 10 years ago. Thus, a significant price/performance gain has been achieved. However, this is one example where the step motor outshines the dc motor.

A dc motor system could easily be designed to provide the same performance, but not at a cost which would allow a complete reader to sell in the OEM marketplace for \$300 to \$500. The step motor and perforated tape reader are ideally matched.

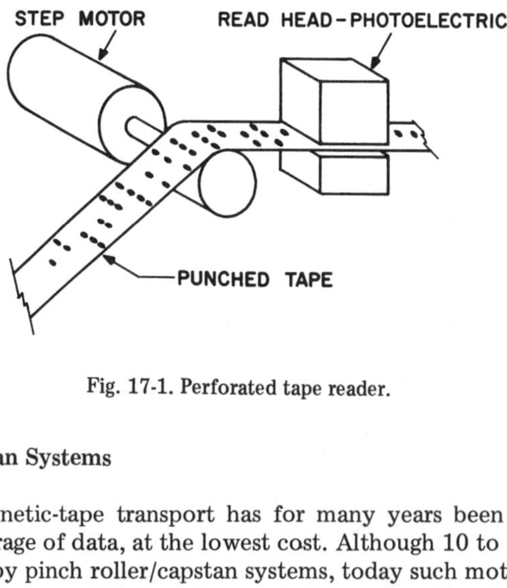

Fig. 17-1. Perforated tape reader.

Magnetic-Tape Capstan Systems

The digital magnetic-tape transport has for many years been the main device used for the bulk storage of data, at the lowest cost. Although 10 to 15 years ago tape motion was created by pinch roller/capstan systems, today such motion is almost universally created by the use of high-performance dc servo motors, in a velocity servo loop (Fig. 17-2).

While at first glance such a system might not seem to be incremental, since it must operate at various speeds from 10 to 200 inches per second, in reality it is really a very flexible type of incremental system.

The data on magnetic tape are written at densities of 200, 556, 800, 1600, and 6250 bytes per inch. These data are formatted into blocks, which theoretically could be as short as 0.01 inch and as long as 2400 feet. The data are written or read a block at a time, with the block lengths being of random length. Thus, the capstan system is incremental, but must respond in a random fashion, as dictated by the information being written onto or read from the tape, while maintaining certain positional accuracy requirements. (Industry standards specify certain inter-block distances between the recorded data blocks; nominally 0.600 inch. Therefore, for worst-case operation, cap-

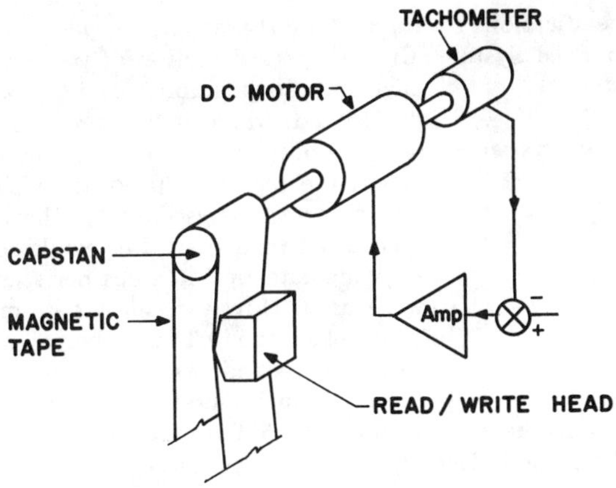

Fig. 17-2. Magnetic-tape capstan drive.

stan systems must be capable of accelerating to speed and decelerating to zero in a nominal distance of 0.200 inch.)

A step motor cannot be used for this application since:

1) A step motor system can only move in multiples of its basic stepping angle, which is not consistent with the random block nature of the data.
2) There is no way of physically synchronizing the tape with the motor motion, as is possible with the sprocket hole that is punched into perforated tape.

The dc-motor-driven capstan system is one example of a more general case of an incremental system than that represented by a step-motor system. The very nature of the step motor (its ability to take a fixed, known step) creates a bounded system which can only perform in multiples of that step. On the other hand, a dc-motor-driven system is not bounded in this sense and can accommodate itself to random requirements of either the work piece itself (the data on the tape in this example) or the various motion commands being fed into the system.

A more detailed example of a capstan system is given in Sec. 17.3.

Printers

The use of various types of printing devices is growing each year in the computer industry. Whereas 10 years ago there were typically only slow terminal devices (the Teletype model 33) and expensive high-performance line printers, today we have a large diversity of equipment.

However, they all use one or more subsystems that are incremental in operation. The paper must be advanced a discrete distance after each line is printed; the printing device must be moved across the paper, either incrementally or continually, but usually in a combination of the two; and the printing ribbon must be moved during printing to expose fresh ribbon to the print hammer device.

Two good examples of modern day terminal printers are the Hytype II by Diablo, and the DECwriter by Digital Equipment Corporation. Both of these printers use step motors for the paper feed system and dc motors with digital encoding devices to achieve incremental motion for the print carriage system.

Typical specifications for the Hytype II are:

Paper Feed System—Creates vertical motion of paper past the printing mechanism: 48 positions per inch, at a rate of 4 inches per second, which is the same as 192 steps per second.

Carriage System——Creates horizontal motion of the printing mechanism across the paper: 120 positions per inch, at a rate of 45 characters per second. If the characters are to be spaced 0.1 inch apart, this would require an equivalent stepping rate of 540 steps per second if a step motor were used. However, since a dc motor is used, it can accomplish the 0.1-inch increments at a rate of 45 steps per second, while also being capable of meeting the 1/120-inch increments when required.

Return Stroke——15 inches in 300 milliseconds.

In each subsystem, the better of the two devices has been used to meet functional requirements.

Disc Systems

The proliferation of printing devices is easily matched by the expansion of disc recording equipment. Devices are available in a wide range of data capacity, access times, data transfer rates, and physical sizes, ranging in price from $400 to tens of thousands of dollars.

However, these units all have one element in common. The recording head or heads must be moved to various points (concentric circles) on the recording discs to provide for proper read/write operation. Table I contains a comparison of certain specifications for three types of disc systems: the floppy-disc drive, the 2314 type of disc memory, and the 3330 disc memory.

Note that as the number of tracks per surface increases, the task of positioning the head becomes more difficult, not with respect to positioning speed, since the track-to-track motion time is a constant 10 milliseconds for decreasing distances traversed, but with respect to positioning accuracy. The dead space between tracks on a floppy-disc drive (0.0068 inch) is more than twice the width of a track itself on a 3330 system. Thus, this class of equipment gives an excellent example of the use of various incremental systems in one type of product.

Table I. Comparison of Positioning Parameters for Disc Systems.

Parameter	Floppy Disc	2314	3330
Number of tracks per surface	77	203	411
Track width	0.014	0.007	0.003
Track spacing (center to center)	0.0208	0.01	0.0052
Track separation (between tracks)	0.0068	0.003	0.0022
Access time (msec) (track to track)	10	10-20	10

Floppy Disc: The floppy disc contains a single read/write head on a carriage, with a total weight of 2 or 3 ounces (Fig. 17-3). With only one or two exceptions, all floppy-disc systems use a step motor and lead screw configuration to move the head. The motor has a 15-degree-per-step motion, with a lead screw of proper pitch to give the required 0.0208-inch motion of the head carriage.

The system is open loop, and the total positional accuracy of the head position is the sum of the accuracies contributed by: motor, lead screw, bell nut, carriage, head, concentricity of spin motor, disc mount and disc, and temperature. There is no detenting or holding torque other than that provided by the step motor itself.

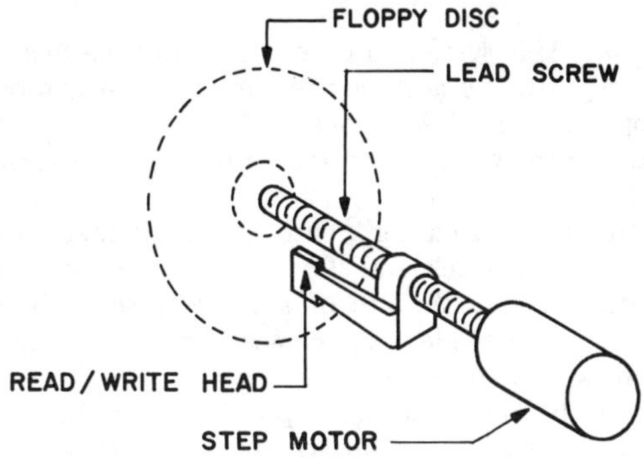

Fig. 17-3. Floppy-disk head positioner.

2314 Disc System: The 2314 type of disc system contains an assembly of 20 read/write heads on a carriage with a total weight of approximately 2 pounds (Fig. 17-4). In addition, its positioning accuracy requirements are 50 percent more difficult in comparison to the floppy disc.

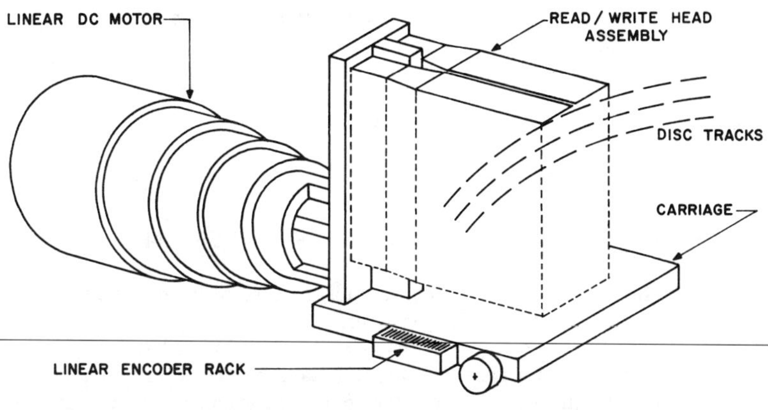

Fig. 17-4. 2314 head motion mechanism.

Therefore, rather than using a step motor, a type of dc motor (known as a linear voice-coil motor) is used, in conjunction with a linear encoder transducer mounted on the structure, in a closed-loop positional system. Thus, some of the tolerances have been included within a closed loop and reduced. The total positional accuracy is now dependent on: encoder rack, carriage, head, concentricity of spin motor, disc mount and disc, temperature, and loop gain and stability.

Some 2314 systems use a mechanical detent after null position has been achieved, and some use an electronic lock to maintain null. In addition, a special alignment disc, with purposely displaced eccentric tracks is used to "split the difference" and achieve best head alignment.

3330 Disc System: This system also contains multiple heads and an additional 50 percent increase in positional accuracy requirements in comparison to the 2314. To achieve proper operation in this system, additional elements are included within the closed loop. This is accomplished by recording a special servo track on the disc (Fig. 17-5).

A linear dc motor is also used here, with the result that the system both moves incrementally from track to track and "locks onto" the servo track when finally achieving null position, through the use of a separate servo track read head. Now, total positional accuracy is dependent upon: accuracy of recorded servo track, servo track pickup, gain and stability of loop.

Thus, we see how for one class of equipment, the requirement for greater and greater accuracy has led to the use of various incremental systems, each proper for its particular application, producing the right cost/performance trade-off.

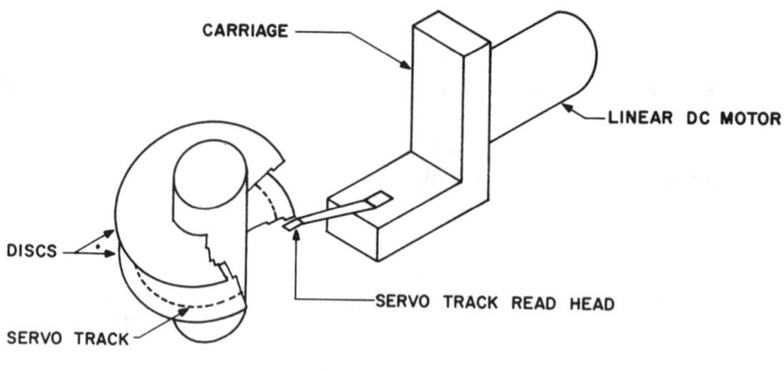

Fig. 17-5. 3330 servo track.

17.3 DESIGN EXAMPLE 1—DUAL-MODE CAPSTAN SYSTEM

An example of the use of a dc motor in a capstan system is the design created for a digital-tape transport for the Poseiden program.

The tape transport was to be used to replace perforated tape equipment, and therefore had to operate in a true incremental fashion, character by character, but at a higher rate, to make use of existing software and interface hardware. In addition, it was required to operate in a normal, block-by-block mode, as discussed previously.

The specifications for this system were:

Recording density: 200 bytes per inch

Stepping rate: 300 steps per second maximum
Normal speed: 45 inches per second

A capstan drive assembly was designed consisting of a PMI U9M2 motor, an encoder disc, and an analog tachometer (Fig. 17-6).

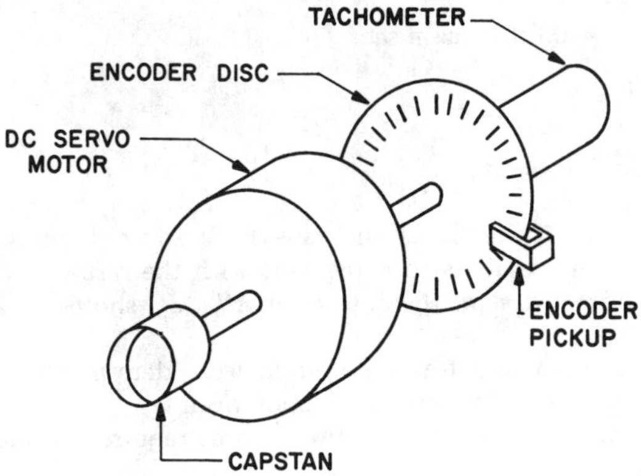

Fig. 17-6. Dual-mode capstan assembly.

Specifications for the U9M2 motor are:

K_t = 3.2 oz-in/ampere

K_e = 2.36 volts/1000 RPM

J_m = 0.0035 oz-in-sec^2

R_a = 0.42 ohms at 25 degrees C

I_a max. at stall = 8.9 amperes

The encoder disc was chosen to have 1500 slots per revolution.
Since the capstan peripheral motion per step is:

$$1/200 = 0.005 \text{ inch}$$

and the capstan angular motion per step is

$$2\pi/1500 = \pi/750 \text{ radians/step}$$

it results in a capstan radius of:

$$(0.005)(750/\pi) = 3.75/\pi \text{ inch}$$

and a capstan circumference of

$$(2\pi)(3.75/\pi) = 7.5 \text{ inch}$$

The various component inertias are:

$$J_{disc} = 0.001 \text{ oz-in-sec}^2$$

$$J_{tachometer} = 0.0005 \text{ oz-in-sec}^2$$

$$J_{capstan} = 0.0016 \text{ oz-in-sec}^2$$

Therefore,

$$J_{total} = 0.0066 \text{ oz-in-sec}^2$$

Note that the moment of inertia of the capstan is approximately 1/4 of the total moment of inertia. This result is in agreement with the result in Chapter 15 which states that under optimal design, the capstan inertia, J_c, should be between 1/4 and 1/5 of the total inertia.

Acceleration requirements for this system were determined by conditions for both the stepping mode and block-by-block operation.

For the stepping mode, a velocity profile was required as shown in Fig. 17-7. In this case the peak velocity is

$$\dot{\theta} = \frac{3 \text{ in/sec}}{7.5/2\pi \text{ in}} = 2.514 \text{ rad/sec}$$

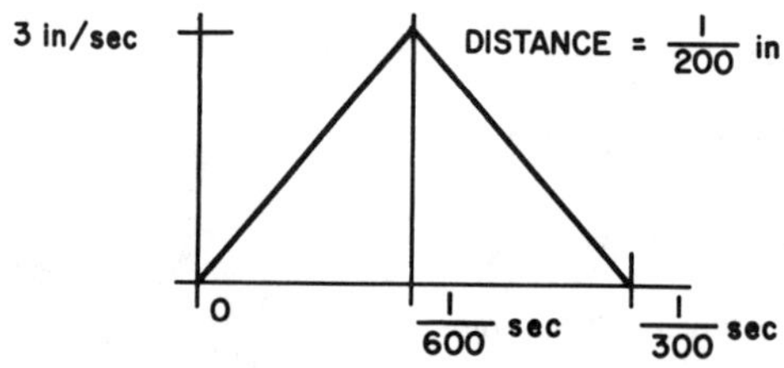

Fig. 17-7. Step mode velocity profile.

Therefore, the acceleration is:

$$\ddot{\theta} = \frac{2.514}{1/600} = 1508 \text{ rad/sec}^2$$

and the required torque is:

$$T_g = J\ddot{\theta} = 0.0066 \times 1508$$

$$= 9.95 \text{ oz-in}$$

The required current is:

$$I_a = \frac{9.95 \text{ oz-in}}{3.2 \text{ oz-in/amp}}$$

$$= 3.11 \text{ amperes}$$

For the normal run mode, the required acceleration profile is shown in Fig. 17-8. Since:

$$(\frac{1}{2}t)(45 \text{ in/sec}) = 0.3 \text{ inches}$$

$$t = 13.3 \text{ msec}$$

therefore, the maximum velocity is:

$$\dot{\theta} = \frac{45 \text{ in/sec}}{7.5/2\pi \text{ in}} = 37.7 \text{ rad/sec}$$

and the acceleration is:

$$\ddot{\theta} = \frac{37.7}{0.0133} = 2828 \text{ rad/sec}^2$$

The torque and current are:

$$T_g = 18.7 \text{ oz-in}$$

$$I_a = 5.8 \text{ amperes}$$

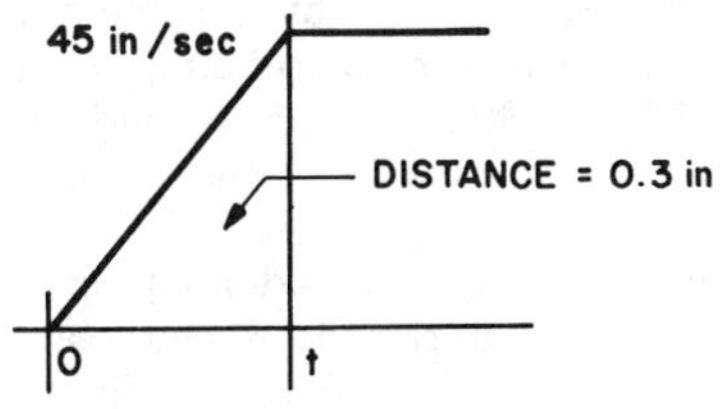

Fig. 17-8. Block mode acceleration profile.

Thus, the motor operates well below its maximum current rating for both modes of operation, and requires no external cooling.

Figure 17-9 shows a block diagram of the system. In the incremental write-mode operation, stepping action is determined by the disc in conjunction with control commands to the logic. The write strobe is activated by the trailing edge of the pulse from

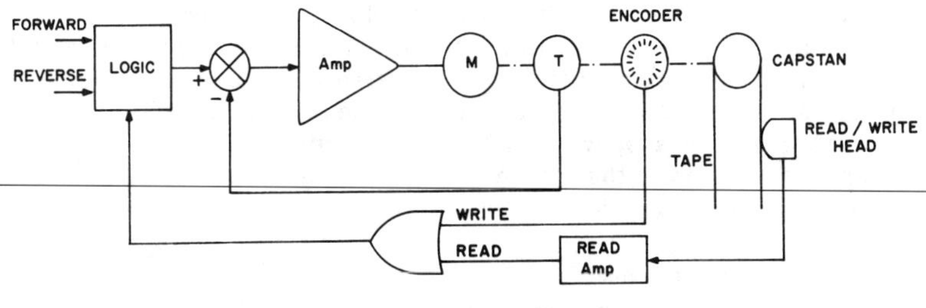

Fig. 17-9. Dual-mode capstan block diagram.

the disc, thus assuring that the accuracy of data location on the tape is determined by the disc accuracy and will be independent of system velocity fluctuations.

In the incremental read-mode operation, stepping action is determined by the read clock pulse from recorded information. The read clock pulse is "substituted" for the disc pulses and thus synchronizes the motor operation with the data on the tape.

The system zero-speed position in the incremental mode is approximately half way between adjacent data bytes.

17.4 DESIGN EXAMPLE 2—LINE PRINTER PAPER-FEED SYSTEM

Another example of the use of a dc motor for achieving incremental motion is in the paper-feed system of a high-speed line printer. Unlike the tape-drive capstan system, a printer paper-feed system is a system requiring a known, nonvarying motion for each step.

For the system under discussion, it was required that operation be possible in either 6, 8, or 10 lines per inch, with spacing accuracy of ±0.0005 inch, at rates as high as 1500 lines per minute.

Three methods of instrumenting the system were explored: a clutch-brake system, a step motor system, and the use of a dc motor.

The clutch-brake system was rejected first due to poor performance on previous designs with respect to life, repeatability, and accuracy over various duty cycle conditions.

In order for a step motor to meet the specifications, the control circuitry would have to drive it for 20 steps to achieve a displacement of 1/6 inch, 15 steps for 1/8 inch, and 12 steps for 1/10 inch (each step being 1/120 inch). It would have to do this at average rates as high as 1333 steps per second with an output torque of about 300 oz-in. Such a motor was not available, and it was decided to use the dc motor approach.

In addition to meeting the spacing and stepping rate requirements, the system must also provide holding torque in the zero-speed (standstill) condition, and provide a method by which the operator can make fine adjustments in the relative position between the forms being printed and the printer hammer assembly.

The first of these requirements was met by using a one-way passive clutch and a steady-state "reverse" current in the motor during no motion times. The clutch consists of an inner race attached to the motor shaft and an outer race attached to the frame of the machine. The "sense" of the clutch is such to allow the motor to rotate in only one direction, namely the direction that will advance paper through the machine. By applying a steady-state current in the motor, in opposition to the clutch, tension is produced in the paper-drive belt, and the paper itself. Also, when stopping, small velocity fluctuations around zero velocity are isolated from the paper. This is a more straightforward and economical approach, as compared to a complete bi-directional positional system, with its inherent stability problems.

The second requirement was easily met, once a dc motor approach was taken, by adding an operator-selectable variable-speed reference signal at the input of the system. This allows the designer, after deactivating the one-way clutch, to run the paper either forward or reverse, at slow speed, to line up forms.

A simplified diagram of the system is shown in Fig. 17-10.

Highlights of the system design are as follows:

J_{total} = 0.053 oz-in-sec^2

(including all pulleys, shafts, motor, etc.)

Motor: PMI type U12M4 with U6T tachometer

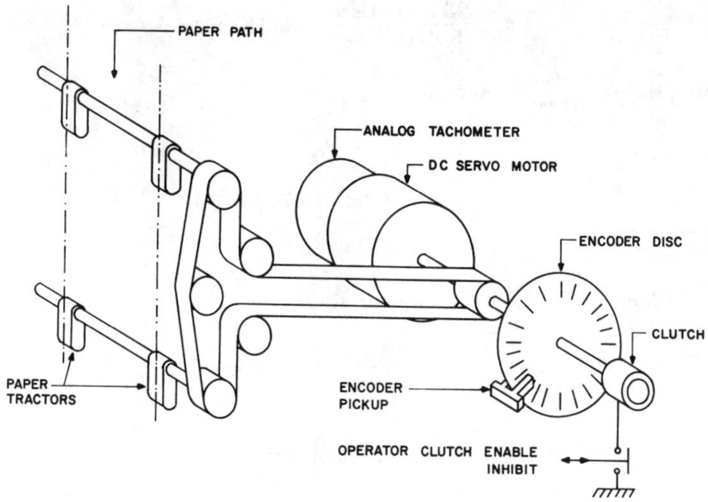

Fig. 17-10. Line printer paper feed.

In designing the encoder disc for this system, the dimensions and inertias of the pulleys, shafts, and tractor mechanism were considered, in order to create a design that could be manufactured economically. An encoder wheel with many slots is expensive and would result in large pulleys to convert a small angular rotation into 1/6, 1/8, or 1/10 inch of linear motion. On the other hand, an encoder with a small number of slots would result in pulleys and shafts of impractical small diameters.

The final encoder design consists of a disc with slots every 15 degrees ($\pi/12$ radians) for the 6-lines-per-inch mode, every 11.25 degrees ($\pi/16$ radians) for the 8-lines-per-inch mode, and every 9 degrees ($\pi/20$ radians) for the 10-lines-per-inch mode.

Since, for a fixed-line motion rate (1500 lines per minute), the 6-lines-per-inch mode results in the motor spending the largest amount of time in motion, it represents a worst-case condition, and, therefore, the design parameters for this system will be based on this requirement.

Since an angular rotation of 12 degrees ($\pi/12$ radians) results in a linear motion of 1/6 inch, the diameter of the motor output pulley, D, is

$$1/6 = (\pi/12)(D/2)$$

$$D = 1.273 \text{ inches}$$

It was decided to have the system reach a steady-state velocity for a period of time before being required to stop. This would aid in achieving accurate, repeatable displacements. A velocity of 15 in/second was chosen, with an acceleration time of 5 milliseconds, as shown in Fig. 17-11. Therefore, since

$$15 \text{ in/sec} = 23.56 \text{ rad/sec}$$

$$\text{displacement during acceleration} = (1/2)(5 \times 10^{-3})(23.56) = 0.0589 \text{ rad}$$

and

$$\ddot{\theta} = \frac{23.56 \text{ rad/sec}}{5 \times 10^{-3} \text{ sec}} = 4712 \text{ rad/sec}^2$$

Therefore, since $T_{friction} = 60$ oz-in

$$\begin{aligned} T_g &= J\ddot{\theta} + T_{friction} \\ &= 0.053 \times 4712 + 60 \\ &= 310 \text{ oz-in} \end{aligned}$$

For the U12M4, $K_t = 15.6$ oz-in/ampere, therefore

$$I_a = \frac{310}{15.6} = 20 \text{ amperes}$$

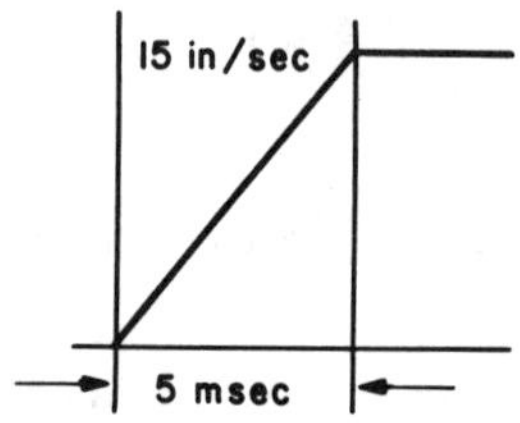

Fig. 17-11. System acceleration profile.

Since this system has considerable friction, the deceleration time is appreciably less than the acceleration time (Fig. 17-12).

$$T = 15.6 \times 20 + 60 = 372 \text{ oz-in}$$

Therefore,

$$\ddot{\theta} = \frac{372}{0.053} = 7019 \text{ rad/sec}^2$$

and

$$t = \frac{23.56}{7019} = 3.36 \text{ msec}$$

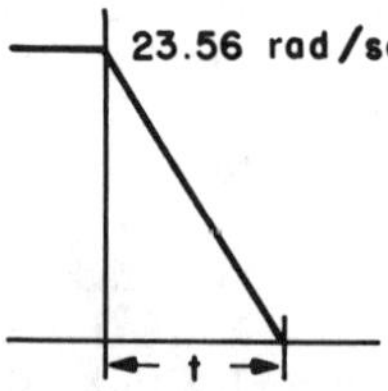

Fig. 17-12. System deceleration profile.

A complete velocity profile for one step is shown in Fig. 17-13, where

t_1 = 5 msec

t_3 = 3.4 msec

t_4 = 2 msec (an arbitrary settle time)

t_2 = 6.9 msec for 6 lines/inch
= 4.13 msec for 8 lines/inch
= 2.47 msec for 10 lines/inch

d_1 = 0.0589 radians

d_3 = 0.040 radians

d_2 = 0.163 radians for 6 lines/inch
= 0.097 radians for 8 lines/inch
= 0.058 radians for 10 lines/inch

with

t_{total} = 17.3 msec for 6 lines/inch
= 14.53 msec for 8 lines/inch
= 12.87 msec for 10 lines/inch

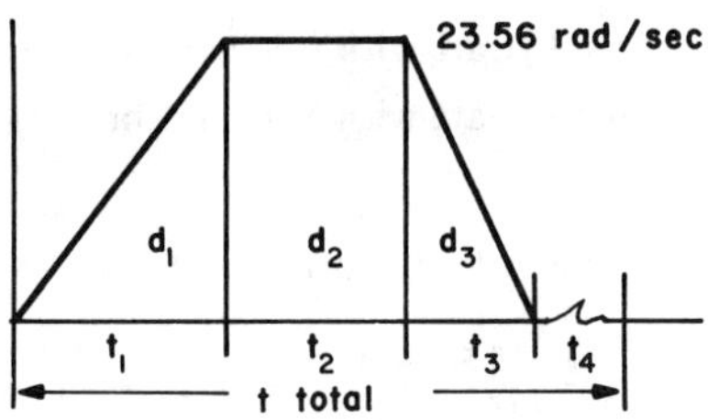

Fig. 17-13. Total velocity profile for one step.

Since the maximum stall current rating of the U12M4 is 7.5 amperes, it was required that the need for cooling be determined. A typical current waveform would be as shown in Fig. 17-14.

Using

$$I_{rms} = \sqrt{\frac{1}{t}\int_0^T i^2(t)dt}$$

it can be shown that

$$I_{rms} = 10.2 \text{ amperes}$$

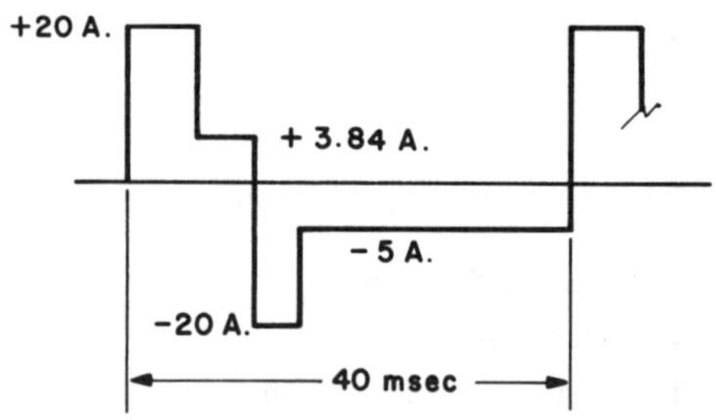

Fig. 17-14. Motor current waveform for one step.

For the U12M4,

$$R_a = 0.75 \text{ ohms at 25 degree C}$$
$$= 1.11 \text{ ohms at 150 degree C}$$

$$P_{max} = \frac{150 \text{ degree C} - T_{ambient}}{\theta_T}$$

where θ_T, the thermal resistance, is given by:

$$\theta_T = 0.8 \text{ degree C/watt with 0.4 lbs/min}$$
$$= 0.5 \text{ degree C/watt with 0.8 lbs/min}$$
$$= 0.3 \text{ degree C/watt with 2.0 lbs/min}$$

for forced-air cooling.

Assuming a maximum ambient temperature of 45 degrees C,

$$\theta_{T(required)} = \frac{150 - 45}{(10.2)^2(1.11)} = 1 \text{ degree C/watt}$$

Thus, forced-air cooling at a rate of about 0.4 lbs/min is needed.

In the final implementation of this design, two double-track encoder discs are actually used. One contains tracks for 6 and 8 lines per inch and the other contains tracks for 8 and 10 lines per inch. The particular line spacing to be used is under operator front-panel control.

In addition, the printer contains control circuitry to allow the system to accelerate to 30- and 75-inches-per-second speeds, when performing a paper-slew operation. In either case, when the paper moves to a position 6 lines ahead of where it will finally stop, decelerate circuits become enabled and reduce paper speed to 15 inches per second at a point one line away from the final stopped position.

17.5 CONCLUSION

The peripheral equipment area of the computer industry has provided a wide range of applications for incremental motion devices and subsystems. Many of the more common devices used today to create this type of motion were either created especially for, or had their design optimized for, the highly-sophisticated applications in peripheral equipment. This will most likely continue, with even more stress placed on response time, positional accuracy, and long, maintenance-free life.

The approach in which a dc motor is used to achieve incremental motion will find continued and broadening application in this industry. In a number of applications, the dc motor is the device to use, together with a tachometer, and in many cases, a digital encoder device, since it allows the designer a large degree of freedom in tailoring the system characteristics to meet specification. It is the only approach that can operate both incrementally and proportionately, and will allow the stepping action to be partly or completely controlled by the requirements of the load.

REFERENCES

1. F. G. Moritz, "6 Ways To Drive Tape," *Control Engineering*, March 1968.
2. *A Capstan Drive System For Digital Tape Transports*, Printed Motors, Inc., Syosset, New York.
3. A. S. Audeh and A. N. Deodhar, "Speed Control: Four Methods Compared," *Control Engineering*, pp. 53-57, December 1970.
4. C. J. Brown and J. T. Ma, "Time-Optimal Control of a Moving-Coil Linear Actuator," *IBM Journal of Research and Development*, September 1968.
5. *OEM Manual, SA800/801 Diskette Storage Drive*, Shugart Associates, Sunnyvale, California.
6. *Technical Manual, RPR02660-1 Disc Drive*, Memorex Corporation, Santa Clara, California.

18 High-Performance Phase-Locked Incremental Drive For Microfilm Moving Documents

J. TAL and D. C. SCHLICK

18.1 INTRODUCTION

Microfilm systems are designed to produce reduced-size images of documents for the purpose of storage and retrieval. The trend toward increasingly widespread use of microfilming techniques is itself uncovering new and varied applications for this versatile documenting process, spurring the system designer to constantly search out methods for improving the microfilming procedure.

The main limitation of current microfilm systems is their processing speed. The objective of the system design presented herein is to increase the processing speed, thereby boosting the system's throughput rate.

The design approach taken to achieve this objective represents a fundamental departure from previous system designs. In current systems, the documents advance into position, come to a complete halt to be photographed, and then move out of position to make room for the next document. The output rate is thus a function of the acceleration and deceleration rates of the documents as they move through the system. The new approach allows the documents to advance nonstop through the system, thereby removing that speed limitation. This is achieved by moving the lens at a velocity proportional to that of the documents, thus maintaining a relative fixed-position relationship between the document and the camera when both are in motion.

While this approach allows a significant increase in the processing rate, it also places very stringent requirements on the accuracy of the lens motion. The lens must maintain a velocity proportional to that of the document, even during variations in the document advancement rate. The lens is mounted on a shuttle which moves in synchronization with the document by means of a phase-locked servo system (PLS). The required resolution demands that the shuttle velocity be synchronized to within a tolerance of ±1.5 percent. An additional demand is placed on the system; the lens shuttle must attain the desired velocity within 10 milliseconds.

The system described in this chapter produces images reduced from the original by a ratio of 51:1. It allows an increase in the throughput rate to a maximum of 240 documents per minute. The system design is described in the following section. A special emphasis has been placed on design problems and their solutions.

18.2 MOTION REQUIREMENTS AND DESIGN APPROACH

The motion requirements are as follows. The lens shuttle must move forward upon command and synchronize with the document velocity within 10 milliseconds. This velocity must then be maintained within a tolerance of ±1.5 percent. Upon completion of the forwarding motion, the lens shuttle must return to its initial position within 30 milliseconds. While the general forward-backward requirements are quite common in incremental motion drives, the additional requirements of speed synchronization and accuracy are unique and add considerably to the complexity of the design.

Due to the complex motion requirements, a multimode drive system was designed. The system is represented by the block diagram in Fig. 18-1. As may be observed from the block diagram, the system has three basic modes of operation: the forward-control mode, the return mode, and the home-lock mode.

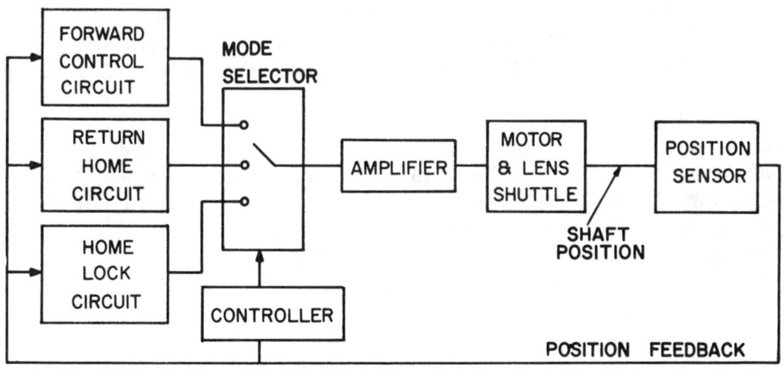

Fig. 18-1. Functional block diagram of the system.

When the document to be filmed reaches the document-present sensor, a command is sent which initiates the forward mode by switching the amplifier input to the forward-control circuit. The forward-control circuit incorporates the phase detector and the low-pass filter of the PLS. In addition, it has two rate detectors which provide the effect of analog velocity feedback. As the motion is initiated, the velocity feedback plays a dominant role, and the motor is accelerated at a high rate. As the desired velocity is approached, the PLS takes over the fine control.

The return mode is initiated when the document has cleared the document-present sensor. The return-home circuit incorporates a counter, a D/A converter, and a rate detector. The counter provides information about the distance from home in digital form. This information is then converted to analog form by the D/A converter.

This constitutes a position signal, which, along with the rate signal, is used to form a position-control loop. The position-control loop returns the lens shuttle home.

When the home position is reached, the system goes into the home-lock mode. This is a position-control mode whose function is to hold the lens shuttle at the desired home position.

A detailed description of the three operation modes is given in the following sections.

18.3 FORWARD MODE

In the forward mode, the system is required to attain the desired velocity, as determined by the reference signal, within 10 milliseconds. This velocity must be maintained within a tolerance of ±1.5 percent to meet the resolution requirements. We will use a PLS with a wide bandwidth in order to obtain the necessary fast response and to satisfy the resolution requirements.

An important point to consider is the fact that the lens shuttle is driven by a crank. As a consequence, the ratio between the motor angular velocity and the linear velocity is not constant, but varies as a sinusoidal function of the motor angular position. Therefore, in order to provide a constant linear velocity, the motor velocity must follow a curve of $1/\cos\theta$ shape, where θ, the angular position of the motor, varies over 57.5 degrees for a 9.5-inch document. This is further complicated by the fact that the position sensor must be mounted on the motor shaft due to size limitations. Thus, we are faced with the necessity of sensing the shuttle linear position from the motor angular position when the two are not proportional. This problem was solved by designing a special timing disk in which the spaces between the slots are distributed sinusoidally to compensate for the position variation. This mechanical arrangement results in a linear relationship between the velocity of the lens shuttle and the frequency of the timing disk signal.

The phase detector we have selected is a phase-frequency detector (PFD) because of the additional information it provides on velocity difference. The PFD is a three-state digital circuit which senses the leading edges of the input signals and changes states accordingly. A complete description of the PFD model and operation is given in [2]. We have chosen to implement the PFD with dual D flip-flops and associated gating, as shown in Fig. 18-2. This has been found to be superior to integrated-circuit detectors.

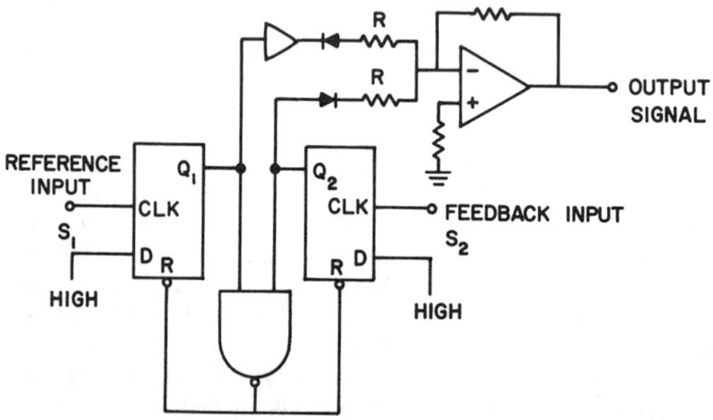

Fig. 18-2. Phase-frequency detector. The input S_1 is the reference signal; S_2 is the feedback signal.

To achieve the desired bandwidth without a tachometer, a positive-current feedback and an equivalent-velocity feedback were used. The positive-current feedback, in addition to negative-voltage feedback, was used to improve the loop bandwidth. A detailed description of this method is found in [6].

In order to form a velocity feedback, it is necessary to generate an analog signal which is proportional to the frequency of a given square wave. This is produced by an FM demodulator which was built as a phase-locked loop (PLL) employing the CD4046 integrated circuit. The details of the design and operation of such a PLL can be found in [8].

A bias signal proportional to the speed of the document-drive roll is derived from the reference signal using an FM demodulator identical to the one mentioned above. The bias signal is used as a velocity command, and the output of the first FM demodulator is used as a velocity feedback. This velocity feedback increases the loop bandwidth to the desired level. A block diagram of the loop in this mode is shown in Fig. 18-3.

The accuracy of the center frequency of the PLL when it is used as an FM demodulator is a function of the supply voltage. At a supply voltage of 5 volts, the accuracy is given as ±30 percent. Thus, the velocity feedback and the bias signal can differ by as much as 60 percent when both document and lens shuttle velocities are matched correctly. This large error must be corrected by the PLS, but the generation of a correction term adds a burden on the loop. This condition was improved by operating the FM demodulator at a supply voltage of 15 volts, where the velocity and bias signals can differ by only 20 percent.

The complete system is capable of reaching the command velocity and locking to the reference signal in 8 to 10 milliseconds. It was observed that the acceleration time can be further shortened to about 5 milliseconds.

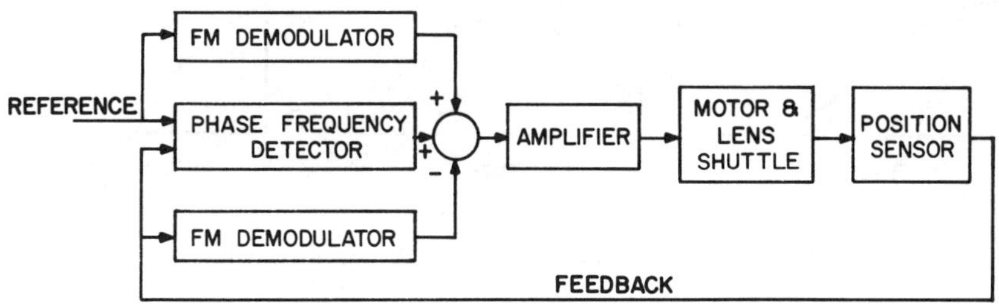

Fig. 18-3. Block diagram of the system in the forward mode.

18.4 RETURN-HOME MODE

The motion requirements in this mode are specified in terms of both position and velocity. The lens shuttle must return to the home position, and its velocity must decrease to nearly zero by the time it arrives there. Furthermore, the system must satisfy this position/velocity requirement for the full range of documents varying in length from 4.25 inches to 9.0 inches. In order to accomplish this, both position and velocity feedback signals are needed.

The position information is derived from a counter; it counts up the number of feedback pulses during the forward mode, and counts down when the return-home mode is initiated. The resulting count at any given time is directly proportional to the distance from home. This count is used as an input for a digital-to-analog converter, providing a position feedback, P(t).

The rate signal, R(t), is derived by a PLL FM demodulator, as described in the previous section. This signal is combined with the position signal to produce a control signal of the form

$$V_{in} = \alpha P(t) - \beta R(t) \tag{18-1}$$

where α and β are constants adjusted by the input impedances between the signals and the summing amplifier. The block diagram of the system in this mode is given in Fig. 18-4, and the relationship between P(t), R(t), and V_{in} is illustrated in Fig. 18-5. It can be seen that the control signal, V_{in}, will accelerate the motor during the first half of the distance and decelerate it during the second half, forming a nearly triangular waveform.

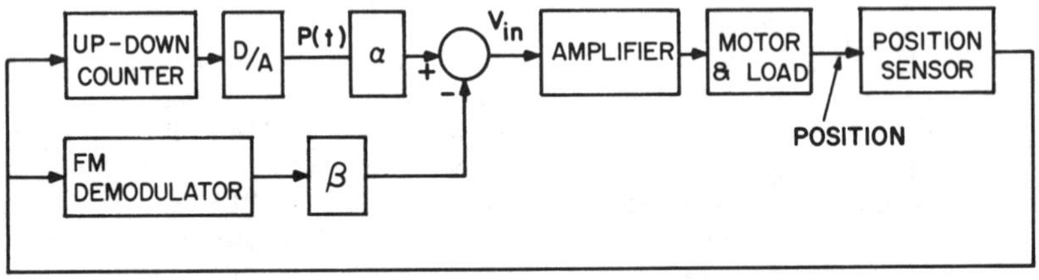

Fig. 18-4. Block diagram of the system in the return-home mode.

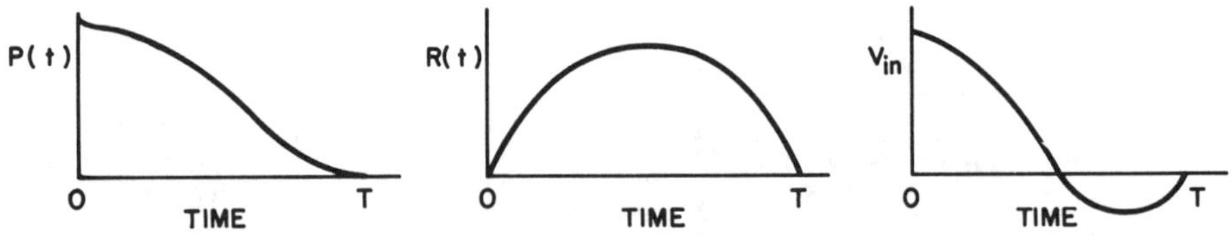

Fig. 18-5. The position, the rate, and the control signal.

18.5 HOME-POSITION LOCK

When the output of the D/A reaches zero, the home-lock mode is initiated. The purpose of this mode is to insure that the system remains at the precise home position. This is achieved by forming a position-control loop around the home position.

The position signal is obtained by sensing the location of the edge of a fan-shaped timing disk. The output of the home position sensor varies linearly between ±5 volts over a 9×10^{-3}-inch interval. This small interval guanantees the home position accuracy; however, the high gain of the sensor output versus position has a detrimental effect on the stability of the loop. In order to appreciate this difficulty, consider the block diagram of the position-control loop as represented by Fig. 18-6.

Let us first suppose that there is no compensation, G(s) = 1; and let the motor and the load be represented by the simplest transfer function, a pure integrator, as shown in Fig. 18-6. Due to the pole at the origin and the two additional poles, the system will become unstable if the loop gain is increased to a high value, as is desired in this case.

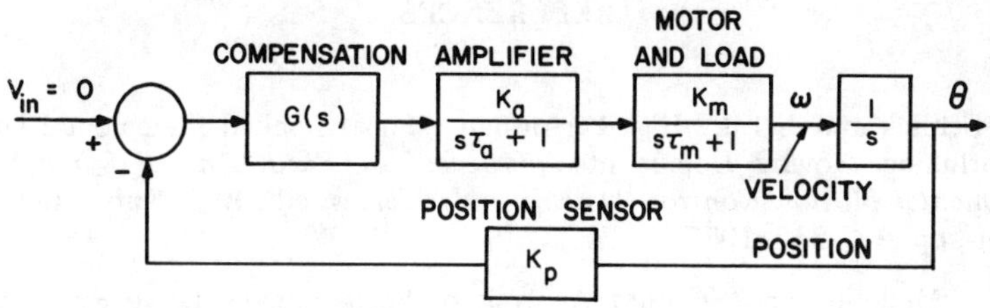

Fig. 18-6. Block diagram of the system as a position-control loop in the home-lock mode.

In order to stabilize the system while maintaining a sufficiently high gain, we must add a compensation network. We have selected a lead-lag network for its simplicity and for the wide bandwidth it provides. This network is illustrated in Fig. 18-7. The values of the network components are:

$$R_1 = 15\ \text{k}\Omega$$

$$R_2 = 2.2\ \text{M}\Omega$$

$$C = 0.0015\ \mu\text{F}$$

The effect of adding R_1 and C is represented by the transfer function G(s).

$$G(s) = \frac{1 + sC(R_1 + R_2)}{1 + sCR_1} = \frac{1 + 3.32 \times 10^{-3}\,s}{1 + 2.25 \times 10^{-5}\,s}$$

This transfer function has a unity gain at low frequencies, a zero (lead) at $\omega = 300$ rad/sec, and a pole (lag) at $\omega = 45{,}000$ rad/sec. This network results in a stable system with sufficient gain.

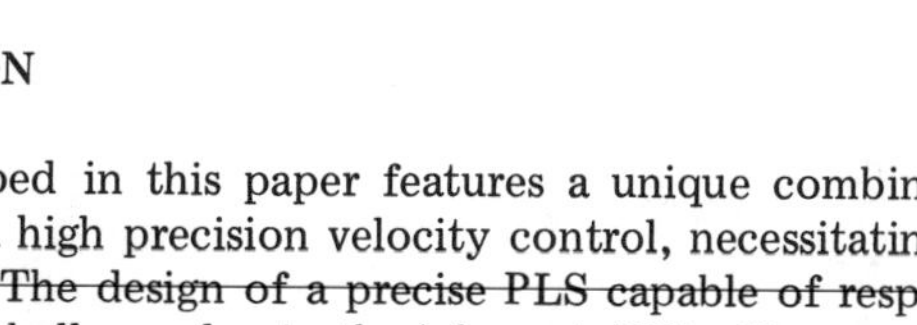

Fig. 18-7. Lead-lag compensation network.

18.6 CONCLUSION

The system described in this paper features a unique combination of incremental motion drive and a high precision velocity control, necessitating the use of a phase-locked servo drive. The design of a precise PLS capable of responding in 10 milliseconds was indeed a challenge due to the inherent difficulties encountered in the design requirements of this system. The idea of deriving the velocity feedback by an FM demodulator was quite helpful in achieving the design goal.

A system was built according to the design described in this chapter, and the test results satisfied the motion requirements.

REFERENCES

1. D. C. Schlick and J. Tal, "High-Performance Phase-Locked Incremental Drive for Microfilming Moving Documents," *Proceedings, Sixth Annual Symposium on Incremental Motion Control Systems and Devices*, ed., B. C. Kuo, University of Illinois, pp. 237-242, 1977.

2. J. Tal, "Modeling and Stability Analysis of Phase-Locked Servo Systems," *Proceedings, Sixth Annual Symposium on Incremental Motion Control Systems and Devices*, ed., B. C. Kuo, University of Illinois, pp. 229-235, 1977.

3. J. Tal, "Phase-Locked Servo Systems—A New Approach to Speed Control," *Proceedings, Fourth Annual Symposium on Incremental Motion Control Systems and Devices*, ed., B. C. Kuo, University of Illinois, pp. AA-1—AA-5, 1975.

4. *DC Motors, Speed Controls, Servo Systems*, 3rd ed., Engineering Handbook, Electro-Craft Corporation, 1975.

5. D. H. Smithgall, "A Phase-Locked Loop Motor Control System," *IEEE Trans. IECI*, Vol. IECI-22, No. 4, pp. 487-490, Nov. 1975.

6. D. T. Phan and V. H. Malakian, "The Design of A Velocity Control System Without A Tachometer—The EMF Feedback Approach," *Proceedings, Sixth Annual Symposium on Incremental Motion Control Systems and Devices*, ed., B. C. Kuo, University of Illinois, pp. 243-259, 1977.

7. C. A. Sharpe, "A 3-State Phase Detector Can Improve Your Next PLL Design," *EDN*, pp. 55-59, Sept. 1976.

8. RCA Application Note 1CAN-6101. Also found in *RCA CMOS/MOS Integrated Circuits Databook*, pp. 471-478, 1975.

9. V. W. Eveleigh, *Control Systems Design*, McGraw-Hill Book Company, New York, 1972.

Conversion Tables Of Units

1. LENGTH OR DISTANCE

SI or Metric Units

meter	(m)
kilometer	(km)
centimeter	(cm)
millimeter	(mm)
micrometer	(μm)
nanometer	(nm)
Angstrom	(Å)

British Units

mile (statute)	(mi)
yard	(yd)
feet	(ft)
inch	(in)
mil	(mil)

	m	mi	yd	ft	in	mil
1 m =	1	6.21×10^{-4}	1.0936	3.2808	39.369	3.9369×10^{4}
1 mi =	1609.3	1	1760	5280	6.336×10^{4}	6.336×10^{7}
1 yd =	0.9144	5.682×10^{-4}	1	3	36	3.6×10^{4}
1 ft =	0.3048	1.894×10^{-4}	0.3333	1	12	1.2×10^{4}
1 in =	0.0254	1.58×10^{-5}	2.778×10^{-2}	8.333×10^{-2}	1	10^{3}
1 mil =	2.54×10^{-5}	1.58×10^{-8}	2.778×10^{-5}	8.333×10^{-4}	10^{-3}	1

$1 \text{ m} = 100 \text{ cm} = 10^3 \text{ mm} = 10^6\ \mu\text{m} = 10^9 \text{ nm} = 10^{10}$ Å

2. AREA

SI or Metric Units

square kilometer	(km^2)
square meter	(m^2)
square centimeter	(cm^2)
square millimeter	(mm^2)

British Units

square mile	(mi^2)
acre	(acre)
square yard	(yd^2)
square feet	(ft^2)
square inch	(in^2)

	m^2	yd^2	ft^2	in^2
1 m^2 =	1	1.196	10.764	1.55×10^3
1 yd^2 =	0.836	1	9	1.296×10^3
1 ft^2 =	9.29×10^{-2}	0.1111	1	144
1 in^2 =	6.45×10^{-4}	7.72×10^{-4}	6.94×10^{-3}	1

1 acre = 4840 yd^2 = 43560 ft^2
1 mi^2 = 640 acres = 2.59 km^2
1 m^2 = 10^4 cm^2 = 10^6 mm^2

3. VOLUME

SI or Metric Units

cubic meter (m^3)
cubic centimeter (cm^3)
cubic millimeter (mm^3)
liter (l)

British Units

gallon (gal)
pint (pt)
cubic yard (yd^3)
cubic feet (ft^3)
cubic inch (in^3)

	m^3	yd^3	ft^3	in^3
1 m^3 =	1	1.308	35.32	6.102×10^4
1 yd^3 =	0.765	1	27	4.666×10^4
1 ft^3 =	0.0283	0.03704	1	1.728×10^3
1 in^3 =	1.639×10^{-5}	2.14×10^{-5}	5.79×10^{-4}	1

1 gal = 8 pt = 4.545 l = 0.1605 ft^3
1 pt = 0.568 l
1 m^3 = 10^6 cm^3 = 10^9 mm^3

4. LINEAR VELOCITY

SI or Metric Units

kilometer per hour (km/hr)
meter per second (m/sec)
centimeter per second (cm/sec)
millimeter per second (mm/sec)

British Units

miles per hour (mph)
feet per second (ft/sec)
inch per second (in/sec)

	km/hr	m/sec	mph	ft/sec	in/sec
1 km/hr =	1	0.2778	0.6214	0.9113	10.936
1 m/sec =	3.6	1	2.238	3.2808	39.369
1 mph =	1.609	0.4468	1	1.4667	17.6
1 ft/sec =	1.0973	0.3048	0.6818	1	12
1 in/sec =	9.144×10^{-2}	0.0254	0.0568	0.0833	1

5. LINEAR ACCELERATION

SI or Metric Units

meter per second per second (m/sec^2)

British Units

feet per second per second (ft/sec^2)
inch per second per second (in/sec^2)

	m/sec^2	ft/sec^2	in/sec^2
$1\ m/sec^2 =$	1	3.2808	39.37
$1\ ft/sec^2 =$	0.3048	1	12
$1\ in/sec^2 =$	0.0254	0.0833	1

6. PLANE ANGLE

SI or Metric or British Units

radian (rad)
degree (deg, °)
revolution (rev)
arc minute ($\widehat{min}$, ′)
arc second ($\widehat{sec}$, ″)

	rad	deg	rev	$\widehat{min}$	$\widehat{sec}$
1 rad =	1	5.7296	0.1592	343.776	2.0626×10^4
1 deg =	0.1745	1	2.778×10^{-3}	60	3600
1 rev =	2π	360	1	2.16×10^4	1.296×10^6
1′ =	2.909×10^{-3}	1.667×10^{-2}	4.63×10^{-5}	1	60
1″ =	4.848×10^{-5}	2.778×10^{-4}	7.716×10^{-7}	1.667×10^{-2}	1

$\pi = 3.14159...$

7. ANGULAR VELOCITY

SI or Metric or British Units

radian per second (rad/sec)
degree per second (deg/sec)
revolution per minute (RPM)
revolution per second (RPS)

	rad/sec	deg/sec	RPM	RPS
1 rad/sec =	1	5.7296	9.549	0.1592
1 deg/sec =	1.745×10^{-2}	1	0.1667	2.778×10^{-3}
1 RPM =	0.1047	6	1	1.667×10^{-2}
1 RPS =	6.283	360	60	1

8. ANGULAR ACCELERATION

SI or Metric or British Units

radian per second per second (rad/sec^2)
degree per second per second (deg/sec^2)
revolution per minute per second (RPM/sec)
revolution per second per second (RPS/sec)

	rad/sec^2	deg/sec^2	RPM/sec	RPS/sec
1 rad/sec^2 =	1	5.7296	9.549	0.1592
1 deg/sec^2 =	1.745×10^{-2}	1	0.1667	2.778×10^{-3}
1 RPM/sec =	0.1047	6	1	1.667×10^{-2}
1 RPS/sec =	6.283	360	60	1

9. MASS

SI or Metric Units

kilogram (kg)
gram (g)

British Units

pound (lb)
ounce (oz)
slug (slug)

	kg	g	lb	slug	oz
1 kg =	1	10^3	2.2046	6.852×10^{-3}	35.274
1 g =	10^{-3}	1	2.2046×10^{-3}	6.852×10^{-6}	3.5274×10^{-2}
1 lb =	0.4536	453.59	1	0.03108	16
1 slug =	14.5939	1.459×10^5	32.174	1	514.78
1 oz =	2.835×10^{-2}	28.35	0.0625	1.943×10^{-3}	1

10. FORCE OR WEIGHT

SI or Metric Units

Newton (N)
kilopond (kp)
kilogram (kg)
pond (p)
gram (g)
dyne (dyne)

British Units

pound (lb)
ounce (oz)
poundal (poundal)

	N	kp	kg	p	g	lb	oz
1 N =	1	0.102	0.102	102	102	0.2248	3.5968
1 kp =	9.804	1	1	10^3	10^3	2.2046	35.274
1 kg =	9.804	1	1	10^3	10^3	2.2046	35.274
1 p =	9.804×10^{-3}	10^{-3}	10^{-3}	1	1	2.2046×10^{-3}	3.5274×10^{-2}
1 g =	9.804×10^{-3}	10^{-3}	10^{-3}	1	1	2.2046×10^{-3}	3.5274×10^{-2}
1 lb =	4.448	0.4536	0.4536	453.6	453.6	1	16
1 oz =	0.278	2.835×10^{-2}	2.835×10^{-2}	28.35	28.35	0.0625	1

1 N = 10^5 dyne = 7.233 poundal

11. TORQUE

SI or Metric Units

Newton-meter (Nm)
kilopond-meter (kpm)
gram-centimeter (g-cm)
dyne-centimeter (dyne-cm)

British Units

pound-foot (lb-ft)
ounce-inch (oz-in)

	Nm	kpm	lb-ft	oz-in	g-cm	dyne-cm
1 Nm =	1	0.102	0.738	141.61	1.02×10^4	10^7
1 kpm =	9.804	1	7.233	1.389×10^3	10^5	9.807×10^7
1 lb-ft =	1.355	0.1382	1	192	1.383×10^4	1.356×10^7
1 oz-in =	7.062×10^{-3}	7.2×10^{-4}	5.208×10^{-3}	1	72.01	7.062×10^4
1 g-cm =	9.804×10^{-5}	10^{-5}	7.233×10^{-5}	1.389×10^{-2}	1	980.7
1 dyne-cm =	10^{-7}	1.02×10^{-8}	7.376×10^{-8}	1.416×10^{-5}	1.02×10^{-3}	1

12. MOMENT OF INERTIA

SI or Metric Units

kilogram-meter square ($kg\text{-}m^2$)
gram-centimeter square ($gm\text{-}cm^2$)
gram-centimeter-second square ($gm\text{-}cm\text{-}sec^2$)

British Units

pound-foot square ($lb\text{-}ft^2$)
foot-pound-second square ($ft\text{-}lb\text{-}sec^2$)
ounce-inch square ($oz\text{-}in^2$)
ounce-inch-second square ($oz\text{-}in\text{-}sec^2$)
moise (moise)
slug-foot square ($slug\text{-}ft^2$)

	kg-m^2	gm-cm^2	gm-cm-sec^2	lb-ft^2	ft-lb-sec^2	oz-in^2	oz-in-sec^2	moise
1 kg-m^2 =	1	10^7	1.02×10^4	23.73	0.737	5.467×10^4	141.7	1.417×10^5
1 gm-cm^2 =	10^{-7}	1	1.02×10^{-3}	2.373×10^{-6}	7.35×10^{-8}	5.467×10^{-3}	1.417×10^{-5}	1.417×10^{-2}
1 gm-cm-sec^2 =	9.8×10^{-5}	980	1	2.325×10^{-3}	7.207×10^{-5}	5.359	1.389×10^{-2}	13.893
1 lb-ft^2 =	4.214×10^{-2}	4.214×10^5	429.97	1	0.031	2.304×10^3	5.968	5.968×10^3
1 ft-lb-sec^2 =	1.357	1.357×10^7	1.387×10^4	32.2	1	7.435×10^5	192	1.919×10^5
1 oz-in^2 =	1.829×10^{-5}	182.9	0.1866	4.34×10^{-4}	1.345×10^{-6}	1	2.59×10^{-3}	2.59
1 oz-in-sec^2 =	7.057×10^{-3}	7.057×10^4	71.98	0.1675	5.21×10^{-3}	386	1	10^3
1 moise =	7.057×10^{-6}	70.57	7.198×10^{-2}	1.675×10^{-4}	5.21×10^{-6}	0.386	10^{-3}	1

1 moise = 10^{-3} oz-in-sec^2
1 slug-ft^2 = 1 ft-lb-sec^2

13. ENERGY

SI or Metric Units

Joule (J)
Newton-meter (Nm)
dyne-centimeter (erg) (dyne-cm)

British Units

BTU
foot-pound (ft-lb)

	J	Nm	erg	BTU	ft-lb
1 J =	1	1	10^7	9.478×10^{-4}	0.738
1 Nm =	1	1	10^7	9.478×10^{-4}	0.738
1 erg =	10^{-7}	10^{-7}	1	9.478×10^{-11}	7.38×10^{-8}
1 BTU =	1.055×10^3	1.055×10^3	1.055×10^{10}	1	778.2
1 ft-lb =	1.356	1.356	1.356×10^7	1.285×10^{-3}	1

14. POWER

SI or Metric Units

Watt (W)
kilowatt (kW)
Joules per second (=kW) (J/sec)

British Units

foot-pound per second (ft-lb/sec)
BTU per second (BTU/sec)
Horsepower (hp)

	W	kW	ft-lb/sec	BTU/sec	hp
1 W =	1	10^{-3}	0.7376	1.341	1.341×10^{-3}
1 kW =	10^3	1	737.6	0.948	1.341
1 ft-lb/sec =	1.36	1.36×10^{-3}	1	1.285×10^{-3}	1.818×10^{-3}
1 BTU/sec =	1.055×10^3	1.055	778.2	1	1.415
1 hp =	745.7	0.7457	550	0.7068	1

Index